T0293668

# Recent Developments in Acoustics and Ultrasonics

# Recent Developments in Acoustics and Ultrasonics

Edited by Brock James

**States Academic Press,**
109 South 5th Street,
Brooklyn, NY 11249, USA

Visit us on the World Wide Web at:
www.statesacademicpress.com



ISBN: 978-1-63989-745-2



**Cataloging-in-Publication Data**

Recent developments in acoustics and ultrasonics / edited by Brock James.
p. cm.
Includes bibliographical references and index.
ISBN 978-1-63989-745-2
1. Sound. 2. Ultrasonics. 3. Physics. I. James, Brock.
QC225.15 .R43 2023
534--dc23

# Table of Contents

# Preface

Acoustics is the branch of physics concerned with the study of mechanical waves in solids, liquids, and gases. It covers topics such as ultrasound, vibration, sound and infrasound. Acoustics are applied for ensuring the proper transmission of sound. Ultrasonics refers to the study of sound waves or vibrations with frequencies greater than the human audible range. These ultrasound waves are commonly used by species such as bats and whales, which can communicate and navigate using ultrasounds. Ultrasonics can be utilized to develop techniques for measuring the location of internal organs and studying their functions. It is also used for ranging and navigating, and nondestructive testing of materials. This book elucidates the concepts and innovative models around prospective developments with respect to acoustics and ultrasonics. It presents researches and studies performed by experts across the globe. This book, with its detailed analyses and data, will prove immensely beneficial to professionals and students involved in this area at various levels.

This book is a comprehensive compilation of works of different researchers from varied parts of the world. It includes valuable experiences of the researchers with the sole objective of providing the readers (learners) with a proper knowledge of the concerned field. This book will be beneficial in evoking inspiration and enhancing the knowledge of the interested readers.

In the end, I would like to extend my heartiest thanks to the authors who worked with great determination on their chapters. I also appreciate the publisher's support in the course of the book. I would also like to deeply acknowledge my family who stood by me as a source of inspiration during the project.

**Editor**

# 1

# Acoustic Radiation by 3D Vortex Rings in Air

**Fedor V. Shugaev *, Dmitri Y. Cherkasov and Oxana A. Solenaya**

Faculty of Physics, M.V. Lomonosov Moscow State University, GSP-1, Leninskiye Gory, Moscow 119991, Russia; E-Mails: dre21@yandex.ru (D.Y.C.); solarka@list.ru (O.A.S.)

* Author to whom correspondence should be addressed; E-Mail: shugaev@phys.msu.ru

Academic Editor: Luís M. B. C. Campos

---

**Abstract:** Acoustic radiation emitted by three-dimensional (3D) vortex rings in air has been investigated on the basis of the unsteady Navier–Stokes equations. Power series expansions of the unknown functions with respect to the initial vorticity which is supposed to be small are used. In such a manner the system of the Navier–Stokes equations is reduced to a parabolic system with constant coefficients at high derivatives. The initial value problem is as follows. The vorticity is defined inside a toroid at $t = 0$. Other gas parameters are assumed to be constant throughout the whole space at $t = 0$. The solution is expressed by multiple integrals which are evaluated with the aid of the Korobov grids. Density oscillations are analyzed. The results show that the frequency band depends on the initial size of the vortex ring and its helicity. The presented data may be applied to the study of a flow in a wake region behind an aerodynamic body.

**Keywords:** vortex ring; acoustic radiation; Navier–Stokes equations

---

## 1. Introduction

Vortical structures (vortex rings and cylindrical vortices) play an important role in the sound radiation of gaseous flows. These structures convert a fraction of their rotational energy into sound waves [1]. The case of the three-dimensional (3D) vortex is of special interest. In this case we have three velocity components ($v_r, v_\varphi, v_\theta$), with $r, \varphi, \theta$ being spherical coordinates. The aim of the present investigation is to determine the frequency band of acoustic radiation emitted by 3D vortex rings in air.

The analysis is based on the unsteady Navier–Stokes equations. A new method for solving the equations is set forth. It uses power series expansion of the unknown functions with respect to the initial vorticity which is supposed to be small. By applying this procedure the Navier–Stokes equations are reduced to a parabolic system with constant coefficients. As a result we get the solution in the form of a power series, with multiple integrals being appropriate coefficients. The first term of the series is the main one that determines the properties of acoustic radiation at small vorticity. General questions of sound generation by vortex structures were considered in [2,3]. Acoustic radiation by a solitary vortex ring in incompressible and weakly compressible fluid was analyzed in [4]. There are investigations of the acoustic radiation during the interaction of two vortex rings [5,6]. Linear problems of sound production by vortices are considered in [1]. The influence of vortex rings on properties of turbulent flows is confirmed by numerous experiments [7,8] and by computations [9,10]. Nevertheless, further investigations are necessary in the field of acoustic radiation by a solitary vortex ring in viscous heat-conducting gas. The topic of this paper is the study of the frequency band of acoustic radiation by a 3D solitary vortex ring in air and the evolution of the radiation.

## 2. Governing Equations

We use the Helmholtz decomposition of the velocity field as a potential part and a solenoidal one:

$$v(x,t) = -\frac{\nabla}{4\pi}\int_{R^3}\frac{s(\xi,t)}{|x-\xi|}d\xi + \frac{\nabla}{4\pi}\times\int_{R^3}\frac{\Omega(\xi,t)}{|x-\xi|}d\xi,$$

$$s = \nabla v,\ \Omega = \nabla\times v,\ \nabla = \left\{\frac{\partial}{\partial x_1},\frac{\partial}{\partial x_2},\frac{\partial}{\partial x_3}\right\} \tag{1}$$

Taking into account Equation (1), the Navier–Stokes equations in dimensionless form can be written as follows:

$$\frac{\partial\Omega_i}{\partial t} = \nu\,\Delta\Omega_i + \frac{3}{4}\varepsilon_{ijl}\nu\left(\frac{\partial v_l}{\partial x_m}+\frac{\partial v_m}{\partial x_l}\right)\frac{\partial^2 h}{\partial x_j\partial x_m} - v_j\frac{\partial\Omega_i}{\partial x_j} + \Omega_m\frac{\partial v_i}{\partial x_m} + f_{1i},$$

$$\frac{\partial w}{\partial t} = -v_j\frac{\partial w}{\partial x_j} + s,$$

$$\frac{\partial s}{\partial t} = \frac{1}{\gamma}\exp(h)\Delta w + \frac{4}{3}\nu\,\Delta s - \left(\frac{1}{\gamma}\exp(h) + 0.5\nu s\right)\Delta h - v_j\frac{\partial s}{\partial x_j} + 1.5\nu\frac{\partial v_i}{\partial x_j}\frac{\partial^2 h}{\partial x_i\partial x_j} + f_2, \tag{2}$$

$$\frac{\partial h}{\partial t} = \frac{\gamma}{\mathrm{Pr}}\eta\,\Delta h - (\gamma-1)s - v_j\frac{\partial h}{\partial x_j} + f_3,$$

$$\Delta = \frac{\partial^2}{\partial x_j\partial x_j},\ w = -log\rho,\ h = logT,\ \nu = \mu/\rho,\ \eta = \lambda/\rho,\ i = 1,2,3;\ j = 1,2,3;\ l = 1,2,3;\ m = 1,2,3$$

Here, $\varepsilon_{ijl}$ is the antisymmetrical tensor; $\rho, T, v$ are the dimensionless density, temperature, velocity (divided by $\rho_0, T_0, c_0$, respectively); $\mu, \nu, \lambda, c$ are the viscosity, kinematic viscosity, heat conductivity, and low-frequency sound speed; $\gamma$ is the adiabatic exponent; Pr is the Prandtl number. The functions $f_{1i}, f_2, f_3$ are non-linear terms with respect to the first derivatives over coordinates. Subscript "0" refers to the initial state. The system Equation (2) was made dimensionless by using the characteristic length $l_0 = \nu_0/c_0$ and the characteristic time $t_0 = \nu_0/c_0^2$.

### 2.1. Initial Value Problem

At the initial instant the vorticity has non-zero values only within a gaseous toroid [11]:

$$x_1 = (r_{00}\, u_1 \sin u_2 + R_c)\cos u_3,\; x_2 = (r_{00}\, u_1 \sin u_2 + R_c)\sin u_3,\; x_3 = r_{00} u_1 \cos u_2, \quad 0 < u_1 < 1, \quad 0 < u_2, u_3 < 2\pi, \tag{3}$$

$r_{00}, R_c$ being the initial dimensions of the vortex ring, namely $r_{00}$ is the radius of its cross-section, $R_c$ is the radius of the ring. The problem is considered under the assumption that the dimensionless initial vorticity $\omega_0$ is small: $\omega_0 << 1$. The initial conditions are

$$\Omega_1 = -\omega_0 \sin u_3,\; \Omega_2 = \omega_0 \cos u_3,\; \Omega_3 = \alpha\, \omega_0 \tag{4}$$

inside the initial toroid,

$$w(x,0) = s(x,0) = h(x,0) = 0 \tag{5}$$

in the whole space. The parameter $\alpha$ refers to helicity. The value of $\alpha$ is $\alpha = 0$ (no helicity) and $\alpha = 1$ (there is the presence of helicity).

### 2.2. Solution to the Problem

Equation (2) represents a non-linear parabolic system. We seek the solution of the parabolic system as a power series expansion. Taking into account Equation (1) and the initial conditions, we get:

$$\begin{aligned} \Omega_i(x,t) &= \varepsilon\, \Omega_i^{(1)}(x,t) + \varepsilon^2\, \Omega_i^{(2)}(x,t) + \varepsilon^3\, \Omega_i^{(3)}(x,t) + \ldots \\ v_i(x,t) &= \varepsilon\, v_i^{(1)}(x,t) + \varepsilon^2\, v_i^{(2)}(x,t) + \varepsilon^3\, v_i^{(3)}(x,t) + \ldots \end{aligned} \tag{6}$$

Inserting Equation (6) into Equations (1) and (2) gives

$$\begin{aligned} w(x,t) &= \varepsilon^2 w^{(1)}(x,t) + \varepsilon^3 w^{(2)}(x,t) + \varepsilon^4 w^{(3)}(x,t) + \ldots \\ s(x,t) &= \varepsilon^2 s^{(1)}(x,t) + \varepsilon^3 s^{(2)}(x,t) + \varepsilon^{(4)} s^{(3)}(x,t) + \ldots \\ h(x,t) &= \varepsilon^2 h^{(1)}(x,t) + \varepsilon^3 h^{(2)}(x,t) + \varepsilon^4 h^{(3)}(x,t) + \ldots \end{aligned} \tag{7}$$

We have for the lowest-order functions:

$$\frac{\partial \Omega_i^{(1)}}{\partial t} = \Delta \Omega_i^{(1)},$$

$$\frac{\partial w^{(1)}}{\partial t} = s^{(1)},$$

$$\frac{\partial s^{(1)}}{\partial t} = \frac{1}{\gamma}\Delta w^{(1)} + \frac{4}{3}\Delta s^{(1)} - \frac{1}{\gamma}\Delta h^{(1)} + \psi_3^{(1)}, \tag{8}$$

$$\frac{\partial h^{(1)}}{\partial t} = \frac{\gamma}{\Pr}\Delta h^{(1)} - (\gamma - 1)s^{(1)} + \psi_4^{(1)},$$

$$\psi_3^{(1)} = \frac{\partial v_m^{(1)}}{\partial x_j}\frac{\partial v_j^{(1)}}{\partial x_m}, \psi_4^{(1)} = \frac{1}{2}\gamma(\gamma-1) D_{mj}v^{(1)} D_{mj}v^{(1)}, D_{mj}v^{(1)} = \frac{\partial v_m^{(1)}}{\partial x_j} + \frac{\partial v_j^{(1)}}{\partial x_m}$$

From Equation (1) one deduces:

$$v^{(1)}(x,t) = -\frac{0.03125}{\pi^{5/2}t^{3/2}}\int_{R^3} d\xi \int_0^\infty dr' \int_0^\pi \sin\theta' d\theta' \int_0^{2\pi} d\varphi' \Omega^{(1)}(\xi,0) \times n \exp\left(-\frac{0.25}{t}|x + x' - \xi|^2\right), \tag{9}$$

$$n = \{\sin\theta'\cos\varphi', \sin\theta'\sin\varphi', \cos\theta'\}, x' = \{r'\sin\theta'\cos\varphi', r'\sin\theta'\sin\varphi', r'\cos\theta'\}$$

The system Equation (8) consists of three homogeneous parabolic equations with respect to $\Omega_i^{(1)}$ and a non-homogeneous parabolic subsystem. All equations have constant coefficients at higher derivatives. The solution to the subsystem can be obtained with the aid of the Fourier transform.

The first Equation (8) yields

$$\Omega_i^{(1)}(x,t) = \frac{0.125}{(\pi t)^{3/2}}\int_{R^3} \Omega_i^{(1)}(\xi,0)\exp\left(-\frac{0.25}{t}|x-\xi|^2\right)d\xi \tag{10}$$

Equation (9) allows us to determine the terms $\psi_3^{(1)}, \psi_4^{(1)}$. The Fourier transform of the homogeneous parabolic subsystem in Equation (8) gives

$$\frac{d\tilde{w}^{(1)}}{dt} = \tilde{s}^{(1)},$$

$$\frac{d\tilde{s}^{(1)}}{dt} = -\frac{k^2}{\gamma}\tilde{w}^{(1)} - \frac{4}{3}k^2\tilde{s}^{(1)} + \frac{k^2}{\gamma}\tilde{h}^{(1)}, \tag{11}$$

$$\frac{d\tilde{h}^{(1)}}{dt} = -\frac{\gamma}{\Pr}k^2\tilde{h}^{(1)} - \gamma(\gamma-1)\tilde{s}^{(1)}$$

The wavy line denotes the Fourier transform, $k$ being the wave number.

The characteristic equation of the system in Equation (11) is [12]

$$f^3 + k^2\left(\frac{4}{3} + \frac{\gamma}{\Pr}\right)f^2 + k^2\left(\frac{4\gamma}{3\Pr}k^2 + 1\right)f + \frac{k^4}{\Pr} = 0 \tag{12}$$

At $0 < k < k_*$, $k_* \approx 1$ for air, the roots of Equation (12) are

$$f_1 = \sigma_1(k), \quad f_{2,3} = \sigma_2(k) \pm i\omega_r(k); \quad \sigma_1, \sigma_2 < 0 \tag{13}$$

At $k > k_*$ all roots are real ones and decay rapidly in a very short time, so we do not take this case into account. The dispersion curve $\omega_r(k), 0 < k < k_*$ has two branches. We consider only the branch that refers to smaller values of the attenuation coefficients $\sigma_1, \sigma_2$.

The fundamental solution matrix $A$ of the subsystem in Equation (11) is given by

$$\begin{gathered} A = \left[a_{jl}\right], a_{j1}(k, t-\tau) = c_{j1}(k)\exp\{\sigma_1(k)(t-\tau)\}, \\ a_{j2}(k, t-\tau) = c_{j2}(k)\exp\{\sigma_2(k)(t-\tau)\}\cos\{\omega_r(k)(t-\tau)\}, \\ a_{j3}(k, t-\tau) = c_{j3}(k)\exp\{\sigma_2(k)(t-\tau)\}\sin\{\omega_r(k)(t-\tau)\} \end{gathered} \tag{14}$$

Here $c_{jm}(k)$ must be defined from the initial conditions. Our goal is to investigate the density evolution. We get for the function $w^{(1)}$:

$$w^{(1)}(x,t) = \frac{1}{(2\pi)^{3/2}} \int_0^t d\tau \int_{R^3} dk \int_{R^3} d\xi \exp\{ik(x-\xi)\}\left\{a_{12}\psi_3^{(1)}(\xi,\tau) + a_{13}\psi_4^{(1)}(\xi,\tau)\right\} \tag{15}$$

Let us introduce a new variable:

$$X = \xi - x, \quad X = \{R_3 \sin\theta_3 \cos\varphi_3, R_3 \sin\theta_3 \sin\varphi_3, R_3 \cos\theta_3\} \tag{16}$$

The change of the variables yields:

$$w^{(1)}(x,t) = \left(\frac{2}{\pi}\right)^{1/2} \int_0^t d\tau \int_0^\infty k dk \int_0^\infty \sin\{kR_3\} R_3 dR_3 \int_0^\pi \sin\theta_3 d\theta_3 \int_0^{2\pi} d\varphi_3 \left\{a_{12}\psi_3^{(1)}(x+X,\tau) + a_{13}\psi_4^{(1)}(x+X,\tau)\right\} \tag{17}$$

The density deviation from its initial value can be written as:

$$(\rho_d - \rho_0)/\rho_0 \approx -w \approx -\omega_0^2 w^{(1)} \tag{18}$$

Here the subscript "$d$" denotes a dimensional value.

The function $w^{(1)}$ does not depend on $\omega_0$ and neither does the frequency of density oscillations. The functions $w^{(n)}(x,t)$, $s^{(n)}(x,t)$, $h^{(n)}(x,t)$, $n > 1$ can be obtained analogously. Thus, the Navier–Stokes equations have been reduced to a parabolic system with constant coefficients at the derivatives. As a result, we get a power series in $\omega_0$. The coefficients of the series are known functions of $x,t$ (multiple integrals). The first terms of the series can be used for the analysis of the frequency band of the density oscillations in the case of small vorticity.

### 3. Results and Discussion

Equation (17) was used for the investigation of density evolution. The multiple integral was evaluated with the aid of the Korobov grids [13]. The density field has been studied in the neighborhood of a vortex ring. The parameters of the ring are as follows: $R_c = 0.15$ cm; $r_{00} = 0.03$ cm, the ratio of the radius of the ring cross-section to the ring radius is equal to 0.2. The value $\omega_0 = 0.000045$

was used in computations. The density values were analyzed within the initial domain of the toroid (at the point $x_1 = r_{00}$, $x_2 = x_3 = 0$) as well as outside it (at the point $x_1 = x_2 = 0$, $x_3 = 0.046$ cm).

Figures 1 and 2 show the dependence of the density on time at the point outside the initial toroid for two cases: (i) there is no helicity (Figure 1); (ii) the helicity is present (Figure 2).

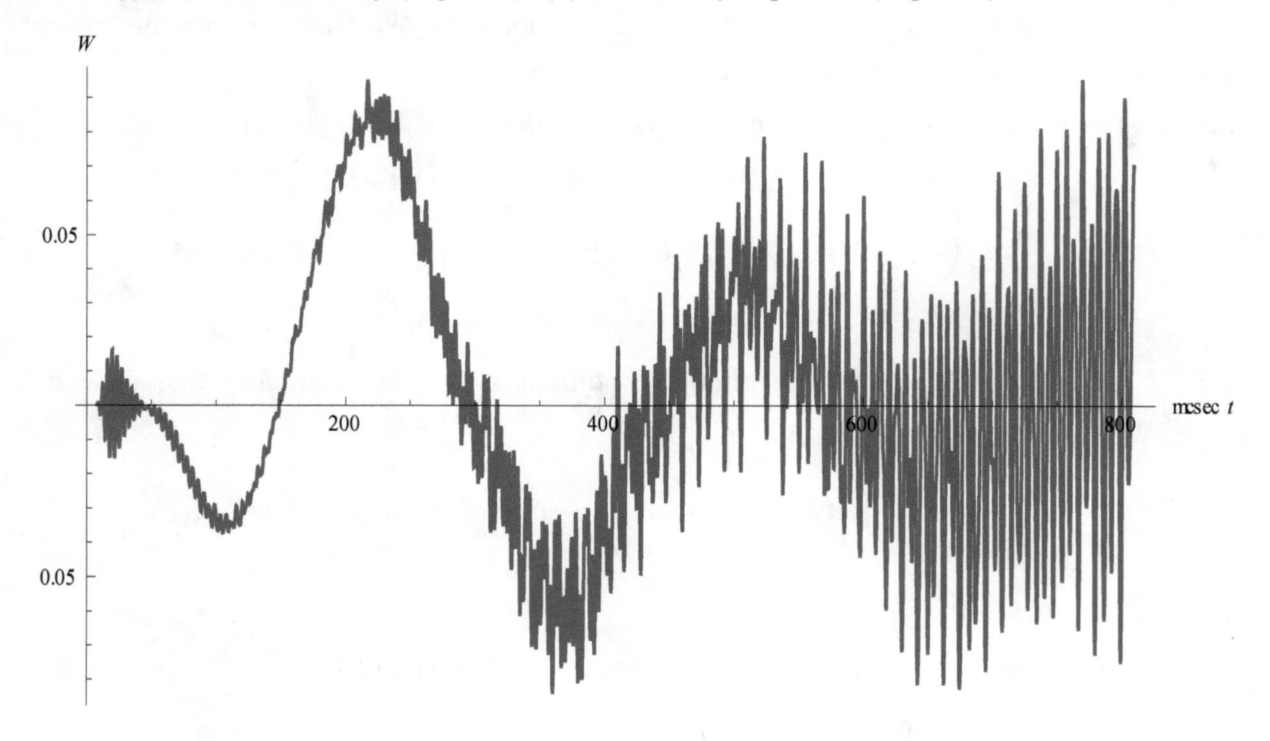


**Figure 1.** Density oscillations for the case of no helicity.

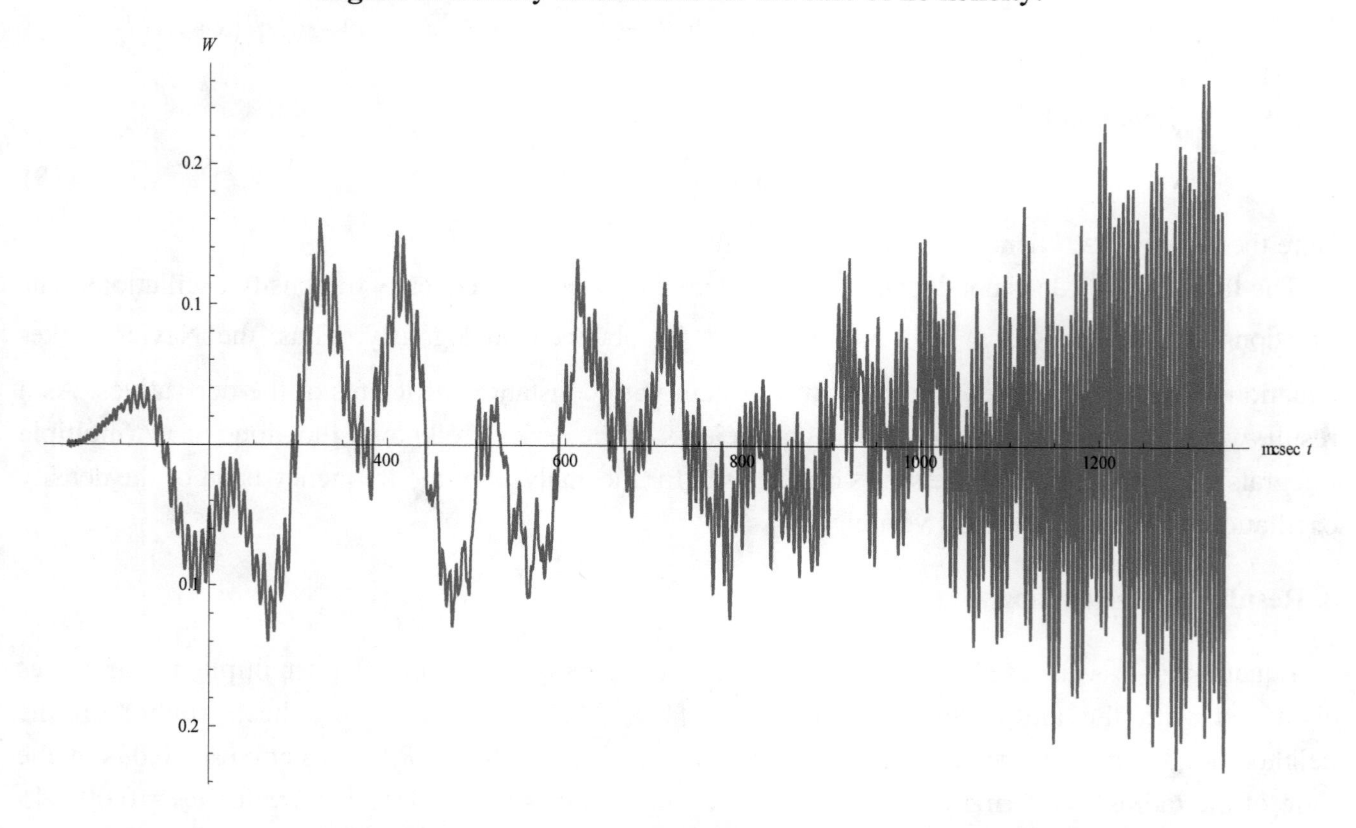


**Figure 2.** Density oscillations for the case of present helicity.

As seen, density oscillations arise. High-frequency oscillations ($f$ = 266 kHz) are modulated by a low-frequency signal ($f$ = 3.3 kHz). First of all, the amplitude of the oscillations increases, then decreases. Later on, the amplitude of the low-frequency oscillations grows and the picture becomes a chaotic one. Figures 3 and 4 show the density oscillations as they decay.

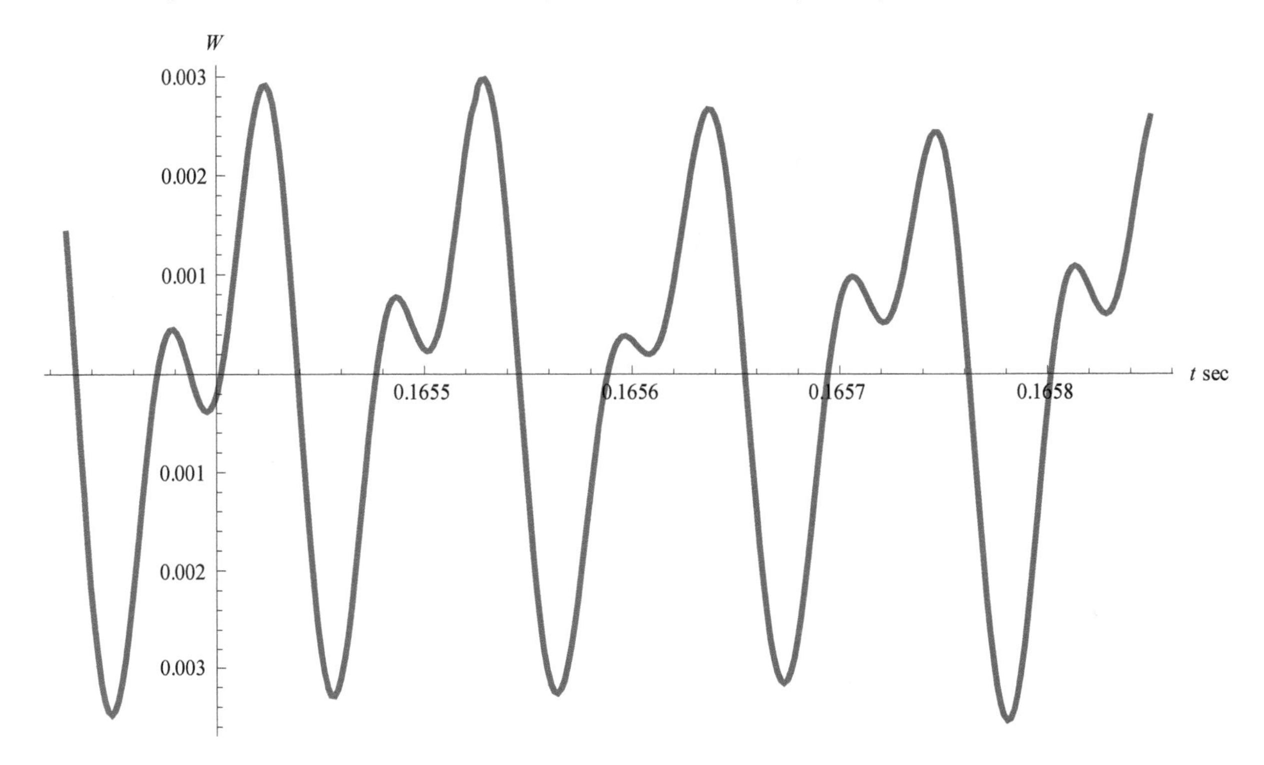


**Figure 3.** Density oscillations at the final stage of the process (no helicity).

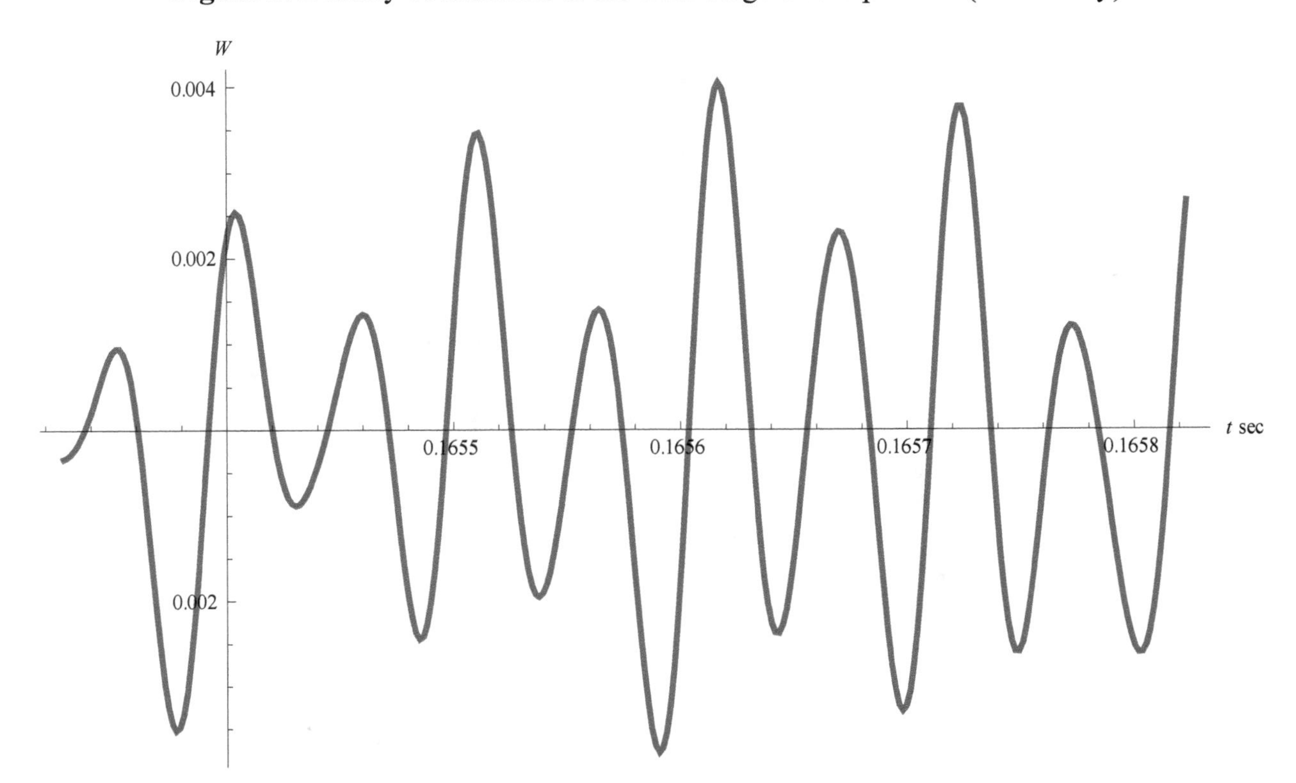


**Figure 4.** Density oscillations at the final stage of the process (helicity is present).

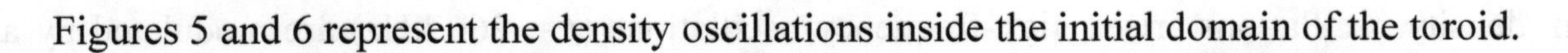

Figures 5 and 6 represent the density oscillations inside the initial domain of the toroid.

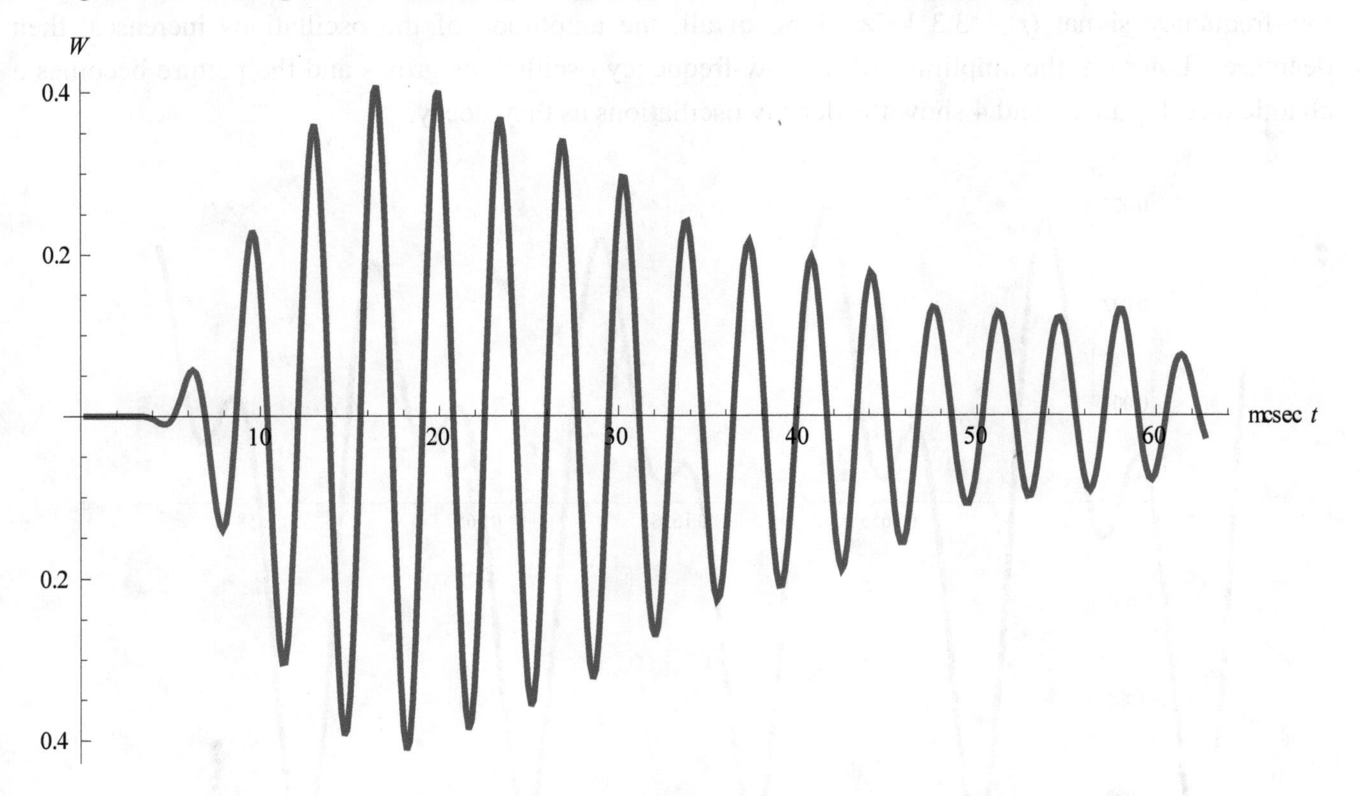


**Figure 5.** Density oscillations inside the initial vortex ring (no helicity).

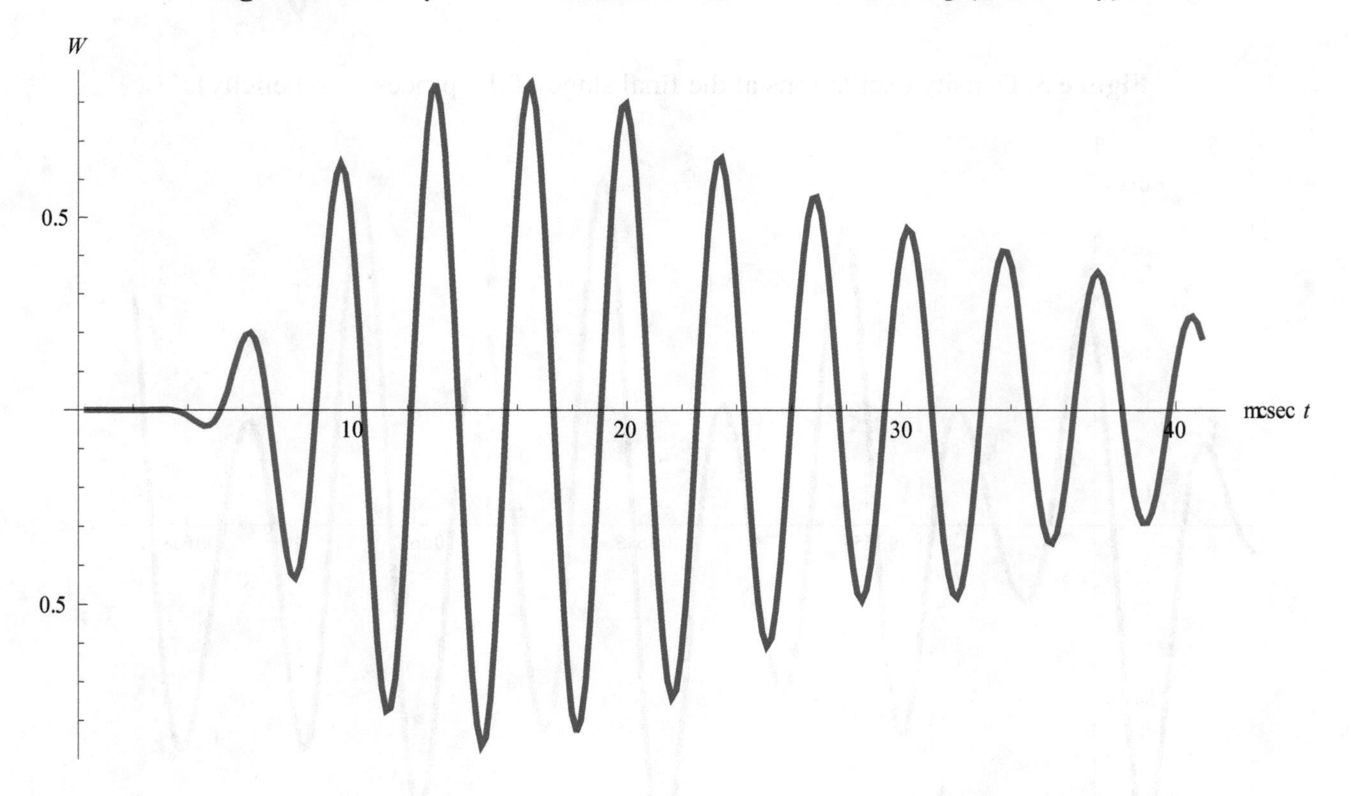


**Figure 6.** Density oscillations inside the initial vortex ring (helicity is present).

One can see that the high-frequency component of the oscillations does not depend on helicity. The presence of helicity has an effect only on the amplitude of the oscillations increasing it. The value of high frequency remains constant during the whole process excluding the final stage of the process (see Figures 3 and 4). The value of high frequency corresponds satisfactorily to the experimental data in [14]. In these experiments an unstable vortex ring appeared behind a shock wave reflected from a concave body. The radius of the ring was equal to 0.1 cm, the frequency of oscillations was 170–220 kHz.

## 4. Conclusions

A new method is set forth which allows us to investigate the frequency of acoustic radiations by 3D vortex rings. As shown, helicity has no effect on the frequency of the high-frequency component. The results may be of interest for aeroacoustics.

## Acknowledgments

The authors are grateful to Dr. O. A. Azarova for fruitful discussions.

## Author Contributions

Fedor V. Shugaev proposed the topic of the present investigation; Dmitri Y. Cherkasov performed the computations for the case of no helicity; Oxana A. Solenaya performed the computations for the case of present helicity.

## References

1. Howe, M.S. *Theory of Vortex Sound*. Cambridge University Press: Cambridge, UK, 2003.
2. Lighthill, M.J. On sound generated aerodynamically. *Proc. Roy. Soc. A* **1952**, *211*, 564–587.
3. Powell, A. Theory of vortex sound. *J. Acoust. Soc. Am.* **1964**, *36*, 177–195.
4. Kopiev, V.F.; Chernyshev, S.A. Vortex ring eigen oscillations as a source of sound. *J. Fluid Mech.* **1997**, *341*, 19–57.
5. Verzicco, R.; Iafrati, A.; Riccardi, G.; Fatica, M. Analysis of the sound generated by the pairing of two axisymmetric corotating vortex rings. *J. Sound Vibr.* **1997**, *200*, 347–358.
6. Inoue, O. Sound generation by the leapfrogging between two coaxial vortex rings. *Phys. Fluids* **2002**, *14*, 3361–3364.
7. Maxworthy, T.J. Turbulent vortex rings. *J. Fluid Mech.* **1974**, *64*, 227–240.
8. Maxworthy, T.J. Some experimental studies of vortex rings. *J. Fluid Mech.* **1977**, *81*, 465–495.
9. Liu, C.; Yan, Y.; Lu, P. Physics of turbulent generation and sustenance in a boundary layer. *Comput. Fluids* **2014**, *102*, 353–384.
10. Yan, Y.; Chen, C.; Fu, H.; Liu, C. DNS study on Λ-vortex and vortex ring formation in flow transition at Mach number 0.5. *J. Turbul.* **2014**, *15*, 1–21.

11. Morton, T.S. The velocity field within a ring with a large elliptic cross-section. *J. Fluid Mech.* **2004**, *503*, 247–271.
12. Truesdell, C. Precise theory of the absorption and dispersion of forced plane infinitesimal waves according to the Navier–Stokes equations. *J. Ration. Mech. Anal.* **1953**, *2*, 643–721.
13. Korobov, N.M. *Theoretical and Numerical Methods in Approximate Analysis*. Moscow State University: Moscow, Russian, 1963.
14. Shugaev, F.V.; Shtemenko, L.S. *Propagation and Reflection of Shock Waves*. World Scientific Publishing: Singapore; Hackensack, NJ, USA; London, UK; Hong Kong, China, 1998.

# 2

# A Hybrid Sender- and Receiver-Initiated Protocol Scheme in Underwater Acoustic Sensor Networks

**Jae-Won Lee [1] and Ho-Shin Cho [2,*]**

[1] Underwater Communication/Detection Research Center, Kyungpook National University, Daegu 702-701, Korea; E-Mail: jwlee@ee.knu.ac.kr

[2] College of IT Engineering, Kyungpook National University, Daegu 702-701, Korea

* Author to whom correspondence should be addressed; E-Mail: hscho@ee.knu.ac.kr

Academic Editor: Gerhard Lindner

---

**Abstract:** In this paper, we propose a method for sharing the handshakes of control packets among multiple nodes, which we call a hybrid sender- and receiver-initiated (HSR) protocol scheme. Handshake-sharing can be achieved by inviting neighbors to join the current handshake and by allowing them to send their data packets without requiring extra handshakes. Thus, HSR can reduce the signaling overhead involved in control packet exchanges during handshakes, as well as resolve the spatial unfairness problem between nodes. From an operational perspective, HSR resembles the well-known handshake-sharing scheme referred to as the medium access control (MAC) protocol using reverse opportunistic packet appending (ROPA). However, in ROPA the waiting time is not controllable for the receiver's neighbors and thus unexpected collisions may occur at the receiver due to hidden neighbors, whereas the proposed scheme allows all nodes to avoid hidden-node-induced collisions according to an elaborately calculated waiting time. Our computer simulations demonstrated that HSR outperforms ROPA with respect to both the throughput and delay by around 9.65% and 11.36%, respectively.

**Keywords:** handshake-based MAC protocol; handshake-sharing; hidden-node problem; long propagation delay; receiver-initiated protocol; sender-initiated protocol; sensor networks; spatial unfairness problem; underwater acoustics

---

## 1. Introduction

Underwater acoustic sensor networks (UWSNs) comprise variable numbers of sensor nodes, which are deployed to perform a variety range of applications, such as oceanic research, oil spill monitoring, submarine detection, offshore exploration, and assisted navigation [1,2]. To communicate between nodes, UWSNs typically employ acoustic signals, which have poor channel conditions compared with terrestrial radio signals. In particular, the speed of sound under water is about 1500 m/s, which is lower than the $3 \times 10^8$ m/s propagation speed of a radio signal, resulting in significantly longer propagation delays. In addition, the underwater acoustic channel is only capable of low data transmission rates due to its extremely limited bandwidth. Consequently, the medium access control (MAC) protocols designed for terrestrial radio channels with high data rates and negligible propagation delays may not work correctly in the underwater channel [3]. Therefore, it is essential to design a new MAC protocol to adapt to the poor underwater channel conditions.

Current research into underwater MAC protocols has focused mainly on handshake-based protocols, which are known to work properly during long-distance delivery via multi-hop relaying [4,5]. The multiple-access collision avoidance (MACA) protocol is a popular terrestrial handshake-based MAC protocol, which uses the request-to-send (RTS)/clear-to-send (CTS) handshake to reserve the shared channel [6]. In MACA, prior to data transmission the sender and receiver exchange RTS and CTS between the sender and receiver. Thus, any neighbors that overhear a control packet can delay their transmission to avoid possible collisions, which is called as a hidden-node problem. In order to apply MACA in an underwater environment, MACA for underwater (MACA-U) was proposed by revising the state transition rules to consider the long propagation delay [7]. However, the simple RTS/CTS exchange fails to fully address the hidden-node problem due to the long propagation delay in the underwater acoustic channel [8]. In addition, the performance of MACA-U is severely constrained by the long propagation delay because of the increased time required for control packet exchanges. Moreover, the long propagation delay also causes a spatial unfairness problem where the channel always becomes clear earlier at nodes closer to the receiver, and thus nodes located far from the receiver might never capture the channel.

Most conventional protocols have improved the channel utilization by sending multiple packets at once in a packet-train form [9–11]. As another approach for channel utilization improvement, a MAC protocol using reverse opportunistic packet appending (ROPA) has permitted multiple nodes to participate in a common handshake [12], which we name "handshake-sharing" in this paper. In general, the handshake-sharing approach is known to be more efficient than the packet-train methods, because the packet-train is only originated from a single sender. In ROPA, an initiating sender polls its neighbors to join the handshake using RTS, which may cause hidden-node collisions at the receiver-side. Such collisions may severely degrade the performance in terms of throughput and latency. Thus, as a solution for the hidden-node collisions, conventional protocols have inserted a specific amount of waiting time into the schedules of both the sender and receiver, but it may still cause additional latency [13–15]. In this paper, we propose a new handshake-sharing approach where a receiver polls its neighbors to join the handshake using CTS. Such a receiver-initiated polling not only improves channel utilization, but also addresses the hidden-node problem.

Based on these observations, we propose an underwater MAC protocol called a hybrid sender- and receiver-initiated (HSR) protocol scheme, which addresses the channel utilization issue and the hidden-node problem. First, to overcome the channel utilization problem, HSR permits multiple nodes to cut into an ongoing communication procedure that involves a handshake between a given sender and a receiver, thereby allowing data packets to be sent without additional handshakes. The method for sharing an ongoing handshaking is similar to that of ROPA. However, in contrast to ROPA, the receiver is the same for both the sender and neighbors in HSR, and thus the transmission schedules of all the nodes can be controlled by the same receiver to avoid the hidden-node problem. HSR combines both sender- and receiver-initiated approaches from the aspect of who initiates communication. The sender begins a handshake by transmitting an RTS to a receiver as part of the typical sender-initiated communication. On the other hand, the receiver invites neighbors to participate in the ongoing communication using three-way signaling via polling/request/grant as part of the receiver-initiated communication [11]. In addition, this participatory process gives neighbors the opportunity to transmit their data packets evenly, which also can address the spatial unfairness problem.

The remainder of this paper is organized as follows: we first review the relevant literature in Section 2. Section 3 presents the problem statements. In Section 4, we explain the proposed protocol design, including our system description. In Section 5, we present our simulation results and discuss them in detail. Finally, we give our conclusions in Section 6.

## 2. Related Work

Considerable research efforts have been made to overcome the effects of long propagation delay, such as low channel utilization, the hidden-node problem, and the spatial unfairness problem, which are also the main concerns of HSR. First, to improve channel utilization, a MACA-based MAC protocol with a packet-train to multiple neighbors (MACA-MN) was proposed wherein multiple packets could be sent to multiple neighbors in every handshake round [9]. Moreover, MACA-U with packet trains (MACA-UPT) was introduced [10] to enhance channel utilization by allowing a sender to transmit a train of data packets for each handshake. Another method for improving channel utilization was presented [8], which is called the adaptive propagation-delay-tolerant collision-avoidance protocol (APCAP). This protocol allows a transmitting node to perform other actions while waiting for the CTS to return, called MAC level pipelining. Nevertheless, APCAP has two constraints: the complexity of MAC level pipelining and time synchronization. Recently, a bidirectional concurrent MAC (BiC-MAC) protocol was proposed, where a sender-receiver pair can exchange bidirectional data transmissions in each handshake simultaneously, which results in channel utilization improvement [10]; however, a bidirectional data packet exchange only occurs when the receiver has data packets destined for the sender. If the receiver does not have any data packets to return, BiC-MAC behaves in a similar manner to MACA-UPT.

In a second attempt to overcome the hidden-node problem, so-called slotted floor acquisition multiple access (Slotted-FAMA) was proposed, which combines carrier sensing and RTS/CTS handshake mechanisms [13]. Slotted-FAMA exploits time slotting to prevent the hidden-node collisions by aligning all of the packet transmissions into slots, but which requires an excessive length of time slot resulting in a low throughput performance. Unlike Slotted-FAMA, the distance-aware collision avoidance protocol

(DACAP) [14] is a non-synchronized protocol that minimizes the duration of handshake by taking advantage of the receiver's tolerance to interference when the two nodes are closer than the maximum transmission range. Moreover, DACAP waits for a certain amount of time before data transmission to avoid possible hidden-node collisions. This waiting time is determined based on a trade-off between throughput and collision avoidance.

Finally, to address the spatial unfairness problem, the spatially fair MAC (SF-MAC) protocol was proposed, which also considers the hidden-node problem [15]. To avoid the hidden-node collisions, SF-MAC delays the CTS transmission after receiving the RTS. In addition, the receiver captures the RTSs of all contenders and determines the earliest RTS transmitter, thereby achieving a fair transmission. However, SF-MAC has a long and fixed RTS collection period and the receiver can only obtain a data packet from one of the potential senders, which severely degrades channel utilization. An efficient handshaking mechanism (EHM) was also proposed to solve the spatial unfairness problem [16]. In EHM, a receiver delays its reply to the RTS that arrives first for a specific amount of time. During this period, if any other RTS that departed earlier than the first-arrival RTS should arrive late due to the long distance, then the receiver also takes the late-arriving RTS into consideration when creating CTS so the long-distance node has an equal transmission opportunity. However, EHM does not consider the control packet collisions caused by hidden nodes, which may significantly decrease the overall network throughput.

## 3. Problem Statements

### *3.1. Hidden-Node Problem in UWSNs*

Conventional handshake-based protocols attempt to reserve the channel by exchanging RTS/CTS control packets, which are probably overheard by neighbors. The neighbors are then aware that the channel will be reserved and they remain in the sleep mode by stopping any transmissions until the occupied channel is released. In this manner, any possible collisions caused by neighboring hidden nodes may be avoided.

However, in the underwater acoustic channel, a new type of hidden-node problem is introduced due to the long propagation delay. In Figure 1, nodes A and D, *i.e.*, the neighbors in this example, may detect a channel reservation too late by overhearing the RTS or CTS after completing the transmission of their control packets, P1 and P2. Thus, the early departure of packets without recognizing channel reservation may cause collisions at the sender (node B) and receiver (node C), as shown by the solid arrows. In this scenario, nodes A and D are hidden from nodes B and C. Therefore, unlike a terrestrial radio channel where a hidden node is located beyond the signal's coverage so its existence is not recognized, a hidden node in an underwater acoustic channel may also occur due to the long propagation delay even when it is located within the region covered.

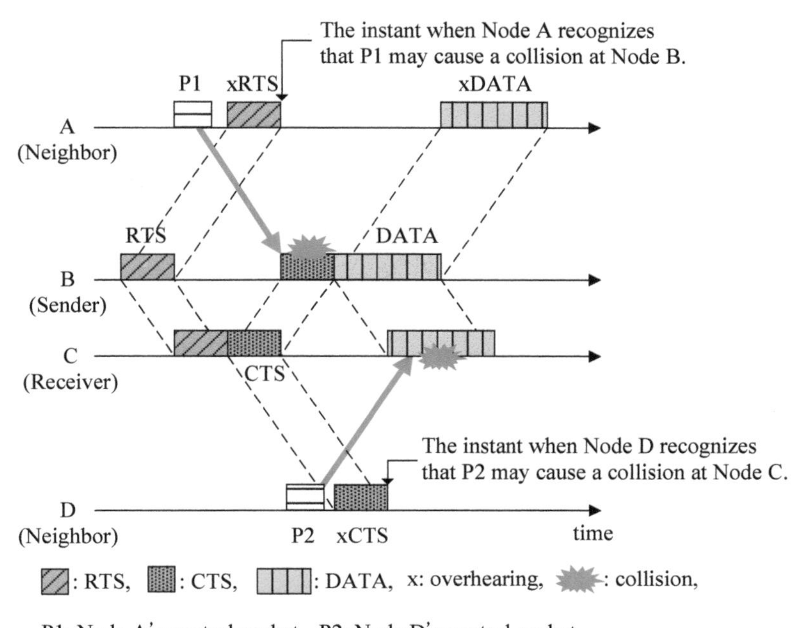


**Figure 1.** Illustration of the hidden-node problem in UWSNs.

### *3.2. Spatial Unfairness Problem in UWSNs*

The "first come, first served" method has been used as the most reasonable approach to ensure fairness among communicating nodes that share a medium, where the requests for the medium accessed by nodes are processed in the order of their arrival. However, due to the long propagation delay in the underwater acoustic channel, a request for channel reservation occurs earlier, but it may arrive later at a long-distance node compared with those nearby so it cannot be served first. This is called the spatial unfairness problem.

In Figure 2, nodes A and B transmit a request packet to node C to reserve the channel. The starting time of node A $t_A$ is before $t_B$ for node B, but the request from node A arrives later than that from node B because node A is located farther than node B from node C. Furthermore, provided that node B has data to transmit, it can preoccupy the channel every time in advance of node A. This causes a severe unfairness problem.

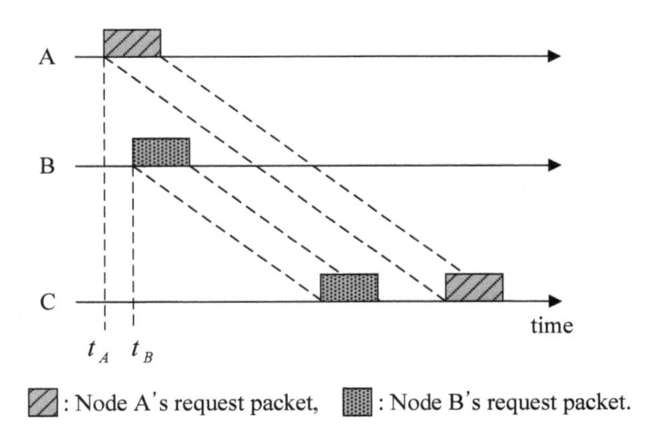


**Figure 2.** Illustration of the spatial unfairness problem in UWSNs.

## 4. Proposed HSR Protocol

### *4.1. System Description*

In HSR, handshake-sharing can be achieved using multi-way polling/request/grant signaling between the receiver and its neighbors. A CTS plays an additional role in polling whether the neighbors have any data packets destined for the receiver. After receiving the CTS, the neighbors who want to participate in the current handshake (defined as "participants") notify their participation by transmitting a control packet called a request-to-participate (RTP). After collecting RTPs, the receiver grants participation by broadcasting a control packet called a clear-to-participate (CTP), which contains the transmission schedules of the sender and participants to avoid collisions at the receiver. We refer to the sender and participants who are granted participation by the receiver as "granted participants".

#### 4.1.1. Assumptions

A multi-hop acoustic network is considered where all of the nodes reach their respective destinations in a multi-hop manner using an omnidirectional and half-duplex underwater acoustic modem. We assume that each node acquires the inter-nodal propagation delay for one-hop distance neighbors through the network initialization stage by measuring the round-trip time or by exchanging some information with its neighboring nodes [17]. The nodes also maintain a list of neighbors to allow the establishment of a bi-directional communication link if necessary. A routing table is maintained in the nodes to facilitate multi-hop relay.

#### 4.1.2. Time Duration Parameters in HSR

In HSR, three parameters related to the time duration are defined, *i.e.*, Delay, Silence, and Waiting. First, Delay (denoted by $D^X(i)$) is an interval during which a node $i$ delays packet $X$'s transmission to avoid collisions. Silence is an interval during which nodes who overhear the control packets associated with channel reservation by others remain quiet and do not communicate. To express the length of the Silence, another time duration called Waiting is defined as an interval between the transmission of a control packet and the reception of the corresponding response, which could be a control packet or a data packet.

Figure 3 shows an example of a RTS/CTS handshaking procedure to illustrate how to obtain the Waiting and Silence. CTS is the response to RTS, so the Waiting of node $i$ for response to RTS is given by:

$$W^{RTS}(i) = 2\tau_{i,j} + T_{control} \tag{1}$$

where $\tau_{i,j}$ is the propagation delay between nodes $i$ and $j$, and $T_{control}$ is the transmission time for the control packet, which is the same for all control packets. We denote $W^X(i)$ as the Waiting of node $i$ to obtain a response to packet $X$. Similarly, since the data from node $i$ correspond to the response to CTS of node $j$, then the Waiting of node $j$ for response to CTS is given by:

$$W^{CTS}(j) = 2\tau_{i,j} + T_{DATA} \tag{2}$$

where $T_{DATA}$ is the transmission time of a data packet. The Waiting is specified in the associated control packets and it is overheard by neighbors so they can estimate how long they should remain in Silence in order to avoid collisions. In Figure 3, the values of $W^{RTS}(i)$ and $W^{CTS}(j)$ are specified in the RTS and CTS, respectively. After overhearing the CTS, the neighbor (node $N$) calculates the Silence by:

$$S(N) = W^{CTS}(j) - 2\tau_{j,N} \quad (3)$$

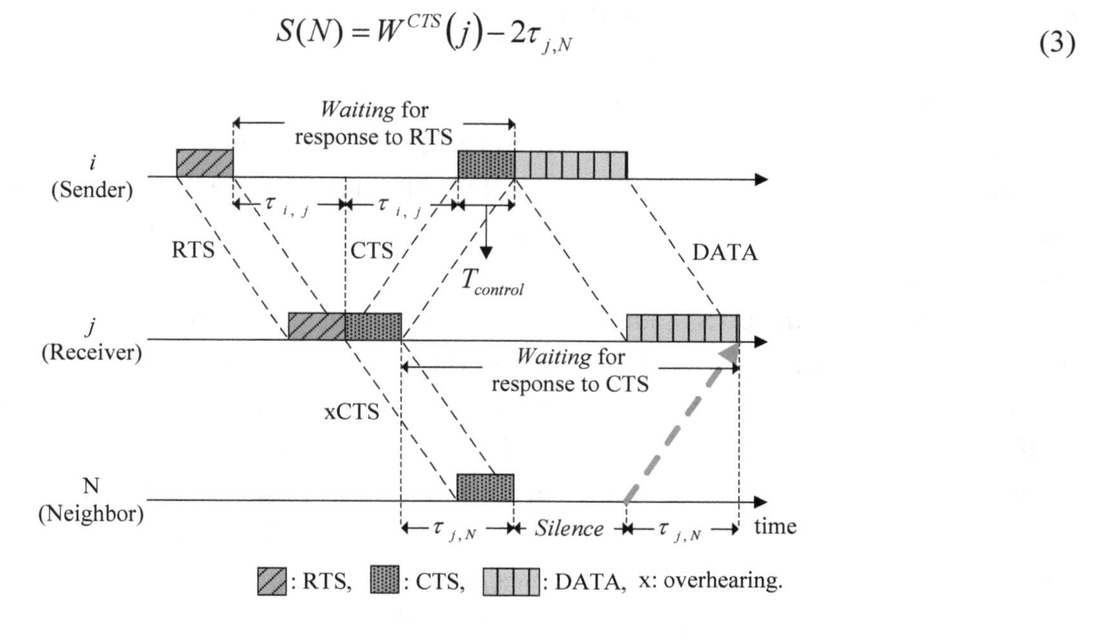


**Figure 3.** Illustration of the Waiting and Silence.

Similarly, all of the neighbors can calculate how long they must maintain the Silence. If a node in Silence overhears another control packet of neighbor, it checks and extends the Silence if necessary.

## *4.2. Fundamental Operation and Features*

Figure 4 illustrates the operation of the HSR protocol. Nodes S and R denote a sender and a receiver, respectively. It is assumed that nodes P1, P2, and P3 are neighbors of node R within one-hop distance. Node S starts the handshake by transmitting RTS to node R. Node R specifies the following information in the RTS: (1) the address of the destination; (2) the batch size of node S, *i.e.*, the number of data packets to be transmitted, $B_{size,S}$; and (3) the Waiting of node S for response to RTS, $W^{RTS}(S)$.

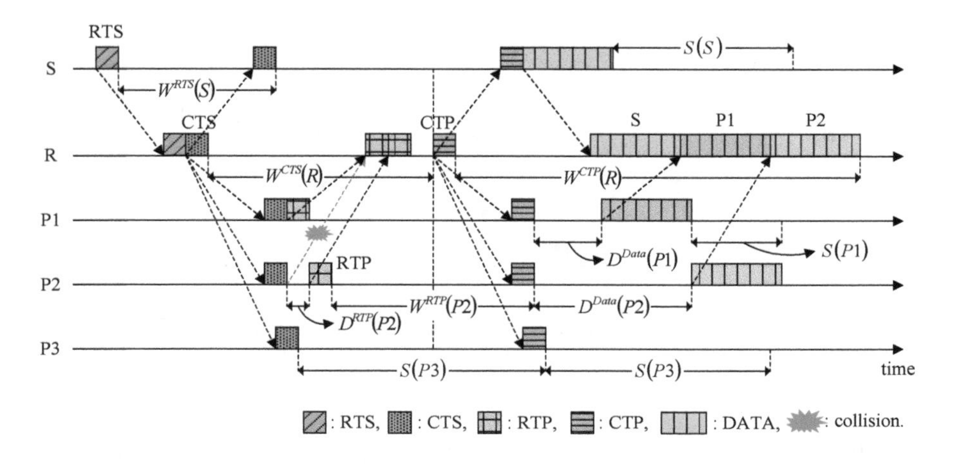


**Figure 4.** Operation of the HSR protocol.

After receiving the RTS, node R broadcasts a CTS, which is used to respond to the RTS but also to invite the neighbors (e.g., nodes P1, P2, and P3) to join the current handshake to send their data packets to the receiver R if they have any. After receiving the CTS, a neighbor with data packets destined for node R participates in the current handshake by transmitting an RTP to node R, and it then waits for a response. As shown in Figure 4, the RTPs from different participants might collide with each other, so node R computes the Delay for all of the participants' RTP transmissions to ensure that collisions do not occur and it specifies these values in the CTS. Thus, the CTS contains two parameters: $W^{CTS}(R)$ and $D^{RTP}(i)$. The methods for calculating $W^{X}(i)$ and $D^{X}(i)$ are explained in Section 4.3. The RTP of participant *i* includes the same information as RTS: (1) address of the destination; (2) $B_{size,i}$; (3) $W^{RTP}(i)$. On the other hand, a neighbor with no data packets destined for node R remains silent during the Silence (as in the case of node P3).

After transmitting the CTS, node R collects RTPs during the Waiting, $W^{CTS}(R)$, and then broadcasts the CTP. In this case, $W^{CTS}(R)$ may induce slightly additional handshake negotiation times for a low offered load condition, but its contribution to the performance, such as collecting the neighbors' notifications of participation and avoiding the collisions caused by hidden nodes, are high enough to ignore the demerit. As shown in Figure 5a, node N is hidden from node S and it may transmit a control packet (e.g., RTS) before overhearing CTS, which causes a collision at node R, as shown by the solid arrow. However, in HSR, $W^{CTS}(R)$ continues until the arrival of all of the control packets that departed from the neighbors earlier than the CTS was overheard, as shown by the dotted arrow in Figure 5b. This method allows the HSR to avoid possible collisions caused by hidden nodes.

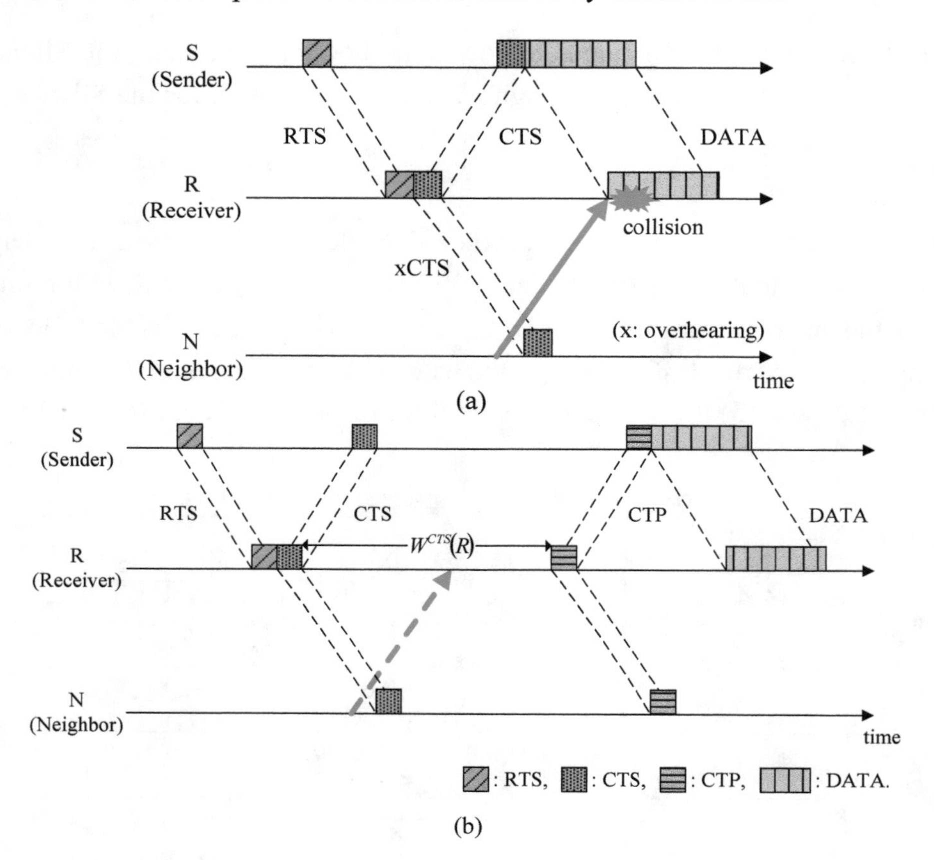


**Figure 5.** Solution to the hidden-node problem: (**a**) Conventional RTS/CTS handshaking protocol; and (**b**) HSR protocol.

In addition, the HSR addresses the spatial unfairness problem caused by different inter-node distances and the long propagation delay, where it provides an additional opportunity to participate in the ongoing communication. For example, as shown in Figure 6, nodes S and P transmit their RTSs to node R. Because it is closer to node R, node S captures the channel even if it sends the RTS later than node P, thereby leading to the spatial unfairness problem. However, in HSR, node P still has a chance to share the channel by transmitting the RTP, as shown in Figure 6. Thus, all of the neighbors of the receiver have the same opportunity to capture the channel, regardless of their locations.

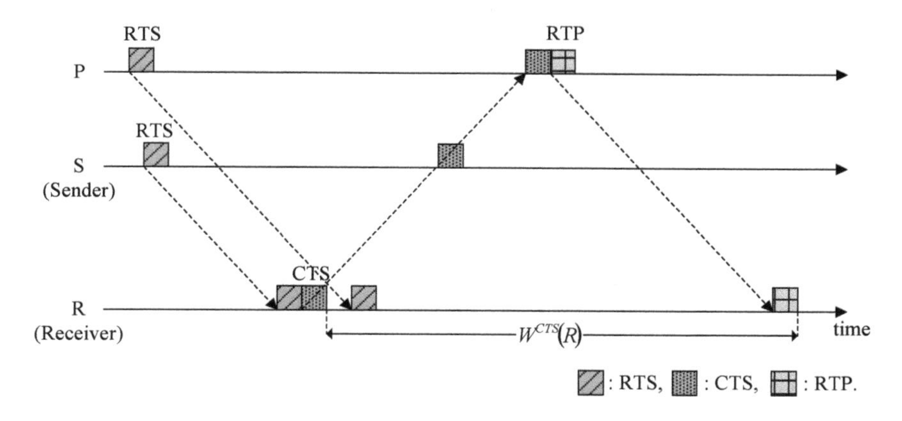


**Figure 6.** Solution to the spatial unfairness problem.

Back to Figure 4, after collecting all of the RTPs, node R calculates the data transmission times for the granted participants so the data packets arrive at the receiver sequentially in a similar manner to a packet-train [9]. The data transmission time for node $i$ is translated into $D^{Data}(i)$ and sent with $W^{CTP}(R)$, where it is included in the CTS. After transmitting the CTP, node R waits for the data packets from the granted participants.

### *4.3. Calculating the Time Duration Parameters*

#### 4.3.1. Delay for the RTP and Data Transmission

In HSR, two types of collisions can occur at a receiver: (1) RTP collisions and (2) data packet collisions. RTP collisions occur when more than two participants at similar distances from the receiver send an RTP immediately after receiving the CTS. Thus, in order to avoid RTP collisions, it is necessary to force all of the participants to delay their RTP transmissions by a specific amount of time, which is denoted by $D^{RTP}$.

Figure 7 shows an example of how to calculate the Delay for RTP transmission, where nodes $(k-1)$, $k$, and $(k+1)$ are participants, and their RTPs are assumed to arrive in the order named. The condition that node $k$'s RTP is scheduled to arrive after node $(k-1)$'s RTP is given by:

$$2\tau_{R,k-1} + T_{control} + D^{RTP}(k-1) \leq 2\tau_{R,k} + D^{RTP}(k) \tag{4}$$

Thus, the Delay for node $k$'s RTP transmission is chosen as the minimum value that satisfies the condition of Equation (4), as follows:

$$D^{RTP}(k) = D^{RTP}(k-1) + T_{control} - 2(\tau_{R,k} - \tau_{R,k-1}) \tag{5}$$

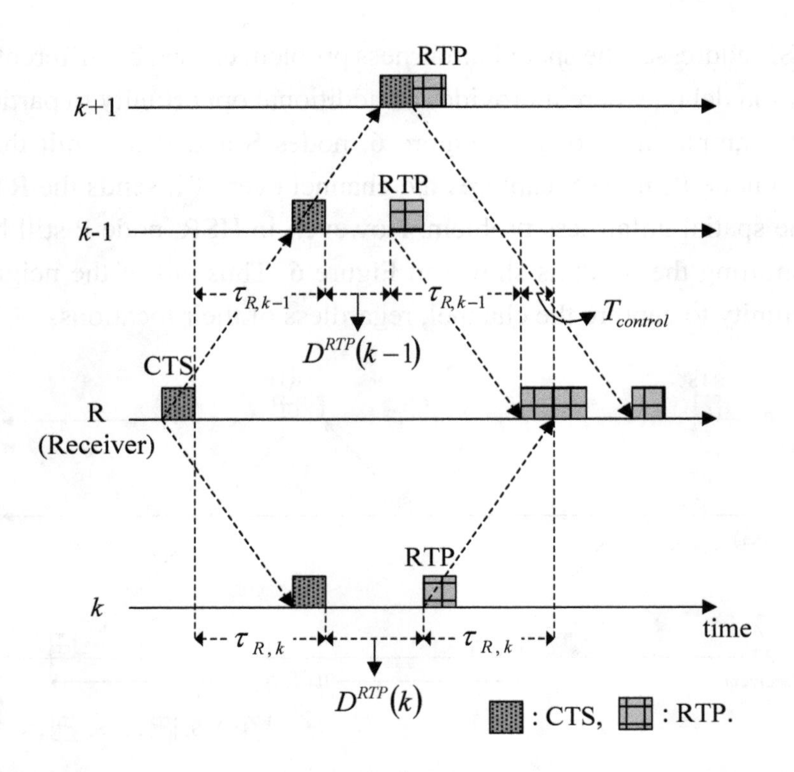


**Figure 7.** Determination of the Delay for RTP transmission.

However, in the case where a Delay is not required, such as node ($k$ + 1), Equation (5) yields a negative value. Thus, the general expression for the Delay for RTP transmission is given by:

$$D^{RTP}(i) = \max\left(0,\ D^{RTP}(i-1) + T_{control} - 2\left(\tau_{R,i} - \tau_{R,i-1}\right)\right) \tag{6}$$

$D^{RTP}(i)$ is calculated for all of the participants and written in the CTS by the receiver, and the participants then adjust the RTP transmission time accordingly.

Data packet collision is also avoided in a similar manner to RTP collision. After receiving the RTPs, the receiver then calculates the Delay for data transmission by the granted participants so the data packets arrive sequentially at the receiver in a packet-train form, as shown in Figure 8, where it is assumed that the data packets of node $k$ arrive after those of node ($k$ − *1*). For this packet-train arrival process, the following relationship should be satisfied:

$$2\tau_{R,k-1} + B_{size,k-1} \cdot T_{DATA} + D^{Data}(k-1) \leq 2\tau_{R,k} + D^{Data}(k) \tag{7}$$

Thus, from Equation (7), the Delay for node $k$'s data transmission is chosen as:

$$D^{Data}(k) = D^{Data}(k-1) + B_{size,k-1} \cdot T_{DATA} - 2\left(\tau_{R,k} - \tau_{R,k-1}\right) \tag{8}$$

and the general expression for granted participant $i$ is:

$$D^{Data}(i) = \max\left(0,\ D^{Data}(i-1) + B_{size,i-1} \cdot T_{DATA} - 2\left(\tau_{R,i} - \tau_{R,i-1}\right)\right) \tag{9}$$

$D^{Data}(i)$ is carried by the CTP and the granted participants then determine their data transmission times accordingly.

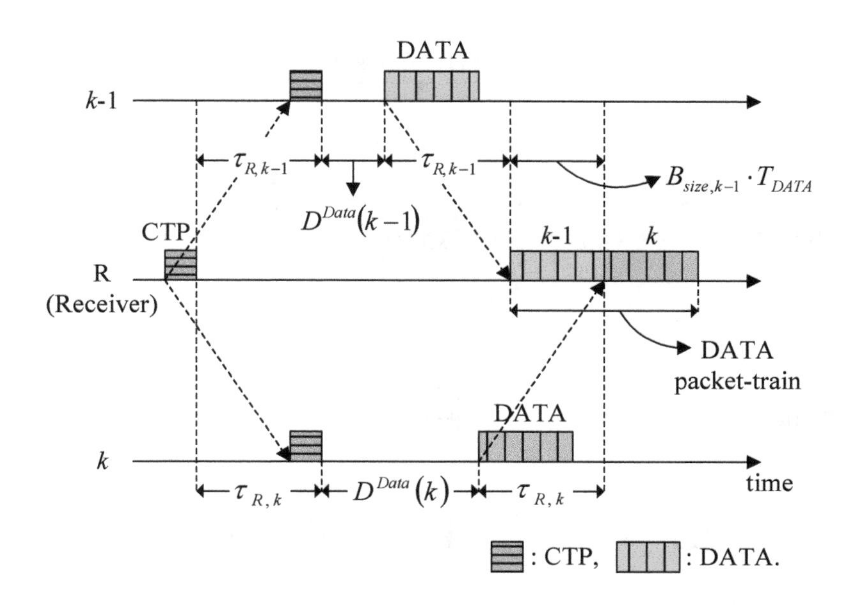


**Figure 8.** Determination of the Delay for data transmission.

4.3.2. Waiting for Responses to RTS, CTS, RTP, and CTP

The Waiting has a different value according to the type of control packet (*i.e.*, RTS, CTS, RTP, and CTP). The Waiting for response to RTS is given in Equation (1). As shown in Figure 9a, the Waiting for response to CTS (corresponding to ① in Figure 9a is given by:

$$W^{CTS}(R) = 2\tau_{R,L} + D^{RTP}(L) + T_{control} \tag{10}$$

where node $L$ denotes the participant farthest from node R (receiver). The Waiting for response to CTP (corresponding to ② in Figure 9a, during which the data packet receptions are all completed as a response to CTP, is obtained by:

$$W^{CTP}(R) = 2\tau_{R,L} + D^{Data}(L) + B_{Size,L} \cdot T_{DATA} \tag{11}$$

where node $L$ is the granted participant most distant from node R.

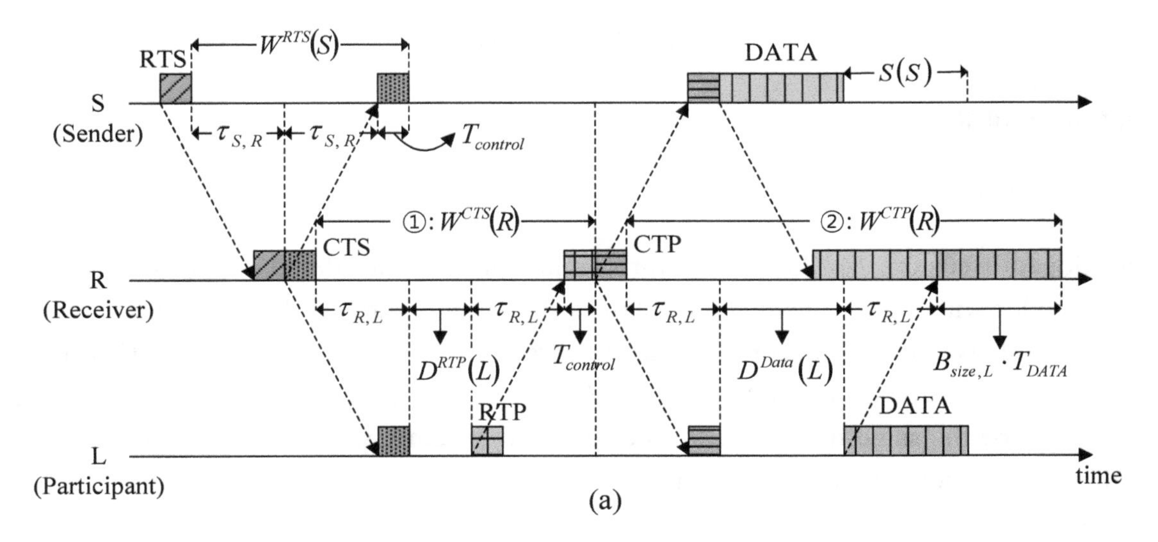


**Figure 9.** *Cont.*

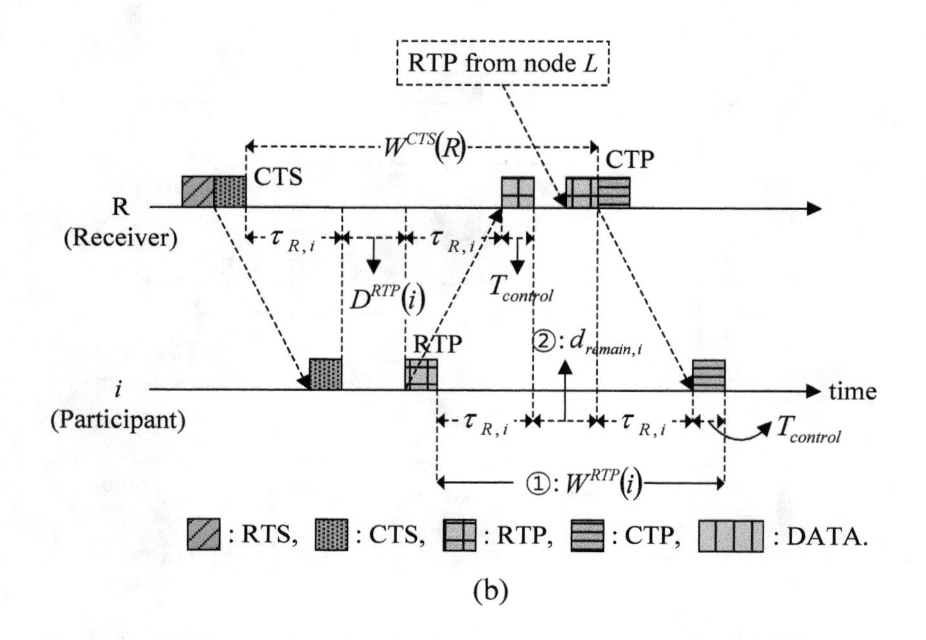


**Figure 9.** Determination of the Waiting for responses: (**a**) Associated with CTS and CTP; and (**b**) Associated with RTP.

The Waiting of participant $i$ for response to RTP (corresponding to ① in Figure 9b, during which CTP reception is completed as a response to RTP, is given by:

$$W^{RTP}(i) = 2\tau_{R,i} + T_{control} + d_{remain,i} \tag{12}$$

where $d_{remain,i}$ is the residual Waiting of node R for response to CTS at the time when participant $i$'s RTP is received (corresponding to ② in Figure 9b. Utilizing the values of $W^{CTS}(R)$ and $D^{RTP}(i)$, which are specified in CTS, participant $i$ calculates $d_{remain,i}$ as:

$$d_{remain,i} = W^{CTS}(R) - \left(2\tau_{R,i} + D^{RTP}(i) + T_{control}\right) \tag{13}$$

After substituting Equation (13) for Equation (12), the Waiting of participant $i$ for response to RTP is given by:

$$W^{RTP}(i) = W^{CTS}(R) - D^{RTP}(i) \tag{14}$$

## 5. Simulations and Results

### *5.1. Simulation Model*

We developed a custom MATLAB network simulator and considered a multi-hop topology that has 36 static-nodes with a grid spacing of 1000 m as shown in Figure 10. To reflect more real situation, the location of node is randomly generated to deviate from a grid intersection point by 10% at maximum of the grid spacing in both vertical and horizontal directions. The transmission power of all of the nodes were assumed to be the same to cover 1.5 times the grid spacing, so the coverage of a node included eight neighbors, as denoted by the dotted circle in Figure 10.

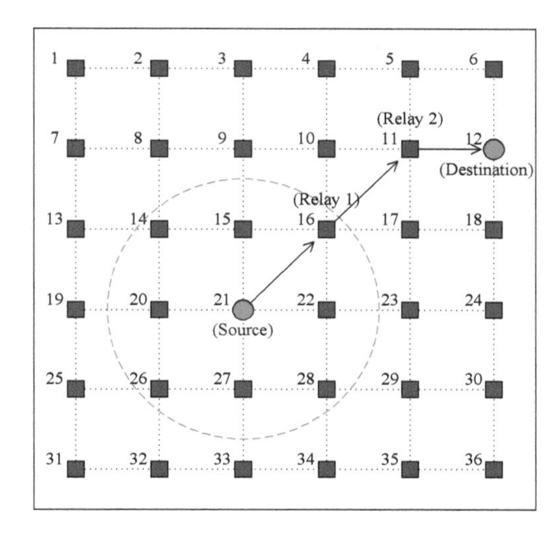


**Figure 10.** Network topology used in the simulations.

Data packet generation for each node followed a Poisson distribution with parameter $\lambda_{node}$ (packets/s) and the destination of each packet was selected randomly with equal probability. In the case of multi-hop transmission, a static routing scheme was employed where every node used a preconfigured routing table [12]. We ignored any channel related packet losses due to noise and multi-path, which only occurred due to packet collisions. For the RTS trials, HSR employed the binary exponential back-off algorithm specified in the IEEE 802.11 standard [18]. Table 1 provides the system parameters and the values used in simulations. The transmission rate was referred to the LinkQuest acoustic modem [19].

**Table 1.** System parameters and values.

| Parameter | Value |
|---|---|
| Transmission Rate | 9600 bps |
| Size of Data Packet | 1200 bits |
| Size of Control Packet | 120 bits |
| Minimum Back-Off Counter | 1 |
| Maximum Back-Off Counter | 64 |
| Capacity of Buffer | 300 packets |
| Speed of Acoustic Wave | 1500 m/s |

*5.2. Simulation Results*

We compared the HSR protocol with conventional handshake-based protocols, *i.e.*, MACA-U [7], MACA-UPT [10], and ROPA [12], with respect to the normalized throughput per node and end-to-end packet delay. The normalized throughput per node is defined by:

$$\gamma = \frac{1}{N} \cdot \frac{\sum_{i=1}^{N} r_i \cdot B_{Data}}{t_{sim}} \tag{15}$$

where $r_i$ is the number of successfully received data packets at node $i$ as a destination; $N$ is the total number of nodes in the network; $B_{Data}$ is a data packet size in bits; and $t_{sim}$ is the total simulation duration. As another performance evaluation metric, the end-to-end packet delay is used, defined as the time spent from initial data generation at a source to successful reception at a destination. Let $\Omega$ be the set of successfully received data packets at a destination. The length of $\Omega$ is given by $N(\Omega)=\sum_{i=1}^{N} r_i$, and each element of $\Omega$ has a different end-to-end packet delay, $t_{delay,j}$, $j=1,2,\cdots,N(\Omega)$. Therefore, the average value of end-to-end packet delay is defined by:

$$\overline{t_{delay}} = \frac{\sum_{j=1}^{N(\Omega)} t_{delay,j}}{N(\Omega)} \tag{16}$$

In terms of the channel occupancy priority, the three priority strategies of foreign-first, dominant-first, and oldest-first described in [20] were applied in order. Foreign-first gives priority to the relayed packets over newly generated packets. Dominant-first gives priority to a larger group of packets, which are commonly destined for a specific node. Oldest-first gives priority to the packets that have traveled a larger number of hops up to the current instant.

5.2.1. Throughput and Delay Analysis

Figure 11 shows the normalized throughput per node and the average end-to-end packet delay performance for various offered loads per node ($\lambda_{node}$). For simplicity, hereafter the terms "normalized throughput per node", "average end-to-end delay", and "offered load per node" are referred to as "throughput", "delay", and "offered load", respectively.

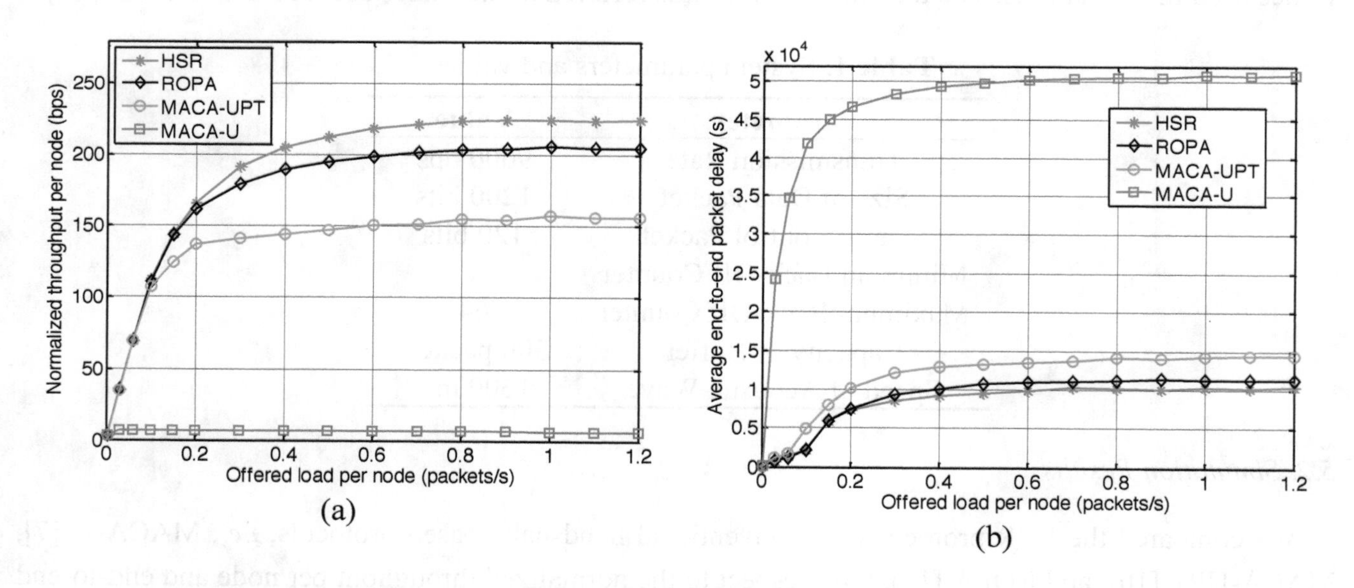


**Figure 11.** Performance comparison between HSR and other underwater MAC protocols: (**a**) Normalized throughput per node; and (**b**) Average end-to-end packet delay.

The results showed that MAC protocols with packet-train approach, such as HSR, ROPA, and MACA-UPT, significantly outperformed the non-packet-train MACA-U protocol in terms of both the throughput and delay, while achieving a stable saturation throughput and delay at high offered load ranges. This is because individual handshake for every single data packet is a very inefficient method, especially in conditions with a long propagation delay such as an underwater channel. Allowing multiple nodes to share an ongoing handshake, HSR and ROPA performed better than MACA-UPT where handshake only involved the sender and receiver.

In the saturation region, a closer analysis of the results for HSR and ROPA showed that HSR performed better in terms of both throughput and delay by around 9.65% and 11.36%, respectively because HSR additionally addresses the hidden-node problem, as mentioned earlier. Figure 12 demonstrates that the number of data packet collisions was lower for HSR compared with that for ROPA and the difference increased with the offered load. The features of the aforementioned MAC protocols are summarized in Table 2.

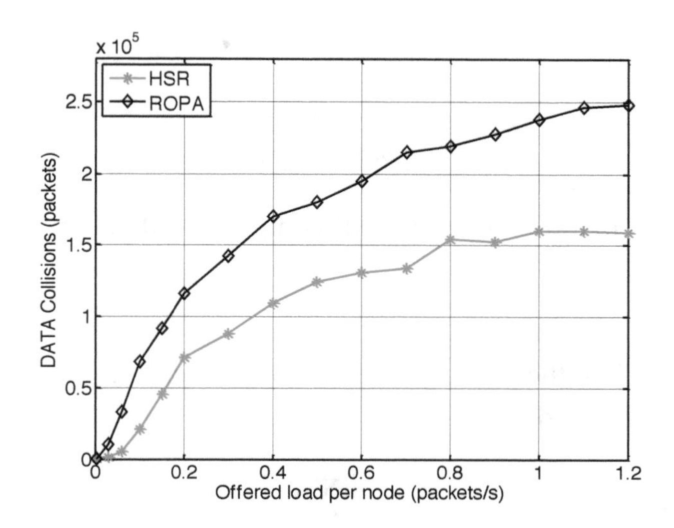


**Figure 12.** Number of data packet collisions with HSR and ROPA.

**Table 2.** Comparison of the features of MAC protocols.

| MAC Protocol \ Feature | Handshake-Sharing | Packet-Train Method | Solving the Hidden-Node Problem |
|---|---|---|---|
| HSR | O | O | O |
| ROPA | O | O | X |
| MACA-UPT | X | O | X |
| MACA-U | X | X | X |

O: used, X: not used.

Figure 13 shows the system throughput for various offered loads, which is defined by:

$$S = \frac{\sum_{i=1}^{N} s_i \cdot B_{Data}}{t_{sim}} \tag{17}$$

where $s_i$ is the total number of successfully received data packets at node $i$, regardless of whether they are relayed or destined. Alternatively, the system throughput can be interpreted as the single-hop performance of the MAC protocol. Thus, with similar reason as the case of throughput, HSR outperformed the other method in terms of the system throughput, which demonstrates that the features of HSR mentioned in Table 2 are also useful in single-hop networks.

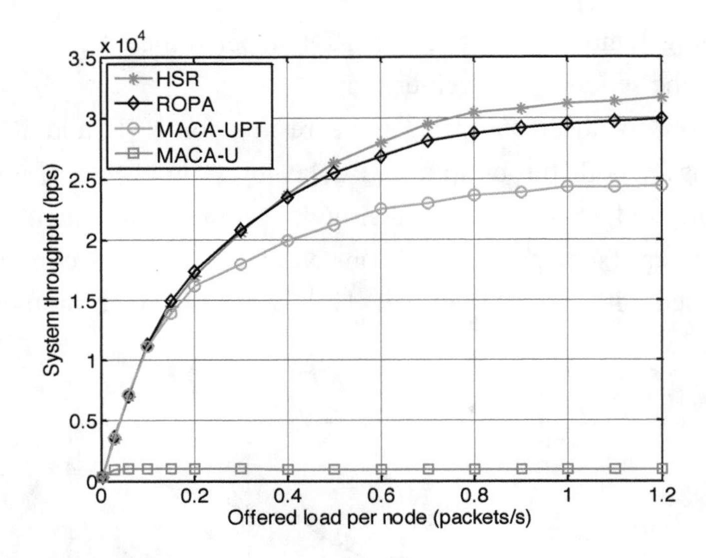


**Figure 13.** System throughput for HSR compared with other MAC protocols.

5.2.2. Analysis of Spatial Fairness

To obtain further insights into the performance of HSR, the spatial fairness is analyzed using Jain's fairness index (FI), which is defined by [21]:

$$FI = \frac{\left(\sum_{i=1}^{N} x_i\right)^2}{N \cdot \sum_{i=1}^{N} x_i^2} \tag{18}$$

where $x_i$ is the throughput of node $i$. The fairness index ranges from 1/N (worst case) to 1 (best case), and the maximum value can be obtained when all of the nodes have the same level of throughput.

Figure 14 shows that the fairness index was highest for HSR, which is because HSR gives all the neighbors of a receiver an equal opportunity to capture the channel regardless of their locations. ROPA also employs a handshake-sharing approach, but HSR offers a solution to the hidden-node problem and this helps to improve the fairness.

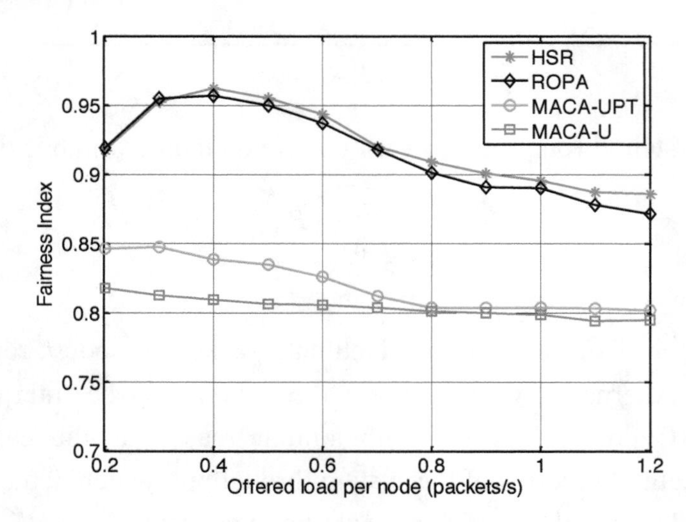


**Figure 14.** Fairness index for HSR compared with other MAC protocols.

## 6. Conclusions

In this paper, we have considered the challenges posed by the long propagation delay in underwater acoustic channels when designing an underwater MAC protocol. Based on these considerations, we have proposed an underwater MAC protocol called the HSR, which employs a packet-train method to improve channel utilization as well as allowing multiple nodes to share an ongoing handshake between a sender and a receiver. This approach also reduces the latency, which is expected to be very high in long propagation delay environments such as those in underwater channels. In addition, HSR addresses the hidden-node problem, which occurs in a different manner compared with terrestrial networks, and this is the main cause of performance degradation in the underwater handshake-based MAC protocol. Addressing the hidden-node problem also improves the fairness between nodes. The results of computer simulations showed that HSR surpasses other popular underwater MAC protocols based on handshake or packet-train methods, such as MACA-U, MUAC-UPT, and ROPA, with respect to both the throughput and delay.

## Acknowledgments

This work was supported by Defense Acquisition Program Administration and Agency for Defense Development under the contract UD130007DD.

## Author Contributions

J.-W.L. developed the HSR protocol and a custom MATLAB network simulator for simulations, and wrote the manuscript. H.-S.C. supervised the research and participated in the data analysis and revision processes.

## References

1. Akyildiz, I.F.; Pompili, D.; Melodia, T. Underwater Acoustic sensor networks: Research challenges. *Ad Hoc Netw.* **2005**, *3*, 257–279.
2. Zhang, B.; Sukhatme, G.S.; Requicha, A.G. Adaptive sampling for marine microorganism monitoring. In Proceedings of the IEEE/RSJ International Conference on Intelligent Robots and System (IROS 2004), Sendai, Japan, 28 September–2 October 2004; pp. 1115–1122.
3. Chen, K.; Ma, M.; Cheng, E.; Yuan, F.; Su, W. A survey on mac protocols for underwater wireless sensor networks. *IEEE Commun. Surv. Tutor.* **2014**, *16*, 1433–1447.
4. Partan, J.; Kurose, J.; Levine, B.N. A survey of practical issues in underwater networks. In Proceedings of the 1st ACM International Workshop on Underwater Networks (WUWNet), Los Angeles, CA, USA, 25 September 2006; pp. 17–24.
5. Chitre, M.; Shahabudeen, S.; Stojanovic, M. Underwater acoustic communication and networks: Recent advances and future challenges. *Mar. Technol. Soc. J.* **2008**, *42*, 103–116.

6. Karn, P. MACA: A new channel access method for packet radio. In Proceedings of the 9th Computer Networking Conference, London, ON, Canada, 22 September 1990; pp. 134–140.
7. Ng, H.H.; Soh, W.S.; Montani, M. MACA-U: A media access protocol for underwater acoustic networks. In Proceedings of the IEEE Global Telecommunications Conference 2008, New Orleans, LO, USA, 30 November–4 December 2008.
8. Guo, X.; Frater, M.R.; Ryan, M.J. Design of a propagation-delay-tolerant mac protocol for underwater acoustic sensor networks. *IEEE J. Ocean. Eng.* **2009**, *34*, 170–180.
9. Chirdchoo, N.; Soh, W.S.; Chua, K.C. MACA-MN: A MACA-based MAC protocol for underwater acoustic networks with packet train for multiple neighbors. In Proceedings of the IEEE Vehicular Technology Conference, Singapore, 11–14 May 2008; pp. 46–50.
10. Ng, H.H.; Soh, W.S.; Montani, M. BiC-MAC: Bidirectional-concurrent mac protocol with packet bursting for underwater acoustic networks. In Proceedings of the MTS/IEEE OCEANS 2010, Seattle, WA, USA, 20–23 September 2010.
11. Chirdchoo, N.; Soh, W.S.; Chua, K. RIPT: A receiver-initiated reservation-based protocol for underwater acoustic networks. *IEEE J. Sel. Areas Commun.* **2008**, *26*, 1744–1753.
12. Ng, H.H.; Soh, W.S.; Montani, M. An Underwater Acoustic MAC protocol using reverse opportunistic packet appending. *Comput. Netw.* **2013**, *57*, 2733–2751.
13. Molins, M.; Stojanovic, M. Slotted FAMA: A MAC protocol for underwater acoustic networks. In Proceedings of the IEEE OCEANS'06 Asia Pacific, Singapore, 16–19 May 2006; pp. 16–19.
14. Peleato, B.; Stojanovic, M. Distance aware collision avoidance protocol for ad-hoc underwater acoustic sensor networks. *IEEE Commun. Lett.* **2007**, *11*, 1025–1027.
15. Liao, W.H.; Huang, C.C. SF-MAC: A spatially fair MAC protocol for underwater acoustic sensor networks. *IEEE Sens. J.* **2012**, *12*, 1686–1694.
16. Lin, W.; Cheng, E.; Yuan, F. EHM: A novel efficient protocol based handshaking mechanism for underwater acoustic sensor networks. *Wirel. Netw.* **2013**, *19*, 1051–1061.
17. Xie, P.; Cui, J.H. R-MAC: An energy-efficient MAC protocol for underwater sensor networks. In Proceedings of the IEEE International Conference on Wireless Algorithms, Systems and Applications 2007 (WASA 2007), Chicago, IL, USA, 1–3 August 2007; pp. 187–198.
18. LAN/MAN (IEEE 802) Standards Committee. *Wireless LAN Media Access Control (MAC) and Physical Layer (PHY) Specifications*; IEEE: New York, NY, USA, 1999.
19. LinkQuest Inc. Available online: http://www.link-quest.com (accessed on 30 August 2014).
20. Lee, J.W.; Cho, H.S. Cascading multi-hop reservation and transmission in underwater acoustic sensor networks. *Sensors* **2014**, *14*, 18390–18409.
21. Syed, A.; Ye, W.; Heidemann, J. Comparison and evaluation of the T-Lohi MAC for underwater acoustic sensor networks. *IEEE J. Sel. Areas Commun.* **2008**, *26*, 1731–1743.

# 3

# An Ultrasonic Sensor System based on a Two-Dimensional State Method for Highway Vehicle Violation Detection Applications

**Jun Liu [1,†], Jiuqiang Han [2,†], Hongqiang Lv [2,*] and Bing Li [2]**

[1] School of Electrical Engineering, Xi'an Jiaotong University, Xi'an 710049, China; E-Mail: jliu1912@gmail.com

[2] School of Electronic and Information Engineering, Xi'an Jiaotong University, Xi'an 710049, China; E-Mails: jqhan@xjtu.edu.cn (J.H.); iacxjtu@163.com (B.L.)

† These authors contributed equally to this work.

* Author to whom correspondence should be addressed; E-Mail: lhqxinghun@163.com

Academic Editor: Felipe Jimenez

**Abstract:** With the continuing growth of highway construction and vehicle use expansion all over the world, highway vehicle traffic rule violation (TRV) detection has become more and more important so as to avoid traffic accidents and injuries in intelligent transportation systems (ITS) and vehicular *ad hoc* networks (VANETs). Since very few works have contributed to solve the TRV detection problem by moving vehicle measurements and surveillance devices, this paper develops a novel parallel ultrasonic sensor system that can be used to identify the TRV behavior of a host vehicle in real-time. Then a two-dimensional state method is proposed, utilizing the spacial state and time sequential states from the data of two parallel ultrasonic sensors to detect and count the highway vehicle violations. Finally, the theoretical TRV identification probability is analyzed, and actual experiments are conducted on different highway segments with various driving speeds, which indicates that the identification accuracy of the proposed method can reach about 90.97%.

**Keywords:** highway vehicle traffic rule violation detection; intelligent transportation systems; two-dimensional state method; ultrasonic sensor system

## 1. Introduction

Since the first Bonn-Cologne highway was built in Germany in 1932, there has been continuous highway growth all over the world, in countries such as Germany, the United States, and China. By the end of 2013, the total mileage of China's highways had reached over 104 thousand kilometers [1], succeeding the United States in having the largest network of highways in the world. At the same time, worldwide vehicle use is also growing very fast, so the highway traffic safety problem has become a significant concern for Intelligent Transportation System (ITS) [2,3]. The driving violations of slower vehicles, especially large heavy trucks, travelling in an improper lane like the "passing lane", might cause seriously negative effects on the highway traffic order, reduce highway traffic efficiency, and become a safety threat for other drivers who have to change lanes more frequently. It has been statistically found that in China, the total number of road traffic crashes, nonfatal injuries, and fatalities increased by 43-fold, 58-fold and 85-fold, respectively from 1951 to 2008 [4]. Most countries have special traffic laws to avoid accidents on highways, for example, China has announced several important national traffic rules and regulations restricting the improper roadway occupation behavior on highways [5,6], in order to reduce the highway TRV induced accidents.

As there are hundreds of thousands kilometers of highways, it is not easy to monitor all vehicles at any time. Traditional TRV detection methods include: ultrasound based systems [3,7–9], capacitive sensor based systems [10], infrared sensors [11], laser and radar sensors [12], traffic video based systems [13], computer vision techniques [2,14], RFID technology [15], *etc.* The former vehicle detection sensors or monitoring cameras were always installed along the highway in fixed positions [16], there might be problems associated with this kind of measurement placement such as: (a) in tunnels or on multiple lane sections of the highways, where sensors might detect vehicles erroneously due to the reception of unnecessary reflected signals; (b) although thousands of sensors have been adopted, they still cannot cover all spots of the highways.

Because the literature has seldom touched the issue of violation detection by moving vehicle measurement or monitoring devices, and most of the past research could not recognize the TRV vehicles in real-time during the whole driving process. This paper proposes a novel ultrasonic sensor system that can perform continuous and reliable TRV detection and counting. The real-time and recorded data from the sensor system can be converted into highway hazard and traffic jam messages, and then be sent to adjacent vehicles (V2V) or roadside infrastructure communication units (V2R, or V2I) through a vehicular *ad hoc* network (VANET) [17,18], so as to improve traffic safety and efficiency. For future implementation of the proposed sensor system, potential challenges will need to be resolved in the fields of efficient medium access control protocol design [19], heterogeneous media provision studies [20], and distributed sensor data fusion algorithms [21], *etc.*, so that safety related and other application messages can be timely and reliably disseminated through vehicular networks.

The structure of the paper is as follows: Section 2 introduces the hardware and software configuration of the ultrasonic sensor system, to be embedded on the host vehicle for highway vehicle TRV detection. The two-dimensional state method to detect the highway vehicle violation is proposed in Section 3, utilizing both the spacial state of the sensors and the past time sequential states being stored. A detailed TRV detection and counting algorithm for the two-dimensional state method is then described, in order to address different driving situations of the vehicles passing-by. Theoretical identification probability for the proposed method is analyzed in Section 4. Real-time experiments on different highway segments with various driving speeds are performed and shown in Section 5, which demonstrates the applicability and high identification accuracy of the proposed method. Finally, the conclusions are given in Section 6.

## 2. Principle of the Ultrasonic Sensor System

After a careful survey and detailed feasibility analysis on different sensor types above, we choose ultrasonic sensors in this study. Ultrasonic sensors have been widely used in ITS and VANET area applications such as vehicle tracking and classification [7,22–25], obstacle detection and mapmaking [26,27], vacant parking slot detection [28], smart traffic signaling [29], ultrasonic ranging and localization [30], *etc.* Ultrasonic sensors are a well accepted technology for distance sensing applications, because of the inexpensive and easy-to-adopt nature, and reliable and stable measurement performance within their measuring range.

Figure 1 shows that a vehicle is driving in the passing lane while three other cars are driving on the carriageway. If the speed of the left vehicle is faster but does not exceed the speed limit of that highway, then it will catch up with the three vehicles to its right and surpass them, which is taken as a normal driving behavior; Otherwise, if the speed of the left vehicle is relatively slower, the other three drivers have to pass it on its right side, then this will be considered an illegal TRV behavior of occupying the passing lane (assuming that the country obeys driving on the right). A real-time vehicle tracking system for this TRV driving situation can be designed, by attaching two ultrasonic sensors to the right side of the vehicle, assisted with a communication device for information transfer. While the vehicle is driving on the highway, the two ultrasonic sensors can detect vehicles consistently whether there are other vehicles overtaking it from the lower speed lane to its right side, then the ultrasonic sensor system will record the situation of the vehicle as a TRV behavior.

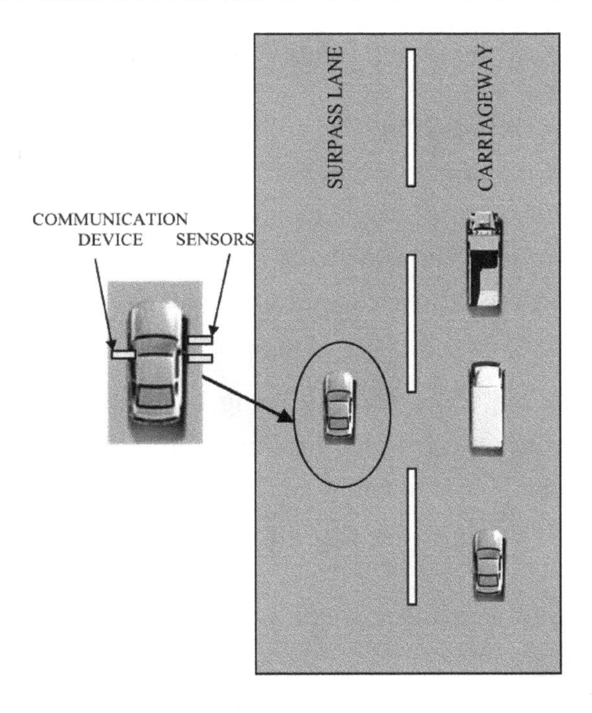


**Figure 1.** Ultrasonic sensor system for highway vehicle TRV identification.

It might be not reliable to detect a passing vehicle by the movement measurement device with only one ultrasonic sensor, because the reflected signal can be influenced by the target height, length, surface flatness, speed, *etc.* Therefore, the ultrasonic sensor system is designed to contain two parallel ultrasonic sensors, and the measurement data of both sensors can act as a complementary source for each other by using proper data processing techniques, in order to deal with problematic measurements.

Figure 2 shows the hardware structure of the ultrasonic sensor system, which is comprised of a central controller, a GPS module, an infrared communication module, and two parallel ultrasonic sensors. The ultrasonic sensors send and receive ultrasonic signals to detect whether there are vehicles within their measuring range, and the GPS module provides the current driving speed of the host vehicle. Then the identification algorithms of the central controller will determine whether the driving situation of the host vehicle is an TRV behavior through measurements, and if it is confirmed as an improper lane-overtaking, the violation counter will increase by 1. The infrared communication module will send the results to the receiver in the highway toll station.

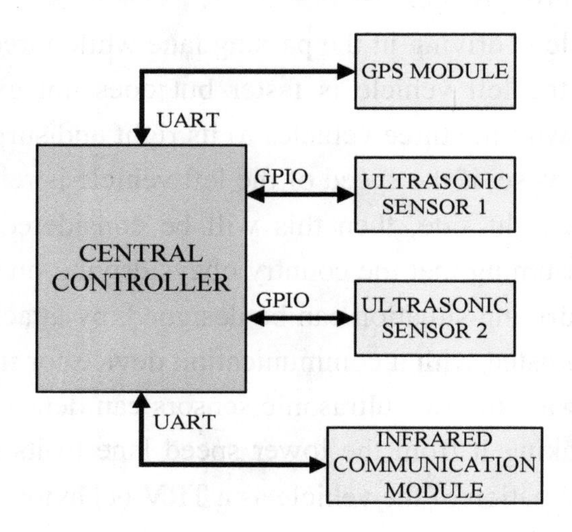


**Figure 2.** The hardware configuration of the ultrasonic sensor system.

In order to meet the target of recognizing the highway TRV driving behavior, the software of the ultrasonic sensor system should include the following functions: system initialization, passing vehicle identification (including both ultrasonic ranging and GPS speed measurement), data display, data storage, and infrared communication. The schematic diagram of the system software can be shown in Figure 3, and each module contains the corresponding drives and interfaces shown in the figure. The detailed functions can be expressed as follows:

(a) System initialization module, does the following things: running the bootstrap, IO interface configuration, flash configuration, timer initialization, loading the system-related parameters including algorithm related parameters and basic information of the host vehicle (license number, owner information, *etc.*). In addition, the initialization module also completes the variable initialization.

(b) Passing vehicle identification module: identifies whether the vehicle is occupying an improper lane, according to the results measured by the ultrasonic sensors and an on-board GPS. (1) GPS speed measurement module, extracts information of speed, time, latitude and longitude coordinates, according to the frame information from the GPS; (2) Ultrasonic measurement module, controls the ultrasonic sensors to transmit and receive the ultrasound waves, and processes the reflected ultrasonic signal to calculate the distance between vehicles and record the signal strength of each measurement.

(c) Data display module, performs the initialization of the LCD, and displays the TRV recognition and counting results of the host vehicle.

(d) Data storage module, runs the flash initialization, flash read-and-write functions, and then be used to store the configuration parameters, the historical measurement data of ultrasonic sensor and vehicle driving status.

(e) Infrared communication module, provides the infrared communication services for the sensor system to highway toll stations, other vehicles, or roadside communication units.

Among all the modules in the second and third rows of Figure 3, the GPS speed measurement module, ultrasonic ranging module and the surpassing vehicle identification module, are the key subsystems of the software design in the sensor system, thereafter the following sections will introduce the identification method of the three modules in detail.

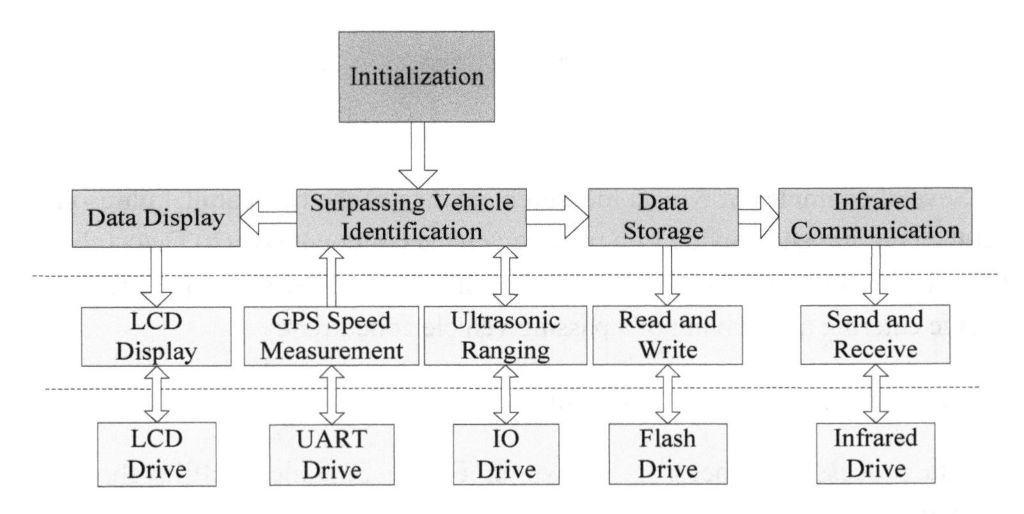


**Figure 3.** Schematic diagram of the system software functions.

## 3. The Two-Dimensional State Method for the Ultrasonic Sensor System

As there are real-time measurement data from two ultrasonic sensors, and the host vehicle of the measurement device is moving, the data processing would be quite important so as to deal with all kinds of measurement data.

In reality, the detection of a passing or to-be-passed vehicle can be rather complicated, as it might be highly related to the driving habits of the driver, vehicle surface condition, relative speed, relative angle between the target and the host vehicle, *etc.* In order to find the relative movement direction of other vehicles according to measurement data from the two sensors, a feasible way is to record the measurement sequences of both sensors as shown in Figure 4, then discriminate the relative motion direction by the logic analysis of the passing or to-be-passed vehicle. In Figure 4, the black line and red line denote the measurement data from ultrasonic sensor 1 and sensor 2, respectively, the red line data of sensor 2 has been shifted slightly down on the vertical axis for clarity, so as to avoid overlapping of the two colored lines. The identification process is easy to conduct when the data flow is clearly distinguishable (such as the situations of Figure 4b–d), but it might not easily make good judgments when there are breakpoints occurring in the measurement results from one sensor (such as the situation of Figure 4a, when the reflecting surface of the target is not flat, or the measured surface has a relatively large angle with the moving direction of the parallel ultrasonic sensors, then the sensors may not be able to detect the target, and thus there will be breakpoints observed by the sensor system). Therefore, a two-dimensional state method is proposed to address this issue, and to increase the identification rate of the ultrasonic sensor system. The two-dimensional state means the spacial state and the past time sequential states of the two ultrasonic sensors.

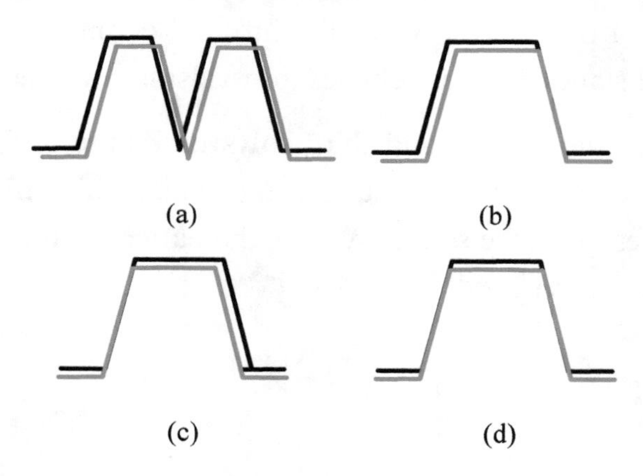


**Figure 4.** Several examples of typical measurement data of the two parallel ultrasonic sensors. (**a**) Detect the entering and leaving of a passing vehicle continuously; (**b**) Detect the leaving of a passing vehicle concurrently; (**c**) Detect the entering of a passing vehicle concurrently; (**d**) Detect the entering and leaving of a passing vehicle concurrently.

### *3.1. Conversion of Binary States of the Ultrasonic Sensors*

The whole data processing process is shown in Figure 5, which fulfills the main function of traffic-rule-violation detection and counting. In the first step of Figure 5, the controller needs to obtain the state $T(n)$ of the ultrasonic sensors. In fact, the direct measurement data is the one-way travelling distance of the ultrasound wave, which is required to be converted into the binary state of $T(n)$. The binary state $X$ of each sensor can be acquired according to the logic rule in Equation (1), which can be 0 or 1, where 0 means that no reflecting signal is received, and 1 means a target is in its measurement scope:

$$X=(d_{meas}>d_{\min})\&(d_{meas}<d_{\max}) \tag{1}$$

where: " & " means the logic "and", $d_{meas}$ is the distance measured by the sensor during this measurement cycle, $d_{\min}$ is the intrinsic blind-area distance of ultrasonic sensors, and $d_{\max}$ is the maximum measured distance or the required effective measurement scope for the ultrasonic sensor system. Then the two sensors' measuring state can be represented by Equation (2):

$$T(n)=((d_1>d_{\min})\&(d_1<d_{\max}))\,|\,((d_2>d_{\min})\&(d_2<d_{\max})) \tag{2}$$

where: "|" denotes the separation between the binary states of the two sensors, $d_1$ and $d_1$ are the measured distances for sensor 1 and sensor 2, respectively.

As it can be seen in Figure 5, after acquiring the ultrasonic sensor state data $T(n)$, the program should determine whether $T(n-1)$ is equal to 00, if $T(n-1) \neq 00$, it means the measurement is an intermediate state, and the program is interrupted to load next inputs; Otherwise, continue to inspect whether $T(n)$ is equal to 00, if $T(n) = 00$, then check whether the vehicle has left, and determine the relative motion to the host vehicle, and update the TRV counting number. If $T(n) \neq 00$, it suggests that the vehicle has just entered the measuring range, then record the measured states and the reflection signal strength of both ultrasonic sensors. The process of Figure 5 denotes only one measurement cycle of the sensor system, it will continue to run when the data of the next measurement cycle are obtained. During the driving process of the host vehicle on the highway, the sensor system runs the vehicle TRV detection and counting algorithm continuously, it records the TRV numbers, and sends any illegal driving behavior of the host vehicle to the highway toll station.

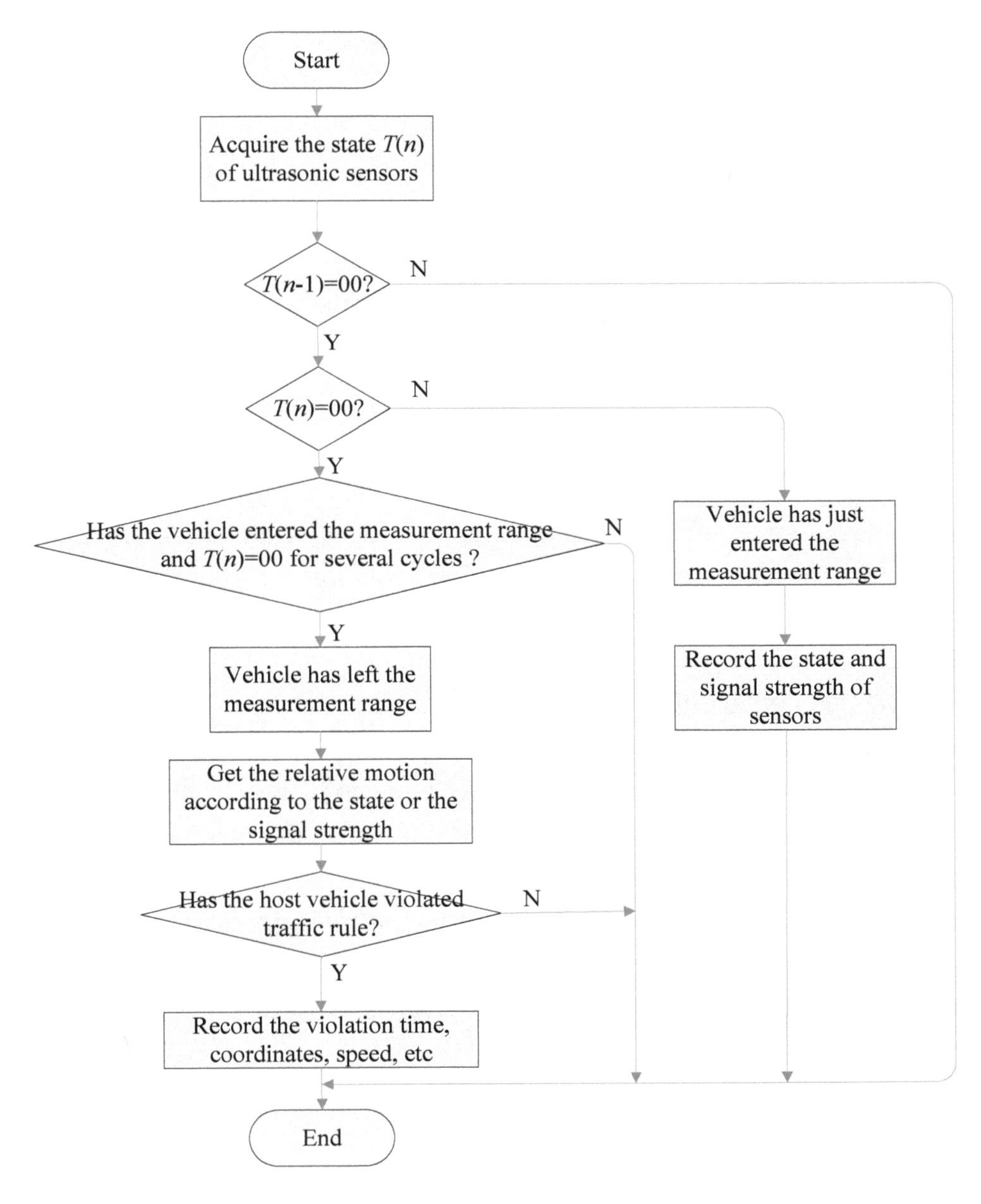


**Figure 5.** Flowchart of the TRV detection and counting algorithm.

### 3.2. Spacial State

In the two-dimensional state method, the first dimension is the spacial state $T(n)$, according to the measured data of the two parallel ultrasonic sensors. The term $T(n)$ in Equation (2) can be simplified as the expression in Equation (3), where the higher digit $X_1$ denotes the sensor measurement result of ultrasonic sensor 1, and the lower digit $X_2$ shows the binary state of sensor 2:

$$T(n) = X_1 X_2 \tag{3}$$

where $n$ denotes the state in time moment $n$, for example the present measurement moment.

If $T(n) = 01$, it means that sensor 2 detects a vehicle in the current moment, and sensor 1 has not detected any vehicle yet. Through this spacial state of the two sensors, this can be used to determine the relative movement direction of a nearby vehicle to the right side. Different values of the lower and higher digits of $T(n)$ directly indicate the relative movement direction. For example, $T(n) = 01$ means that the vehicle approaches sensor 2 earlier, the relative moving direction is from 2 to 1, the vehicle is trying to overtake the host vehicle, hence the TRV behavior is detected, and the violation counter of the host vehicle increases by 1; Otherwise, if $T(n) = 10$, it means that the vehicle reaches sensor 1 first, so the direction for the binary value of 10 indicates the direction is from 1 to 2, the host vehicle is passing other vehicles, hence the violation counter remains unchanged. It is worth noting that, if it is able to identify the direction of relative movement when the vehicle is entering the sensor measuring range, it will not need to count when the vehicle is leaving the ultrasonic range.

### 3.3. Time Sequential State

If the measurement data is clear, the spacial state is sufficient to distinguish the relative motion direction. However, when the measurement has breakpoints, the spacial state parameter only will not be able to point out correctly the passing lane occupation violations. Therefore, another dimensional state should be considered, the complementary time sequences of $T(n-1)$ and $T(n-2)$, together with the present measurement $T(n)$.

If there are other vehicles within the scope of the sensor, but the reflecting surface or the high relative speed of the vehicle, makes the ultrasonic sensors miss the ultrasound reflection occasionally, the measurement sequence will display a certain number of breakpoints. In this kind of situation, the primary concern should be determining whether the vehicle has left completely, which can avoid double TRV counting of the sensor system. Since all vehicles are required to maintain a certain safety distance to the vehicle ahead on the highway, the TRV counter will increase only when the moving object has left for at least two measurement cycles. Otherwise, the counter remains unchanged. The situation with breakpoints can be divided into several conditions, according to the previous state values of $T(n-1)$ at the time moment of $(n-1)$.

#### 3.3.1. $T(n-1) = 00$

For the situation when the present state is $T(n) = 00$, and an object has been detected to be in the measuring range of the sensor system before, then it must decide whether the vehicle has completely driven out of the sensor measurement range. If the vehicle has left, the relative movement direction of the vehicle is from 2 to 1, and the violation counter increases by 1. If $T(n)$ = 00 and no vehicle has been detected in the sensor's measuring scope before, then the counter does not count. These two conditions can be simply shown in Figure 6a,b with binary state plots.

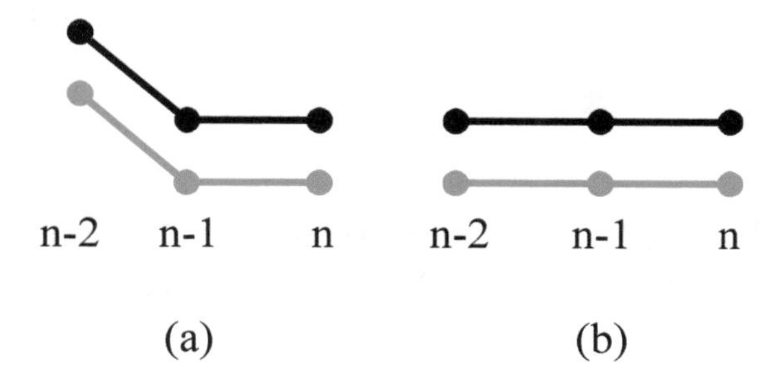


**Figure 6**. $T(n - 1) = 00$ and $T(n) = 00$. (**a**) An vehicle has been detected to be in the measuring range of the sensor system before; (**b**) No vehicle has been detected in the sensor's measuring scope before.

If the present state is $T(n) \neq 00$, Figure 7 shows three different situations of (a) $T(n) = 01$, (b) $T(n) = 10$ and (c) $T(n) = 11$. Figure 7a indicates that sensor 1 (black dotted line) might have two continuous breakpoints and sensor 2 (red dotted line) might have missed one reflecting signal at moment *n*. Figure 7c means that both two sensors experience a breakpoint at moment *n*. Figure 7b indicates there might be another vehicle that is entering the ultrasonic sensor measurement range of sensor 1, or there might be at least two continuous breakpoints for both ultrasonic sensors at $(n-2)$ and $(n-1)$, it will depend on more measurement results to give a correct identification in this situation.

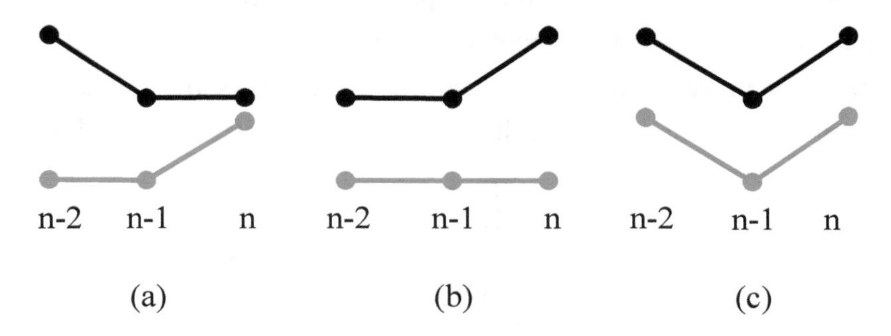


**Figure 7.** $T(n - 1) = 00$ and $T(n) \neq 00$. (**a**) $T(n) = 01$; (**b**) $T(n) = 10$; (**c**) $T(n) = 11$.

3.3.2. $T(n - 1) \neq 00$

For the situation when the present state as $T(n - 1) \neq 00$, three different situations are shown in Figure 8. Figure 8a can be a vehicle that has just left the measuring range, so both sensors display a state with no vehicles for the first time, and it needs to wait for the following measurement cycle so as to avoid the breakpoint judgment. Figure 8b indicates a situation where a new rising edge is being detected by both sensors, the vehicle enters the measuring range of sensor 1 first, and then enters the measuring range of sensor 2. Figure 8c shows three continuous positive values for both sensors, which denotes a moving target is passing through the sensors' coverage, and both sensors detect the moving target at $(n-2)$, $(n-1)$ and *n* time moments.

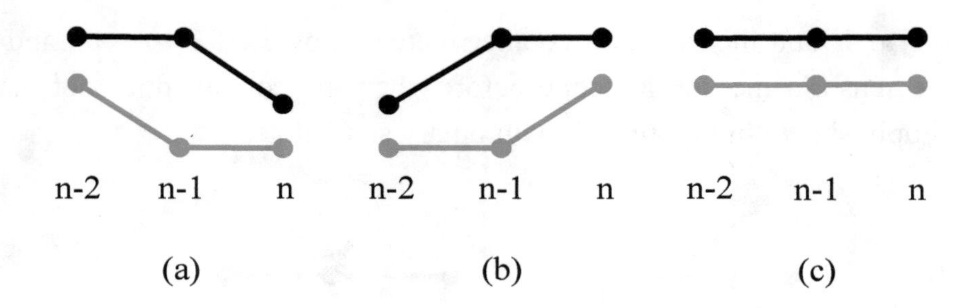


**Figure 8**. $T(n-1) \neq 00$. (**a**) A vehicle has just left the measuring range; (**b**) A vehicle enters the measuring range of sensor 1 and 2 continuously; (**c**) A vehicle is passing through both sensors' measurement range.

### *3.4. Improvement by Measured Reflection Signal Strength*

Typically, the binary spacial and past time sequential states are sufficient to successfully identify the TRV behavior. However, if the vehicle remains in the measuring range of both ultrasonic sensors simultaneously, such as the situation described in Figure 4d and Figure 8c, the binary states will be 11 for $T(n-2)$, $T(n-1)$ and $T(n)$, thus it would be unable to determine the direction of movement of the object. Under this circumstance, decisions will be made according to the measured reflection signal strength. The principle is that, the head and tail part of the moving vehicle is always not flat enough, or has some intersection angle with the ultrasonic sensor plane, as shown in Figure 9, hence the ultrasonic reflection signals by these surfaces are usually weak. For example, although sensor 1 and 2 both detected the moving object at the same moment in Figure 9, the reflection signal of sensor 2 is much bigger than that of sensor 1; therefore, we can set a proper threshold value $\Delta d$ for the signal strength difference between the two sensors, if the signal strength of sensor 2 minus that of sensor 1 is greater than the threshold $\Delta d$, it can be seen that the motion direction is from 2 to 1. Similarly, when a moving object is leaving the measuring range, the signal strength difference will also be applicable.

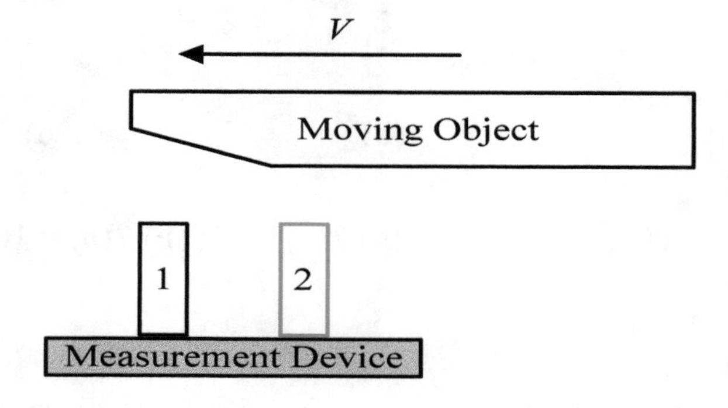


**Figure 9.** Diagram for a moving object with varying reflecting surfaces.

## 4. Theoretical Identification Accuracy Analysis

In the real world, the traffic conditions can be very complicated, the host vehicle might overtake other cars, and other cars might overtake the host vehicle. A simple probability analysis on the TRV identification accuracy of the ultrasonic sensor system is performed, according to the random driving situation of the measurement device and the passing vehicle.

### *4.1. Conditions for Successful Identification of TRV Behavior*

In order to estimate the probability of the TRV identification algorithm, the condition for successful identification of TRV behavior should be analyzed first. Assuming that the distance between the two ultrasonic sensors is $d$, the ultrasonic sending direction is perpendicular to the driving direction of the host vehicle with a time interval of $T$. The relative speed of a passing or to-be-passed vehicle with respect to the host vehicle is assumed to be a constant speed $V$, because the passing time of other vehicles through the measuring unit is typically very short in our tests, with a minimum of less than 1 s, and no more than 10 s for the longest situation, so this constant relative speed assumption is reasonable.

#### 4.1.1. When Other Vehicles Start to Enter the Measurement Range

As it can be seen in Figure 10, another vehicle is approaching the measuring device a measurement cycle $T$ before, the present distance of the head of the vehicle to sensor 2 is $\Delta L_1$, and $0 < \Delta L_1 < VT$.

An ultrasonic measurement cycle $T$ later, the relative travel between the vehicle and the measuring device will be $VT$, then the relative position of vehicle and the sensor system can be as seen in either Figure 11a or b. Figure 11a shows that only ultrasonic sensor 2 detects the passing vehicle, and ultrasonic sensor 1 has not sensed any object at this moment, so the relative motion direction of the vehicle can be easily identified as from 2 to 1. The other vehicle is catching up to the host vehicle, which indicates that the host vehicle is driving in an improper lane; Figure 11b indicates that both ultrasonic sensors detect the passing vehicle, which can also been identified as a TRV of roadway overtaking behavior as Figure 11a does. It can be found in Figure 11 that the inequality relationship of Equation (4) must be satisfied, so as to successfully identify the relative motion direction of the passing vehicle:

$$d + \Delta L_1 > VT \tag{4}$$

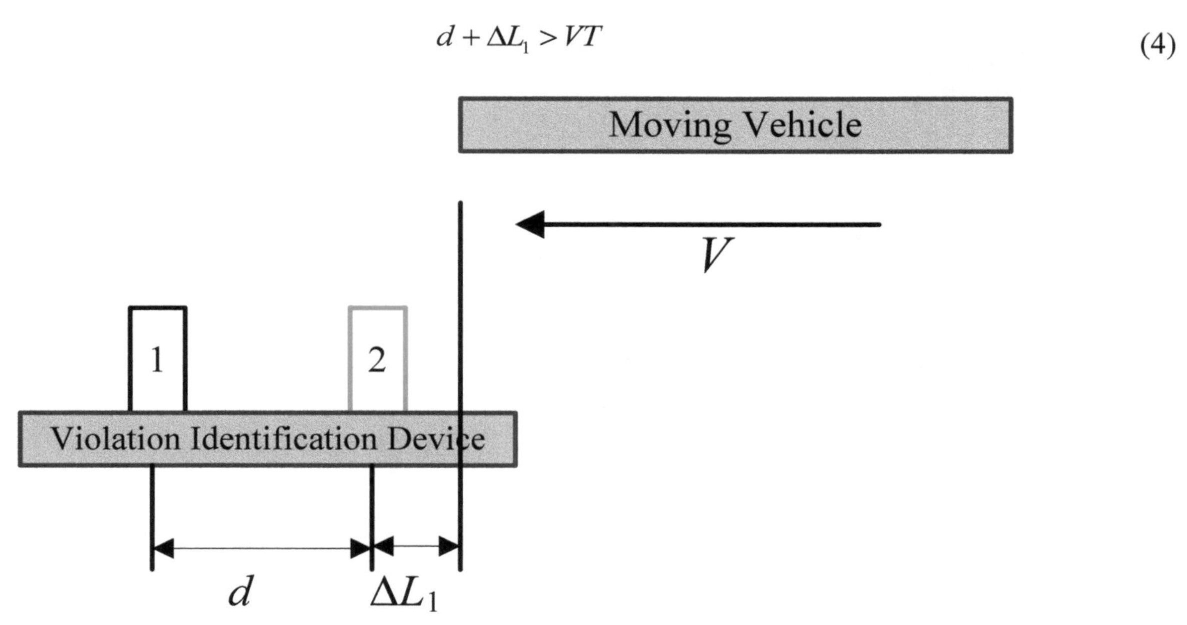


**Figure 10.** A vehicle is about to enter the measurement range of the sensor system.

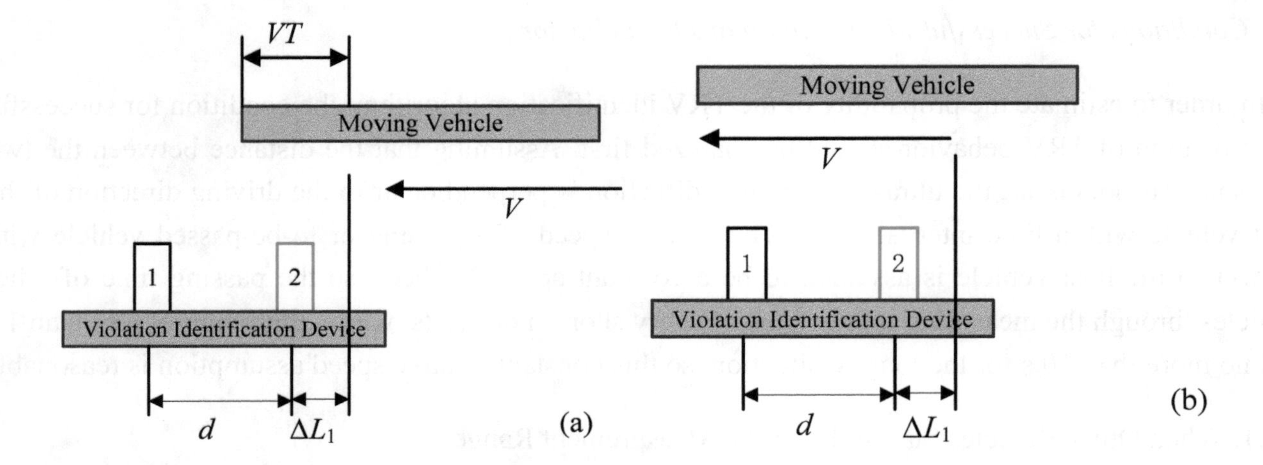


**Figure 11.** Vehicle has just entered the measurement range: (**a**) Detected by sensor 2 only; (**b**) Detected by both sensors.

4.1.2. When Other Vehicles Start to Leave the Measurement Range

Figure 12 shows that another vehicle is leaving the ultrasonic sensors' measuring range, and both ultrasonic sensors can detect the vehicle a measurement cycle before, but at this moment ultrasonic sensor 1 is able to receive its reflecting signal and sensor 2 fails as indicated in Figure 12a, while neither ultrasonic sensor can detect the vehicle in the situation of Figure 12b.

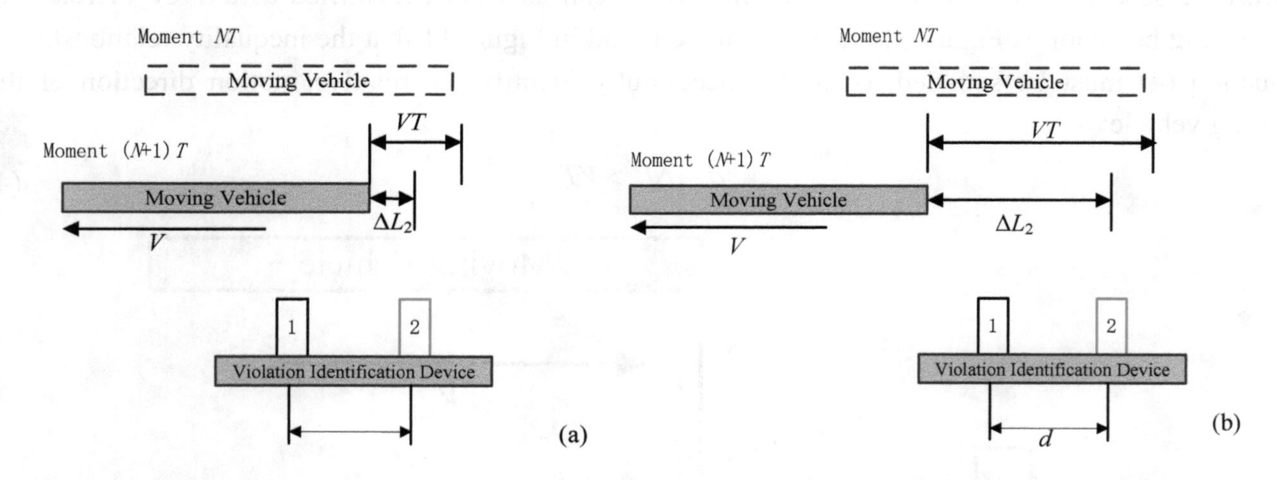


**Figure 12.** Vehicle is leaving the measurement range: (**a**) Detected by sensor 1 only; (**b**) Detected by no sensor.

According to Figure 12a, the inequality relationship of Equation (5) must be satisfied, so as to identify the relative motion direction of the leaving vehicle:

$$0 < \Delta L_2 < d \tag{5}$$

It should be noted that $\Delta L_2 < VT$ must be true in this situation, because both ultrasonic sensors can detect the vehicle a cycle before the state of Figure 12a. The distance $\Delta L_2$ is the distance from the tail of the vehicle to sensor 2, which is obviously related to both the distance $\Delta L_1$ and the vehicle length $L$.

Since Equations (4) and (5) can both be satisfied if $d > VT$, only the $d < VT$ situation should be analyzed further so as to accomplish a successful TRV identification.

From the driving state of Figure 10 to Figure 12a, it must have experienced an integer number of measurement cycles, assuming that there have been $N$ cycles. During this period, the surpassing vehicle has gone through a travel of $L + \Delta L_1 + \Delta L_2$, corresponding to the position of sensor 2:

$$L + \Delta L_1 + \Delta L_2 = NVT \tag{6}$$

where $0 < \Delta L_1$, and $\Delta L_2 < VT$.

Assuming that $L = kVT + \Delta L_3$, the new variable $\Delta L_3$ being introduced, is related only to the vehicle length, and it must satisfy $0 < \Delta L_3 < VT$, and $k < N$, so substituting these relationships into Equation (6) yields:

$$\Delta L_1 + \Delta L_2 + \Delta L_3 = (N - k)\ VT \tag{7}$$

where $\Delta L_1$ relates to the time when the vehicle enters the measurement range, and $\Delta L_3$ relates to the vehicle length, therefore, the two variables are independent of each other, and can be seen as statistically independent random variables.

Because $0 < \Delta L_1 < VT$ and $0 < \Delta L_3 < VT$, then $0 < \Delta L_1 + \Delta L_3 < 2VT$. Thus Equation (7) can be classified into two cases, firstly, $0 < \Delta L_1 + \Delta L_3 < VT$, and $(N - k)$ can only be equal to 1 in this case, otherwise $\Delta L_2$ will be larger than $VT$, therefore, we have:

$$\Delta L_1 + \Delta L_2 + \Delta L_3 = VT, \quad \text{if} \quad 0 < \Delta L_1 + \Delta L_3 < VT \tag{8}$$

Secondly, when $VT < \Delta L_1 + \Delta L_3 < 2VT$, and $(N - k)$ should be equal to 2 in this case, otherwise $\Delta L_2$ will be larger than $VT$, then we have:

$$\Delta L_1 + \Delta L_2 + \Delta L_3 = 2VT, \quad \text{if} \ VT < \Delta L_1 + \Delta L_3 < 2VT \tag{9}$$

Combining Equations (5), (8) and (9), the following inequality relationships must be satisfied in order to identify the relative motion direction of the leaving vehicle:

$$VT - \Delta L_1 - \Delta L_3 < d, \quad \text{if} \ 0 < \Delta L_1 + \Delta L_3 < VT \tag{10}$$

or:

$$2VT - \Delta L_1 - \Delta L_3 < d, \quad \text{if} \quad VT < \Delta L_1 + \Delta L_3 < 2VT \tag{11}$$

According to the analysis above, the parameters should meet the condition of Equation (4) plus Equations (10) or (11) so as to successfully identify a passing or to-be-passed vehicle with the sensor system.

### *4.2. Theoretical Identification Probability for the Proposed Method*

Equations (4), (10) and (11) are all related to parameter $\Delta L_1$, which is an random variable. According to the basic knowledge of probability theory, the summation of probability of identified and unidentified situations for the proposed method should be 1. That is:

$$P_{identified} = 1 - P_{unidentified} \tag{12}$$

Based on the information from Equations (4), (10) and (11), the condition that fails to identify the direction of the vehicle movement, should be the dissatisfaction of "Equation (10) plus Equation (4)" or "Equation (11) plus Equation (4)", which can be expressed as:

$$\begin{cases} VT - \Delta L_1 - \Delta L_3 > d \\ 0 < \Delta L_1 + \Delta L_3 < VT \\ \Delta L_1 + d < VT \end{cases} \tag{13}$$

or:

$$\begin{cases} 2VT - \Delta L_1 - \Delta L_3 > d \\ VT < \Delta L_1 + \Delta L_3 < 2VT \\ \Delta L_1 + d < VT \end{cases} \tag{14}$$

In Equation (13), the first expression $VT - \Delta L_1 - \Delta L_3 > d$ can be rewritten as $\Delta L_1 + \Delta L_3 < VT - d$, together with the other two expressions, $0 < \Delta L_1 + \Delta L_3 < VT$ and $\Delta L_1 + d < VT$, Equation (13) can be simplified as Equation (15):

$$\Delta L_1 + \Delta L_3 < VT - d \tag{15}$$

Similarly in Equation (14), the first expression can be rewritten as $\Delta L_1 + \Delta L_3 < 2VT - d$, which has already been included in the second expression of $\Delta L_1 + \Delta L_3 < 2VT$, then Equation (14) can be simplified as Equation (16):

$$\begin{cases} VT < \Delta L_1 + \Delta L_3 < 2VT - d \\ \Delta L_1 + d < VT \end{cases} \tag{16}$$

Therefore, the probability of identified situations for the proposed method will be:

$$P_{identified} = 1 - P\{\Delta L_1 + \Delta L_3 < VT - d\} - P\{VT < \Delta L_1 + \Delta L_3 < 2VT - d,\ \Delta L_1 < VT - d\} \tag{17}$$

where $0 < \Delta L_1 < VT,\ \Delta L_3 < VT < d$.

As is known from the analysis above, $\Delta L_1$ and $\Delta L_3$ are independent of each other, and they can be seen as obeying a uniform distribution along the range of $(0, VT)$. Then we can calculate the two terms of Equation (17) separately as:

$$\begin{aligned} &P\{\Delta L_1 + \Delta L_3 < VT - d\} \\ &= \int_0^{VT-d} \int_0^{VT-d-\Delta L_1} \frac{1}{(VT)^2} d(\Delta L_1) d(\Delta L_3) = \int_0^{VT-d} \left(\frac{VT - d - \Delta L_1}{(VT)^2}\right) d(\Delta L_1) = \frac{(VT-d)^2}{2(VT)^2} \end{aligned} \tag{18}$$

$$\begin{aligned} &P\{VT < \Delta L_1 + \Delta L_3 < 2VT - d,\ \Delta L_1 < VT - d\} \\ &= \int_0^{VT-d} \int_{VT-\Delta L_1}^{2VT-d-\Delta L_1} \frac{1}{(VT)^2} d(\Delta L_1) d(\Delta L_3) = \int_0^{VT-d} \left(\frac{VT-d}{(VT)^2}\right) d(\Delta L_1) = \frac{(VT-d)^2}{2(VT)^2} \end{aligned} \tag{19}$$

Substituting Equations (18) and (19) into Equation (17), the probability of identified situations can be obtained as:

$$P_{identified} = 1 - \frac{(VT-d)^2}{(VT)^2}, \quad \text{if } d < VT \tag{20}$$

When $d > VT$, the condition (4) and (5) will surely be satisfied, then the probability of identified situations under the condition of $d > VT$ will be:

$$P_{identified} = 1, \quad \text{if } d > VT \tag{21}$$

Equations (20) and (21) give the theoretical probability of the TRV identification algorithm by the proposed ultrasonic sensor system and the two-dimensional state method. In addition, when the driving direction of the passing vehicle has a certain angle $\theta$ with the measurement system, such as the situation being shown in Figure 13, only the ultrasonic reflection that is perpendicular to the vehicle surface can be received by the sensor. Therefore, the measurement distance should be corrected by a certain coefficient, namely, replacing the measurement distance $d$ by $d\cos\theta$ in Equations (20) and (21). The final probability considering the angle between the sensor system and the target vehicle driving direction will be:

$$P_{identified} = \begin{cases} 1 - \frac{(VT - d\cos\theta)^2}{(VT)^2}, & d\cos\theta < VT \\ 1, & d\cos\theta \geq VT \end{cases} \tag{22}$$

It can be concluded from the analysis above and Equation (22) that: (a) The smaller the relative speed $V$, the bigger the probability $P$. (b) The smaller the measurement time interval $T$, the bigger the probability $P$. (c) The greater the spacing $d$ between the two sensors, the bigger the probability $P$. (d) The smaller the angle $\theta$, the greater the probability $P$, and the angle should not be larger than the ultrasound beam angle, otherwise, the ultrasonic sensors will not be able to receive the reflection wave.

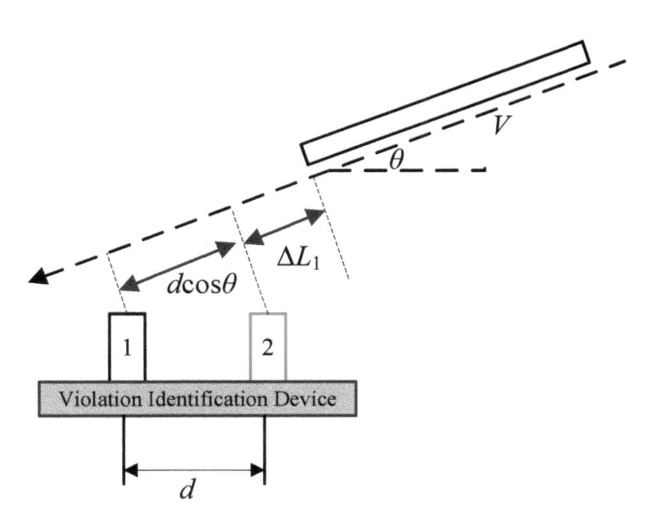


**Figure 13.** Driving direction of other vehicles is not parallel with the violation identification device.

Since the length of the host vehicle is limited and the relative speed $V$ can vary, the theoretical value $P_{identified} = 1$ in the situation of $d > VT$ cannot be easily achieved.

## 5. Experimental Results

According to the hardware and software configuration from Section 2, the physical connection of the ultrasonic sensor with the control board can be seen in Figure 14, the placement of the two sensors is shown in Figure 15, and the developed central controller board and related interfaces for the TRV identification experiments are shown in Figure 16.

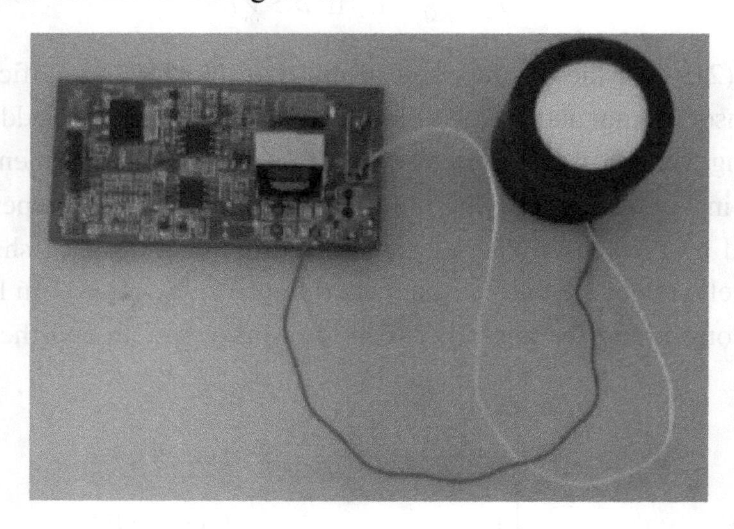

**Figure 14.** The physical connection of the ultrasonic sensor with the control board for TRV identification experiments.

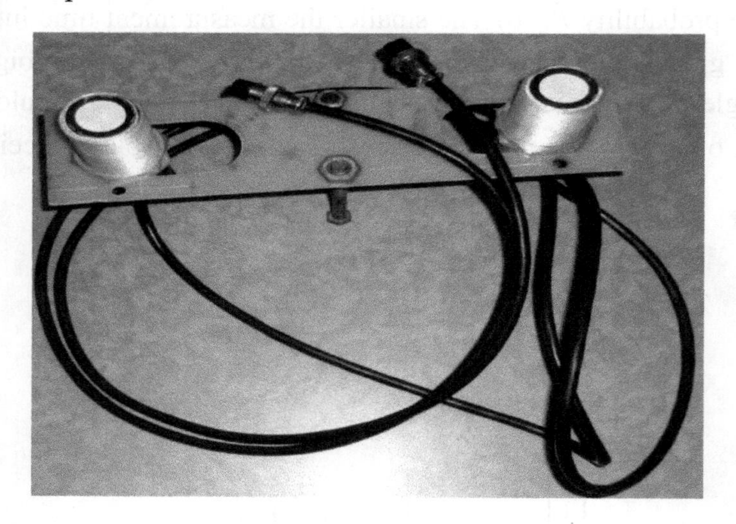

**Figure 15.** The two parallel ultrasonic sensors fixed on one board.

In the ultrasonic measuring devices, the corresponding timer accuracy is 5 μs (frequency resolution is chosen as 100 kHz, since the higher the frequency, the stronger the reflection ability; a typical ultrasound wave has a frequency of more than 20 kHz), assuming the ultrasonic speed in air is about constant at 340 m/s, the distance measuring accuracy will be (340 m/s × 5 μs)/2 = 1.7 mm, thus the distance resolution of the ultrasonic sensor is 1.7 mm. The maximum measuring distance of the ultrasonic sensor is chosen as 3.4 m (approximately the length of one lane on the highway) because it is not reasonable to punish the driver when there are more than two vacant lanes to his/her right side. The ultrasonic sensor can perform a distance measurement every 20 ms, considering the additional calculation time by the TRV detection algorithm, so the actual measuring time interval for the ultrasonic sensors is set as 30 ms. All the sensor parameters are listed in Table 1.

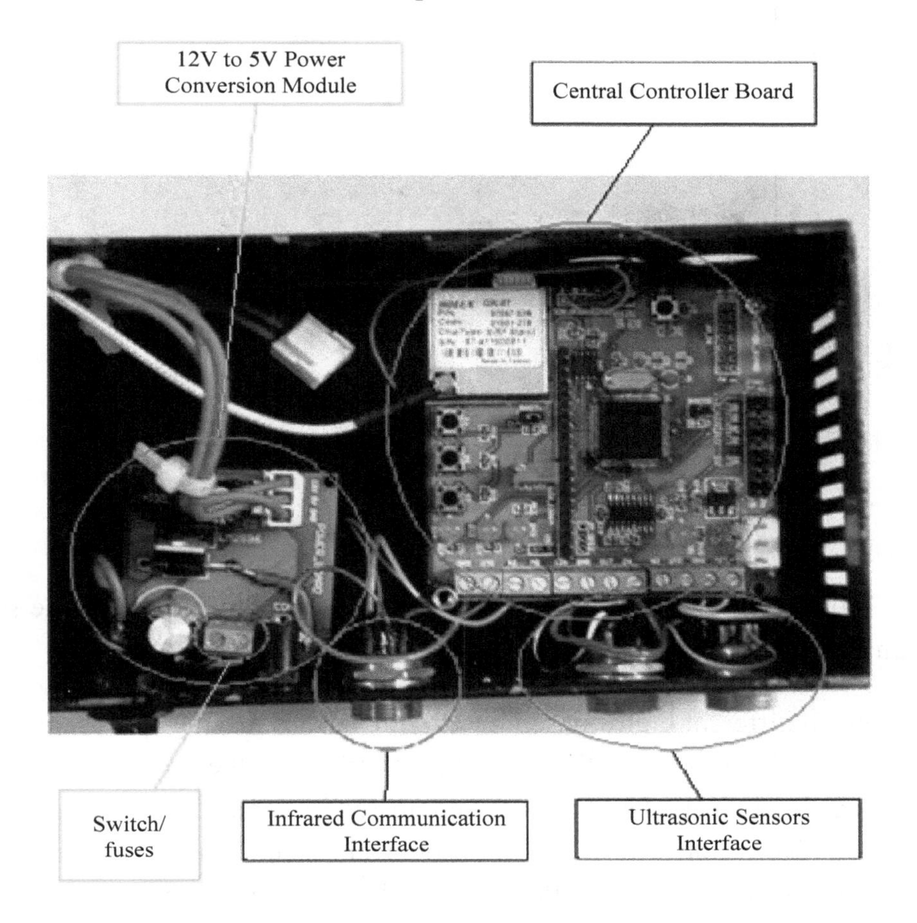


**Figure 16.** The layout of the central controller board and related interfaces.

**Table 1.** Parameters of the ultrasonic sensors.

| Frequency | Minimum Distance | Maximum Distance | Ultrasonic Accuracy | Measurement Interval |
|---|---|---|---|---|
| 100 kHz | 0.35 m | 3.4 m | 1.7 mm | 30 ms |

An experimental ultrasonic distance measurement test has been recorded and transferred via a serial port to the software developed on PC, and the corresponding graphical display is presented in Figure 17.

It can be seen that there are five vehicles being detected by the proposed method within the 27 s test, among which there are three TRV of the host vehicle identified (blue line of sensor 1 lagging the red line of sensor 2), and 2 legal passings of other vehicles (red line lagging the blue line). The upper curve denotes the real-time signal strength measurement data, and the lower curve shows the measured distances of every ultrasonic reflection signal for both sensors. As previously discussed in Section 3, TRV identification can be made according to the binary states converted from the measured distance data from the lower curve, if the binary states are distinguishable. However, the upper signal strength curve must be considered when there are breakpoints detected by the ultrasonic sensors, and this complementary curve can make the identification algorithm more robust.

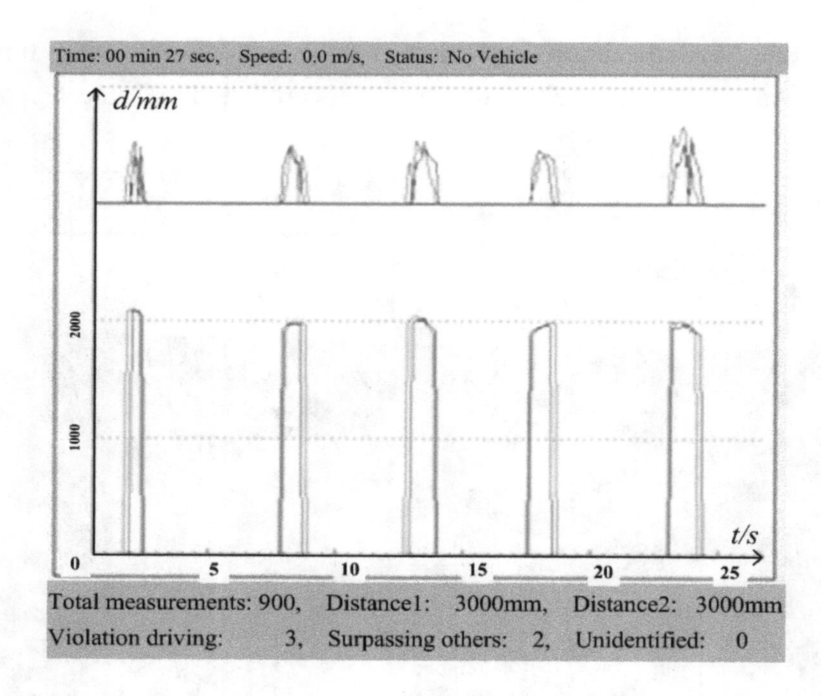


**Figure 17.** Reflection signal strength and real-time distance measured by the sensors and displayed on PC software.

More experiments on different highway segments with various driving speeds, have been conducted on the G65 highway of China, and the TRV counting numbers of the proposed identification method and the counting results by human observation are compared in Tables 2 and 3.

It should be noted that Table 2 uses the ultrasonic distance values only, and Table 3 uses the improved method with the assistance of reflection signal strength difference between the two sensors. It can be seen that the accuracy of Table 3 is comparatively higher than Table 2 when the host vehicle has the same driving speed, because the additional information of the signal strength information can improve the identification accuracy. There is an abnormal count in the first case of Table 3, because a long truck with a very low relative speed and unsmooth surface was passing our host vehicle, and the vehicle has been counted twice during one passing action. To sum up, the error rate of the proposed method can reach 9.03% and 2.91% respectively, without and with the additional signal strength data. In addition, because of the high-frequency and good-reflective characteristics of ultrasonic sensors with the specific measurement range of 0.35–3.4 m, weather conditions, such as rainy, partly-cloudy weather or even night applications, have little influence on the sending and receiving of signals during our tests.

**Table 2.** Experimental results *without* signal strength on 7 June 2014.

| Road Segments | Driving Speed (km/h) | Tested Mileage (km) | No. of System Counter | No. by Human Observation | Error Rate |
|---|---|---|---|---|---|
| Jingyang to Sanyuan | 60 | 14 | 46 | 50 | 8% |
| Sanyuan to Tongchuan | 70 | 28 | 45 | 49 | 8.2% |
| Tongchuan to Sanyuan | 70 | 28 | 32 | 36 | 11.1% |
| Sanyuan to Jingyang | 60 | 14 | 18 | 20 | 10% |
| Total | - | 84 | 141 | 155 | **9.03%** |

**Table 3.** Experimental results *with* signal strength on 13 July 2014.

| Road Segments | Driving Speed (km/h) | Tested Mileage (km) | No. of System Counter | No. by Human Observation | Error Rate |
|---|---|---|---|---|---|
| Caotan to Jingyang | 60 | 16 | 65 | 64 | −1.6% |
| Jingyang to Sanyuan | 70 | 14 | 38 | 40 | 5% |
| Sanyuan to Tongchuan | 70 | 28 | 57 | 58 | 1.7% |
| Tongchuan to Jingyang | 80~100 | 42 | 40 | 44 | 9.1% |
| Total | - | 100 | 200 | 206 | **2.91%** |

If the parameters are assumed as constants of $V = 10$ m/s, $T = 30$ ms, $d = 0.18$ m and $\theta = 0°$, then the theoretical identification rate will be $P = 84\%$ according to the probability analysis in Equation (22) of Section 4. The actual measurement results from Table 2 and Table 3 have a slightly higher detection rate of over 90.97%, because the relative driving speeds are not always a constant for different passing situations, if the relative speed of the passing vehicles with respect to the host vehicle are less than 10 m/s, the detection rate will be higher than the theoretical value calculated using Equation (22).

## 6. Conclusions

The paper aims to track the slower vehicles that are occupying the passing lane of highways for a certain time, which might lead to traffic jams or driving safety problems. A novel ultrasonic sensor system to detect this kind of TRV behavior on a moving measurement device is developed, by monitoring the driving status of other passing vehicles in real-time. Accordingly, a two-dimensional state method is proposed to fulfill the function of TRV detection and counting. The sensor system is comprised of two parallel ultrasonic sensors to scan the passing vehicles, and the distances measured by both sensors are converted into the binary spacial states, and the historical stored measurement data act as the time sequential states, to perform more reliable highway TRV behavior detection. Through the monitoring of the changes of the two-dimensional states, the relative motion direction of other vehicles can be recognized, the theoretical identification rate is analyzed according to the random driving situation of the measurement device and the passing vehicles. Experiments have shown that the proposed ultrasonic sensor system is able to identify the improper TRV driving behavior of the host vehicle to an accuracy of about 90.97%. The proposed ultrasonic sensor system, as well as the TRV detection and counting method will be a significant supplement in intelligent transportation systems (ITS) and vehicular *ad hoc* networks (VANETs), so as to avoid traffic accidents and injuries.

## Acknowledgments

This work was supported in part by China Postdoctoral Science Foundation under Grant 2013M542349, in part by the Fundamental Research Funds for the Central Universities of China under Grant xjj2013026, and in part by the State Key Laboratory of Electrical Insulation and Power Equipment under Grant EIPE14314.

## Author Contributions

All authors have made significantly contributions to the paper. Jun Liu carried out a literature survey and proposed the concept of combining two parallel ultrasonic sensors system and the two-dimensional state method, and wrote the paper as well. The hardware structure was developed by Hongqiang Lv and the software program was coded and tested by Jun Liu under the supervision of Jiuqiang Han. Hongqiang Lv and Bing Li contributed to data collection and modification of the manuscript.

## References

1. Xuerong, L.; Yao, M.; Boxuan, G. *The 2012–2016 Highway Industry Investment Analysis and Outlook Report in China*; China Investigation Consultant: Shenzhen, Guangdong, 2012.
2. Buch, N.; Velastin, S.A.; Orwell, J. A Review of Computer Vision Techniques for the Analysis of Urban Traffic. *IEEE Trans. Intell. Transp. Syst.* **2011**, *12*, 920–939.
3. Agarwal, V.; Murali, N.V.; Chandramouli, C. A Cost-Effective Ultrasonic Sensor-Based Driver-Assistance System for Congested Traffic Conditions. *IEEE Trans. Intell. Transp. Syst.* **2009**, *10*, 486–498.
4. Zhang, X.; Xiang, H.; Jing, R.; Tu, Z. Road Traffic Injuries in the People's Republic of China, 1951–2008. *Traffic Inj. Prev.* **2011**, *12*, 614–620.
5. The Ministry of Public Security of China. *Highway Traffic Management Law of the People's Republic of China*; The Ministry of Public Security of the People's Republic of China: Beijing, China, 1995.
6. The State Council of China. *Implementation Rules of Road Traffic Safety Regulation of the People's Republic of China*; The Ministry of Public Security of the People's Republic of China: Beijing, China, 2004.
7. Jiménez, F.; Naranjo, J.E.; Gómez, O.; Anaya, J.J. Vehicle Tracking for an Evasive Manoeuvres Assistant Using Low-Cost Ultrasonic Sensors. *Sensors* **2014**, *14*, 22689–22705.
8. Hyungjin, K.; Joo-Hyune, L.; Sung-Wook, K.; Jae-In, K.; Cho. D. Ultrasonic Vehicle Detector for Side-Fire Implementation and Extensive Results Including Harsh Conditions. *IEEE Trans. Intell. Transp. Syst.* **2001**, *2*, 127–134.
9. Kohler, P.; Connette, C.; Verl, A. Vehicle Tracking Using Ultrasonic Sensors and Joined Particle Weighting. In Proceedings of the IEEE International Conference on Robotics and Automation (ICRA), Karlsruhe, Germany, 6–10 May 2013; pp. 2900–2905.
10. Schlegl, T.; Bretterklieber, T.; Neumayer, M.; Zangl, H. Combined Capacitive and Ultrasonic Distance Measurement for Automotive Applications. *IEEE Sens. J.* **2011**, *11*, 2636–2642.
11. Hussain, T.M.; Baig, A.M.; Saadawi, T.N.; Ahmed, S.A. Infrared Pyroelectric Sensor for Detection of Vehicular Traffic Using Digital Signal Processing Techniques. *IEEE Trans. Veh. Technol.* **1995**, *44*, 683–689.

12. Martinez, F.J.; Toh, C.K.; Cano, J.C.; Calafate, C.T.; Manzoni, P. Emergency Services in Future Intelligent Transportation Systems Based on Vehicular Communication Networks. *IEEE Intell. Transp. Syst. Mag.* **2010**, *2*, 6–20.
13. Zhang, J.; Gao, T.; Liu, Z. Traffic Video Based Cross Road Violation Detection. In Proceedings of the ICMTMA '09. International Conference on Measuring Technology and Mechatronics Automation, Zhangjiajie, China, 11–12 April 2009; pp. 645–648.
14. Tang-Hsien, C.; Chen-Ju, C. Rear-End Collision Warning System on Account of a Rear-End Monitoring Camera. In Proceedings of the Intelligent Vehicles Symposium, 2009 IEEE, Xi'an, China, 3–5 June 2009; pp. 913–917.
15. Wang, H.; Tang, Y. RFID Technology Applied to Monitor Vehicle in Highway. In Proceedings of the Third International Conference in Digital Manufacturing and Automation (ICDMA), Guilin, China, 31 July–2 August 2012; pp. 736–739.
16. Matsuo, T.; Kaneko, Y.; Matano, M. Introduction of Intelligent Vehicle Detection Sensors. In Proceedings of the IEEE/IEEJ/JSAI International Conference on in Intelligent Transportation Systems, Tokyo, Japan, 5–8 October 1999; pp. 709–713.
17. Tacconi, D.; Miorandi, D.; Carreras, I.; Chiti, F.; Fantacci, R. Using Wireless Sensor Networks to Support Intelligent Transportation Systems. *Ad Hoc Netw.* **2010**, *8*, 462–473.
18. Yousefi, S.; Mousavi, M.S.; Fathy, M. Vehicular Ad Hoc Networks (VANETs): Challenges and Perspectives. In Proceedings of the 6th International Conference on ITS Telecommunications Proceedings, Chengdu, China, 21–23 June 2006, pp. 761–766.
19. Qian, Y.; Lu, K.; Moayeri, N. A Secure VANET MAC Protocol for DSRC Applications. In Proceedings of the IEEE GLOBECOM 2008 Global Telecommunications Conference, New Orleans, LO, USA, 30 November–4 December 2008; pp. 1–5.
20. Zhou, L.; Zhang, Y.; Song, K.; Jing, W.; Vasilakos, A.V. Distributed Media Services in P2P-Based Vehicular Networks. *IEEE Trans. Veh. Technol.* **2011**, *60*, 692–703.
21. Cherfaoui, V.; Denoeux, T.; Cherfi, Z.L. Distributed Data Fusion: Application to Confidence Management in Vehicular Networks. In Proceedings of the 11th International Conference on Information Fusion, Cologne, Germany, 30 June–3 July 2008; pp. 1–8.
22. Jo, Y.; Jung, I. Analysis of Vehicle Detection with WSN-Based Ultrasonic Sensors. *Sensors* **2014**, *14*, 14050–14069.
23. Hsu, L.-Y.; Chen, T.-L. Vehicle Dynamic Prediction Systems with On-Line Identification of Vehicle Parameters and Road Conditions. *Sensors* **2012**, *12*, 15778–15800.
24. Alonso, L.; Milanés, V.; Torre-Ferrero, C.; Godoy, J.; Oria, J.P.; De Pedro, T. Ultrasonic Sensors in Urban Traffic Driving-Aid Systems. *Sensors* **2011**, *11*, 661–673.
25. Hsieh, J.W.; Yu, S.H.; Chen, Y.S.; Hu, W.F. Automatic Traffic Surveillance System for Vehicle Tracking and Classification. *IEEE Trans. Intell. Transp. Syst.* **2006**, *7*, 175–187.
26. Lazarus, S.B.; Ashokaraj, I.; Tsourdos, A.; Zbikowski, R.; Silson, P.M.G.; Aouf, N.; White, B.A. Vehicle Localization Using Sensors Data Fusion Via Integration of Covariance Intersection and Interval Analysis. *IEEE Sens. J.* **2007**, *7*, 1302–1314.

27. Jeon, S.; Kwon, E.; Jung, I. Traffic Measurement on Multiple Drive Lanes with Wireless Ultrasonic Sensors. *Sensors* **2014**, *14*, 22891–22906.
28. Jae, K.S.; Ho, G.J. Sensor Fusion-Based Vacant Parking Slot Detection and Tracking. *IEEE Trans. Intell. Transp. Syst.* **2014**, *15*, 21–36.
29. Dhole, R.N.; Undre, V.S.; Solanki, C.R.; Pawale, S.R. Smart Traffic Signal Using Ultrasonic Sensor. In Proceedings of the 2014 International Conference on Green Computing Communication and Electrical Engineering (ICGCCEE), Coimbatore, India, 6–8 March 2014; pp. 1–4.
30. Gutierrez-Osuna, R.; Janet, J.A.; Luo, R.C. Modeling of Ultrasonic Range Sensors for Localization of Autonomous Mobile Robots. *IEEE Trans. Ind. Electron.* **1998**, *45*, pp. 654–662.

4

# Ultrasonic Transducer Fabricated using Lead-Free BFO-BTO+Mn Piezoelectric 1-3 Composite

**Yan Chen [1,2], Kai Mei [2], Chi-Man Wong [2], Dunmin Lin [3], Helen Lai Wa Chan [2] and Jiyan Dai [1,2,*]**

[1] The Hong Kong Polytechnic University, Shenzhen Research Institute, Shenzhen 518057, China; E-Mail: ap.cheny@connect.polyu.hk

[2] Department of Applied Physics, The Hong Kong Polytechnic University, Hong Kong, China; E-Mails: meikai1990@gmail.com (K.M.); 14902762r@connect.polyu.hk (C.-M.W.); apahlcha@polyu.edu.hk (H.L.W.C.)

[3] College of Chemistry and Materials Science, Sichuan Normal University, Chengdu 610066, China; E-Mail: ddmd222@sicnu.edu.cn

* Author to whom correspondence should be addressed; E-Mail: jiyan.dai@polyu.edu.hk

Academic Editor: Delbert Tesar

---

**Abstract:** Mn-doped 0.7BiFe$O_3$-0.3BaTi$O_3$ (BFO-0.3BTO+Mn 1% mol) lead-free piezoelectric ceramic were fabricated by traditional solid state reaction. The phase structure, microstructure, and ferroelectric properties were investigated. Additionally, lead-free 1–3 composites with 60% volume fraction of BFO-BTO+Mn ceramic were fabricated for ultrasonic transducer applications by a conventional dice-and-fill method. The BFO-BTO+Mn 1-3 composite has a higher electromechanical coupling coefficient ($k_t$ = 46.4%) and lower acoustic impedance ($Z_a$ ~ 18 MRayls) compared with that of the ceramic. Based on this, lead-free piezoelectric ceramic composite, single element ultrasonic transducer with a center frequency of 2.54 MHz has been fabricated and characterized. The single element transducer exhibits good performance with a broad bandwidth of 53%. The insertion loss of the transducer was about 33.5 dB.

**Keywords:** lead-free ceramic; 1-3 composite; ultrasonic transducer

---

## 1. Introduction

Lead-based ceramics, especially $Pb(Zr,Ti)O_3$ (PZT), are the most extensively used piezoelectric ceramics for transducer applications due to their stable and good piezoelectric properties [1–4]. However, environmental problems are caused by preparing lead-based ceramics because of PbO volatility. Recently, lead-free piezoelectric materials with relatively good piezoelectric properties have attracted a great deal of attention owing to environmental conservation [5]. Therefore, the lead-free ceramics were used in various applications, such as ultrasonic transducers [6–9], pyroelectric sensors [10] and actuators [11,12].

BFO-BTO system is proposed to be an important family of high-performance lead-free piezoelectric ceramics because of its good ferroelectric properties [13–17]. Additionally, Mn doping is a common method to enhance the piezoelectric properties and reduce the dielectric loss of the piezoelectric materials [15,16], thus, Mn modified BFO-BTO ceramic with good ferroelectric properties was chosen for transducer applications.

In order to further improve the performance of transducer, such as detecting highly attenuative materials in non-destructive evaluation applications or acquiring high resolution ultrasonic imaging in medical field [18–21]. The 1-3 composite was widely used for further enhancing the acoustic and electrical properties of transducers due to its lower acoustic impedance Z and higher $k_t$ compared to the single phase ceramic [22].

Therefore, in this work, lead-free BFO-BTO+Mn ceramic was characterized, and a single-element ultrasonic transducer was fabricated and characterized using the BFO-BTO+Mn ceramic/epoxy 1-3 composite.

## 2. Experimental Section

### *2.1. Ceramic Characterization*

Mn-doped $BiFeO_3$-$0.3BaTiO_3$ lead-free piezoelectric ceramic were fabricated by a traditional solid state reaction using metal oxides and carbonate powders. The crystalline structure of the ceramic was identified by an X-ray diffraction (XRD) diffractometer (SmartLab, Rigaku Co., Tokyo, Japan). The microstructure was characterized using scanning electron microscopy (SEM, TM3000, HITACHI, Japan). The bulk ceramic density was measured by the Archimedes method. The room temperature polarization-electric field (P-E) hysteresis loops were measured using a modified Sawyer Tower circuit at 100 Hz. The sample was poled under 6 kV/mm at 100 °C for 15 min in a silicon oil bath. The piezoelectric properties of the samples were calculated following the IEEE standards on piezoelectricity [23]. The dielectric properties of the BFO-BTO+Mn ceramic and its 1-3 composite were measured using an impedance analyzer (Agilent 4294A, Santa Clara, CA, USA).

### *2.2. 1-3 Composite Fabrication*

For improving the transducer performance, BFO-BTO+Mn ceramic/epoxy 1-3 composite selected as the active element of the transducer. The composite was fabricated using the traditional dice-and-fill method. A dicing saw with a 50 μm-thick blade was used to dice the sample. The kerf is about 75 μm due to the blade vibration, and the ceramic volume fraction is 60%. The low-viscosity epoxy (Epo-Tek 301, Epoxy Technology, Billerica, MA, USA) was used to fill the kerf. Before the epoxy solidification, the composite sample was vacuumed to remove the bubbles. Then the 1-3 composite was obtained by lapping the excess ceramic and epoxy away. Silver paint (SPI, West Chester, PA, USA) as the electrode was covered on the top and bottom faces of the 1-3 composite disk.

## 3. Results and Discussion

Figure 1 shows the XRD pattern of the BFO-BTO+Mn ceramic. The sample possesses a typical $ABO_3$ perovskite rhombohedral structure and no secondary phase is observed. This suggests that Mn has diffused into the BFO-BTO lattices to form a homogeneous solid solution.

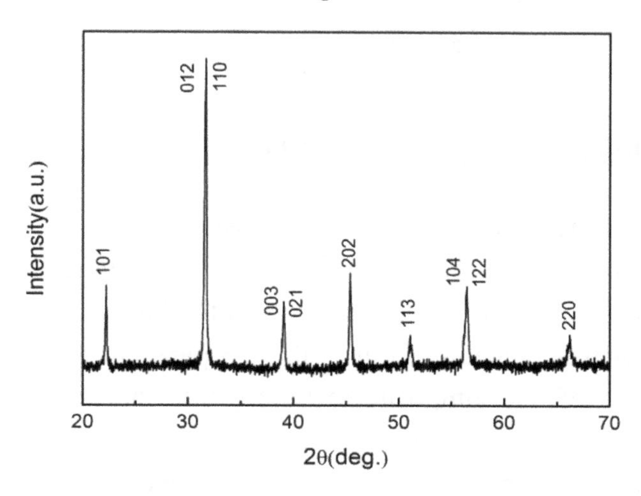


**Figure 1.** XRD pattern of the BFO-BTO+Mn ceramic.

Figure 2 shows the SEM micrograph of the BFO-BTO+Mn ceramic. It can be found that the grain size is about 4 μm and the ceramic is dense and without pores. The density of the ceramic is high with a value of about 7366 $kg/m^3$.

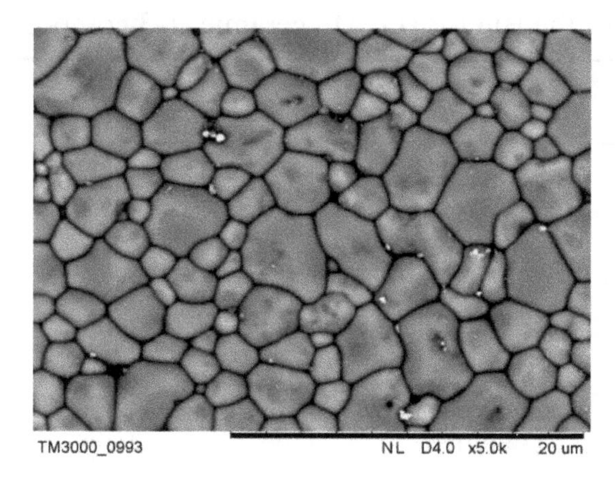


**Figure 2.** SEM micrograph of the BFO-BTO+Mn ceramic.

Figure 3 shows the P-E hysteresis loop of the BFO-BTO+Mn ceramic at room temperature. The remnant polarization $P_r$ value is found to be 32 $\mu C/cm^2$, which is similar to the previous report [16]. The coercive field $E_c$ is about 2 kV/mm. The result suggests that the BFO-BTO+Mn ceramic has good ferroelectric property.

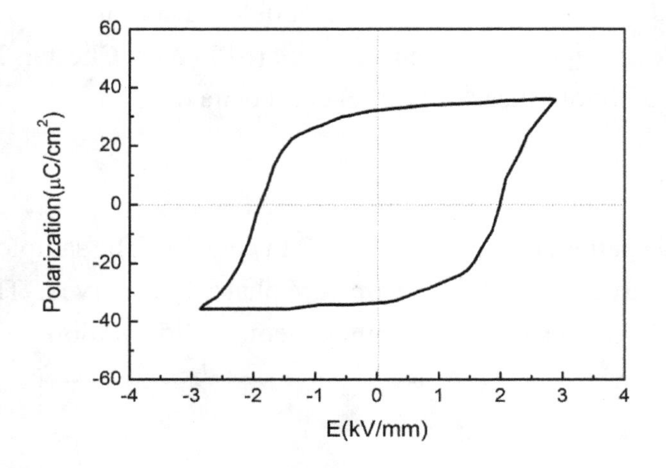


**Figure 3.** P-E loop of the BFO-BTO+Mn ceramic.

Figure 4 shows diagram structure of the 1-3 composite transducer. There are four components including a backing layer, 1-3 piezoelectric composite, matching layers, and metal housing. The 1-3 piezoelectric composite with a diameter of 20.8 mm and a thickness 0.75 mm was mounted in a metal housing. The matching layer was placed in front of the 1-3 piezoelectric composite. The backing layer was molded on the rear side of the composite to reduce the ring-down time of the transducer. The geometry of the transducer is a piston structure. Table 1 shows the properties of the BFO-BTO+Mn ceramic, 1-3 composite and PZT ceramic. It can be seen that the electromechanical coupling coefficient $k_t$ of the 1-3 composite (46.4%) is higher than that of the ceramic (37.5%) and similar to that of PZT ceramic (46%) [24]. Additionally, the acoustic impedance ($Z_a \sim$ 17.76 MRayls) of the 1-3 composite is lower compared to that of the BFO-BTO+Mn ceramic (29.84 MRayls) and the PZT ceramic (32.5 MRayls) [24]. The 1-3 composite, with a higher $k_t$ and lower $Z_a$, is more suitable for transducer applications.

**Table 1.** Properties of the BFO-BTO+Mn ceramic, 1-3 composite and PZT ceramic.

| Material | $k_t$ (%) | $d_{33}$ (pC/N) | ρ (kg/m³) | c (m/s) | $\varepsilon^T$ | Za (MRayls) |
|---|---|---|---|---|---|---|
| BFO-BTO+Mn Ceramic | 37.5 | 82 | 7366 | 4051 | 596 | 29.84 |
| 1-3 composite (60%) | 46.4 | 45 | 4800 | 3700 | 290 | 17.76 |
| PZT ceramic [24] | 46 | 420 | 7700 | 4100 | 1850 | 32.5 |

Single matching layer was designed based on a one-dimensional Krimholtz-Leedom-Matthae (KLM) model software PiezoCAD (Version 3.03 for Windows, Sonic concepts, Wood-inville, WA, USA). The matching layer was fabricated using low-viscosity epoxy (Epo-Tek 301, Epoxy Technology, Billerica,

MA, USA) with aluminum oxide powder. The acoustic impedances of the matching layers are 3.98 MRayls. The thickness of the matching layer was designed to be λ/4 (~0.27 mm), where λ is the wavelength of the acoustic wave transmitting in a matching layer at the resonance frequency. The backing layer is epoxy (Epo-Tek 301, Epoxy Technology, Billerica, MA, USA) loaded with larger-size aluminum oxide powder and polymer micro-bubbles. The acoustic impedance of the backing layer is light, about 5.05 MRayls. The thickness of the backing layer was about 10 mm. The properties of the matching and backing materials are shown in Table 2.

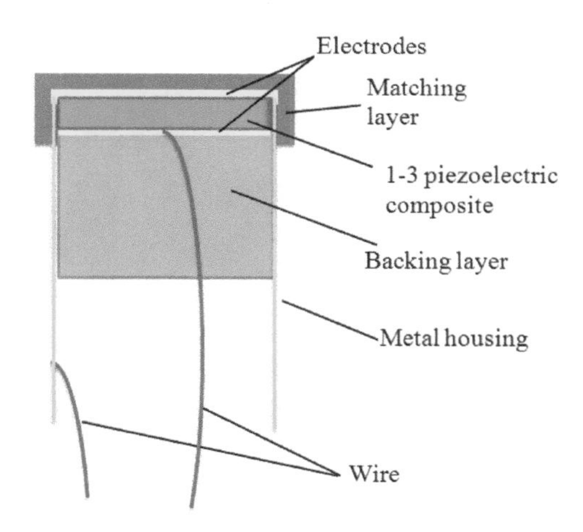


**Figure 4.** Schematic diagram of the 1-3 composite transducer.

**Table 2.** Properties of the matching and backing materials.

| Materials | Use | ρ (kg/m³) | c (m/s) | Za (MRayls) | Thickness (mm) |
|---|---|---|---|---|---|
| Aluminum oxide powder/Epo-tek 301 | Matching layer | 1453 | 2740 | 3.98 | 0.27 |
| Aluminum oxide powder and micro-bubbles/Epo-tek 301 | Backing layer | 1725 | 2930 | 5.05 | 10 |

The square of effective electromechanical coupling coefficient $k_{eff}^2$, which describes the conversion of energy between electrical and mechanical, was calculated as following [23,25]:

$$k_{eff} = \sqrt{1 - \frac{f_s^2}{f_p^2}}$$

where $f_s$ is the frequency of the maximum conductance, and $f_p$ is the frequency of maximum resistance. For this transducer, the values of the $f_s$ and $f_p$ are about 2.62 MHz and 2.93 MHz, respectively. Therefore, the value of $k_{eff}$ for this transducer is 48.5%.

The performance of the transducer was evaluated using a conventional pulse-echo response measurement method. By connecting to an ultrasonic pulser-receiver (Panametrics 5900PR, Olympus, Japan), the transducer was excited by a 1 μJ electrical pulse with 1 kHz repetition and 50 ohms damping. The echo response was captured by the receiving circuit of the pulser-receiver and displayed on an oscilloscope (Infinium 54810A, HP/Agilent, Santa Clara, CA, USA). The frequency domain pulse-echo response was acquired on the oscilloscope by Fast Fourier Transforms (FFT) math feature.

Based on the PiezoCAD software, the modeled pulse-echo waveform and frequency spectrum of the 1-3 composite transducer is shown in Figure 5a. Compared to the modeled results, the experimental results (Figure 5b) show similar characteristics. It is found that the measured center frequency of the transducer (2.54 MHz) agrees well with the modeled result (2.53 MHz). The experimental bandwidth of the transducers is 53%, which matche quite well with the modeled result (54%). This transducer's performance is comparable with that of a PZT transducer (1.88 MHz, 56.4%) and BNKLBT lead-free transducer (1.84 MHz, 63.6%) with a similar frequency and structure [6].

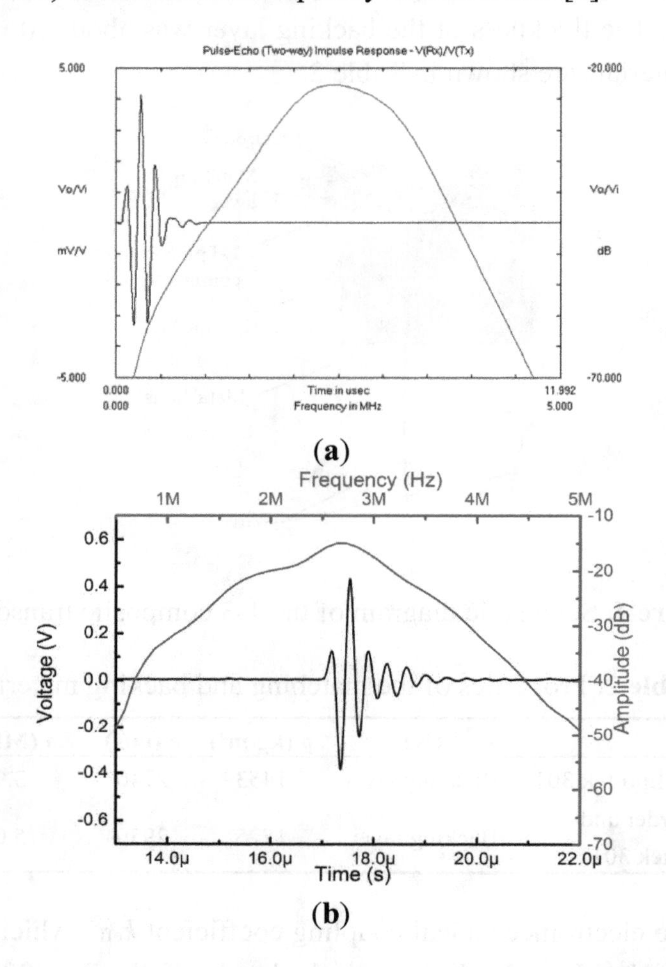


**Figure 5.** (**a**) Modeled and (**b**) measured pulse-echo waveform and frequency spectra of the 1-3 composite transducer.

## 4. Conclusions

The BFO-BTO+Mn/epoxy 1-3 composite ultrasound transducers have been successfully fabricated. The performance of the transducers have been simulated and measured. The measured bandwidth was found to be 53%. The transducers were found to exhibit low insertion loss of ~33.5 dB. The results suggest that the BFO-BTO+Mn lead-free composites have the potential to be used for ultrasonic transducers.

## Acknowledgments

This research was supported by the National key Basic Research Program of China (973 Program) under Grant No. 2013CB632900. Financial support from The Hong Kong Polytechnic University Strategic Importance Plan (No: 1-ZVCG&1-ZV9B).

## Author Contributions

Yan Chen wrote the manuscript and performed the experiments, Kai Mei and Chi-Man Wong performed the experiments, Dunmin Lin, Helen L.W. Chan and Jiyan Dai conceived the project and edited the manuscript.

## References


1. Yamamura, T. Ferroelectric Properties of the $PbZrO_3$-$PbTiO_3$ System. *Jpn. Appl. Phys.* **1996**, *35*, 5104–5108.
2. Lin, D.M.; Xiao, D.Q.; Zhu, J.G.; Yu, P. Piezoelectric and Ferroelectric Properties of $[Bi_{0.5}(Na_{1-x-y}K_xLi_y)_{0.5}]TiO_3$ Lead-Free Piezoelectric Ceramics. *Appl. Phys. Lett.* **2006**, *88*, 062901:1–062901:3.
3. Takenaka, T.; Nagata, H. Current Status and Prospects of Lead-Free Piezoelectric Ceramics. *J. Eur. Ceram. Soc.* **2005**, *25*, 2693–2700.
4. Tani, T.; Kimura, T. Reactive Templated Grain Growth Processing for Lead Free Piezoelectric Ceramics. *Adv. Appl. Ceram.* **2006**, *105*, 55–63.
5. Saito, Y.; Takao, H.; Tani, T.; Nonoyama, T.; Takatori, K.; Homma, T.; Nagaya, T.; Nakamura, M. Lead-Free Piezoceramics. *Nature* **2004**, *432*, 84–87.
6. Edwards, G.C.; Choy, G.C.; Chan, H.L.W.; Scott, D.A.; Batten, A. Lead-Free Transducer for Non-Destructive Evaluation. *Appl. Phys. A* **2007**, *88*, 209–215.
7. Chan, H.L.W.; Choy, S.H.; Chong, C.P.; Li, H.L.; Liu, P.C.K. Bismuth Sodium Titanate Based Lead-Free Ultrasonic Transducer for Microelectronics Wirebonding Applications. *Ceram. Int.* **2008**, *34*, 773–777.
8. Choy, S.H.; Wang, X.X.; Chong, C.P.; Chan, H.L.W.; Liu, P.C.K.; Choy, C.L. $0.90(Bi_{1/2}Na_{1/2})TiO_3$–$0.05(Bi_{1/2}K_{1/2})TiO_3$–$0.05BaTiO_3$ Transducer for Ultrasonic Wire bonding Applications. *Appl. Phys. A* **2006**, *84*, 313–316.
9. Yan, X.; Lam, K.H.; Li, X.; Chen, R.; Ren, W.; Ren, X.; Zhou, Q.; Shung, K.K. Lead-Free Intravascular Ultrasound Transducer Using BZT–50BCT Ceramics. *IEEE Trans. Ultrason. Ferroelectr. Freq. Control* **2013**, *60*, 1272–1276.
10. Barrel, J.; MacKenzie, K.J.D.; Stytsenko, E.; Viviani, M. Development of Pyroelectric Ceramics for High-Temperature Applications. *Mater. Sci. Eng. B* **2009**, *161*, 125–129.
11. Lam, K.H.; Wang, X.X.; Chan, H.L.W. Lead-Free Piezoceramic Cymbal Actuator. *Sens. Actuators A Phys.* **2006**, *125*, 393–397.
12. Lam, K.H.; Lin, D.M.; Kwok, K.W.; Chan, H.L.W. Lead-Free Piezoelectric-Metal-Cavity (PMC) Actuators. *IEEE Trans. Ultrason. Ferroelectr. Freq. Control* **2008**, *55*, 1682–1685.
13. Li, Y.; Jiang, N.; Lam, K.H.; Guo, Y.Q.; Zheng, Q.J.; Li, Q.; Zhou, W.; Wan, Y.; Lin, D. Structure, Ferroelectric, Piezoelectric, and Ferromagnetic Properties of $BiFeO_3$-$BaTiO_3$-$Bi_{0.5}Na_{0.5}TiO_3$ Lead-Free Multiferroic Ceramics. *J. Am. Ceram. Soc.* **2014**, *97*, 3602–3608.

14. Zheng, Q.J.; Luo, L.L.; Lam, K.H.; Jiang, N.; Guo, Y.Q.; Lin, D. Enhanced Ferroelectricity, Piezoelectricity, and Ferromagnetism in Nd-modified $BiFeO_3$-$BaTiO_3$ Lead-Free Ceramics. *J. Appl. Phys.* **2014**, *116*, doi:10.1063/1.4901198.
15. Wan, Y.; Li, Y.; Li, Q.; Zhou, W.; Zheng, Q.J.; Wu, X.C.; Xu, C.G.; Zhu, B.P.; Lin, D. Microstructure, Ferroelectric, Piezoelectric, and Ferromagnetic Properties of Sc-Modified $BiFeO_3$-$BaTiO_3$ Multiferroic Ceramics with $MnO_2$ Addition. *J. Am. Ceram. Soc.* **2014**, *97*, 1809–1818.
16. Leontsevw, S.O.; Eitel, R.E. Dielectric and Piezoelectric Properties in Mn-Modified (1-x)$BiFeO_3$-x$BaTiO_3$ Ceramics. *J. Am. Ceram. Soc.* **2009**, *92*, 2957–2961.
17. Wang, T.H.; Ding, Y.; Tu, C.S.; Yao, Y.D.; Wu, K.T.; Lin, T.C.; Yu, H.H.; Ku, C.S.; Lee, H.Y. Structure, Magnetic, and Dielectric Properties of (1-x)$BiFeO_3$-x$BaTiO_3$ Ceramics. *J. Appl. Phys.* **2011**, *109*, 1–4.
18. Safari, A.; Janas, V.F.; Bandyopadhyay, A. Development of Fine-Scale Piezoelectric Composites for Transducers. *AlChE J.* **2004**, *43*, 2849–2856.
19. Smith, W.A. The Role of Piezocomposites in Ultrasonic Transducers. *IEEE Proc. Ultrason. Symp.* **1989**, 755–766.
20. Gururaja, T.R. Piezoelectrics for Medical Ultrasonic Imaging. *Am. Ceram. Soc. Bull.* **1994**, *73*, 50–55.
21. Kim, K.B.; Hsu, D.K.; Ahn, B.; Kim, Y.G.; Barnard, D.J. Fabrication and Comparison of PMN–PT Single Crystal, PZT and PZT-Based 1–3 Composite Ultrasonic Transducers for NDE Applications. *Ultrasonics* **2010**, *50*, 790–797.
22. Zhou, D.; Lam, K.H.; Chen, Y.; Zhang, Q.H.; Chiu, Y.C.; Luo, H.S.; Dai, J.Y.; Chan, H.L.W. Lead-Free Piezoelectric Single Crystal Based 1–3 Composites for Ultrasonic Transducer Applications. *Sens. Actuators A Phys.* **2012**, *182*, 95–100.
23. IEEE Standard on Piezoelectricity. ANSI/IEEE Standard: New York, NY, USA, 1987.
24. Lam, K.H.; Lin, D.M.; Ni, Y.Q.; Chan, H.L.W. Lead-free Piezoelectric KNN-based Pin Transducer for Structural Monitoring Applications. *Struct. Health Monit.* **2009**, *8*, 283–289.
25. Tressler, J.F. Piezoelectric Transducer Designs for Sonar Applications. In *Piezoelectric and Acoustic Materials for Transducer Applications*, Safari, A., Akdoğan, E.K., Eds.; Springer Science+Business Media: New York, NY, USA, 2008; p. 219.

# 5

# Effect of Ultrasonic Treatment in the Static and Dynamic Mechanical Behavior of AZ91D Mg Alloy

**Helder Puga [1,*], Vitor Carneiro [2,†], Joaquim Barbosa [1,†] and Vanessa Vieira [2,†]**

1 Centre for Micro-Electro Mechanical Systems (CMEMS), University of Minho, Campus of Azurém, 4800-058 Guimarães, Portugal; E-Mail: kim@dem.uminho.pt

2 Department of Mechanical Engineering, University of Minho, Campus of Azurém, 4800-058 Guimarães, Portugal; E-Mails: a53996@alumni.uminho.pt (V.C.); a61938@alumni.uminho.pt (V.V.)

† These authors contributed equally to this work.

* Author to whom correspondence should be addressed; E-Mail: puga@dem.uminho.pt;

Academic Editor: Hugo F. Lopez

---

**Abstract:** The present study evaluates the effect of high-intensity ultrasound (US) in the static and dynamic mechanical behavior of AZ91D by microstructural modification. The characterization of samples revealed that US treatment promoted the refinement of dendrite cell size, reduced the thickness, and changed the β-$Mg_{17}Al_{12}$ intermetallic phase to a globular shape, promoted its uniform distribution along the grain boundaries and reduced the level of porosity. In addition to microstructure refinement, US treatment improved the alloy mechanical properties, namely the ultimate tensile strength (40.7%) and extension (150%) by comparison with values obtained for castings produced without US vibration. Moreover, it is suggested that the internal friction, enhanced by the reduction of grain size, is compensated by the homogenization of the secondary phase and reduction of porosity. It seems that by the use of US treatment, it is possible to enhance static mechanical properties without compromising the damping properties in AZ91D alloys.

**Keywords:** ultrasonic treatment; AZ91D; intermetallic phases; mechanical properties; damping

---

## 1. Introduction

Mg alloys are a promising candidate for applications involving low-weight and high-damping characteristics. Metallic materials have relatively low damping, although Mg alloys, in addition to their very low density, possess the highest damping capacity among metals [1]. However, they are characterized by a relatively low elastic modulus and mechanical strength. This dependence between static and dynamic mechanical properties is a common feature of metals, given that high-damping materials are generally unsuitable in many structural applications [2]. Since, the referred mechanical properties are dependent on the material's microstructure and bulk defects, Mg alloy processing is the main route to manipulate both static and dynamic properties. Enhanced static mechanical properties, *i.e.*, elevated values of yield/ultimate strength and fracture extension, are generally correlated with the microstructure of a given material and the presence of bulk defects. For instance, the values of plastic flow stress ($\sigma_y$) is related to the grain size ($d$) by the Hall-Petch relation [3]:

$$\sigma_y = \sigma_0 + k_y d^{-1/2} \tag{1}$$

where $\sigma_0$ is the friction stress of mobile dislocations and $k_y$ defines the characteristic constant that depends on the amount of impurities and alloying elements [3]. Additionally, the values of ultimate tensile strength and fracture extension can be diminished by the presence of porosities, which are promoters of stress concentration and crack nucleation in bulk materials [4].

As for the dynamic mechanical properties, it is desirable to obtain materials with high damping properties, *i.e.*, a high ability to dissipate elastic strain energy during mechanical vibration or wave propagation [2]. This energy dissipation is promoted by the internal friction generated through inelastic relaxation and/or thermal activation. The presence of point defect relaxation, dislocation mechanisms, interfaces, and their combination with thermal activation processes are the main cause of internal friction. Consequently, they are associated with the morphology of the microstructure, namely by the secondary phases, impurity atoms [5], porosity, grain size, and boundaries [5,6]. This internal friction has a relevant role in the viscoelastic behavior of materials and consequently implies a lag time effect that is generally neglected by the classical Hooke's law formulation. According to Equation (2), the overall strain ($\varepsilon$) is composed by two terms, an elastic strain ($\varepsilon_e$) and an inelastic strain ($\varepsilon_a$) [7]:

$$\varepsilon = \varepsilon_e + \varepsilon_a \tag{2}$$

The instant stress ($\sigma$) and strain values under cyclic loading can be described by Equations (3) and (4) [7]:

$$\sigma = \sigma_0^{\,i\omega t} \tag{3}$$

$$\varepsilon = \varepsilon_0^{\,i(\omega t - \phi)} \tag{4}$$

where $\sigma_0$ and $\varepsilon_0$ are, respectively, the original stress and strain amplitudes, $\omega$ is the circular frequency, $t$ the time, and $\phi$ is the loss angle by which the strain lag is delayed relatively to the stress. These two factors are correlated to the material complex modulus ($E^*$), that can be defined by a dissipated energy part (loss modulus, $E''$) and a stored energy part (storage modulus, $E'$) according to Equation (5). The ratio between the dissipated and stored energies is the internal friction ($Q^{-1}$) of the material, which can be determined by the use of Equation (6) [8]:

$$E^* = \frac{\sigma}{\varepsilon} = E' + iE'' \quad (5)$$

$$Q^{-1} = \frac{E''}{E'} \quad (6)$$

The value of internal friction is known to be associated with the dislocation properties of metals [9] at low temperatures and with thermal energy at high temperatures, according to thermodynamics and Granato-Lücke dislocation theory [10]. According to Hu *et al.* [11], at room temperature, Mg alloys' internal friction is related to dislocations, while at elevated temperatures some crystal defects must be thermally activated to enhance damping.

From the classical work of Granato and Lücke [12], internal friction due to dislocation mechanism is strain dependent, and the logarithmic decrement $\delta$ is characterized by two parts, according to Equation (7) [13]:

$$\delta = \delta_0 + \delta_H(\varepsilon) \quad (7)$$

For low values of strain, the value of internal friction is strain-independent and assumes the value of $\delta_0$. If the strain amplitude values are increase, higher than the material critical strain ($\varepsilon_{cr}$), the internal friction becomes strain-dependent and increases exponentially. The strain-independent internal friction is the result of reversible movements of dislocation and microscopic yielding [14] around the initial pinned positions and is associated to the values obtained by the application of Equation (8) [13]:

$$\delta_0 \sim \rho l^4 \quad (8)$$

where $\rho$ is the dislocation density and $l$ is the mean length of dislocation between the weak pinning points [15]. Dislocation damping can be enhanced by thermal activation. Thermal currents and the stress generated by inhomogenities and individual anisotropic elastic crystals enhance damping in macroscopic polycrystalline metals [16]. This activation, promoted by external thermal energy, is dependent to the temperature, frequency, impurities, and microstructural morphology [17]. It may be observed that the main intrinsic factor related to the static and dynamic mechanical behavior is related with to the microstructure, namely the grain size and boundaries, presence of impurities, secondary phases, and porosity. Thus, in order to optimize material characteristics, it is essential to processes materials to manipulate their microstructure.

The microstructural refinement of Mg alloys can be performed by two basic approaches: (i) thermomechanical treatment through plastic deformation in which dynamic recrystallization leads to the formation of small equiaxed grains [18]; (ii) casting processing where two different routes can be followed: chemical [19] or physical [20–22] treatment. An example of physical processing is the application of ultrasound treatment [23]. Although the effect of refinement by ultrasound in molten metal has been studied [24], this subject has been mainly explored on the effects of ultrasonic vibration in the microstructure and only few works have focused the effect in the dynamic mechanical properties.

This paper presents and discusses the effect of ultrasonic treatment in the microstructure and the static/dynamic mechanical behavior of AZ91D alloy cast in a permanent mold by applying ultrasound indirectly to the melt during the first stages of solidification.

## 2. Experimental Section

The experimental set-up used to perform ultrasonic refinement consisted on a MMM (Multi-frequency Multimode Modulated—MPI, Le Locle, Switzerland) technology US power supply unit, a high power ultrasonic generator (3000 W), a Ø38.1 mm × 92 mm $Ti_6Al_4V$ long acoustic waveguide, an acoustic load composed by a steel die (Ø24 mm × 120 mm long) and the liquid metal, according to Figure 1a. The MMM technology is characterized by synchronously exciting many vibration modes through the coupled harmonics and sub-harmonics in solids and liquid containers, to produce high intensity multimode vibration that are uniform and repeatable, which avoid the creation of stationary and standing waves, so that the whole vibrating system is fully agitated, improving the degassing and refinement process. The equipment is fully controlled through Windows compatible software developed by MPI (Le Locle, Switzerland).

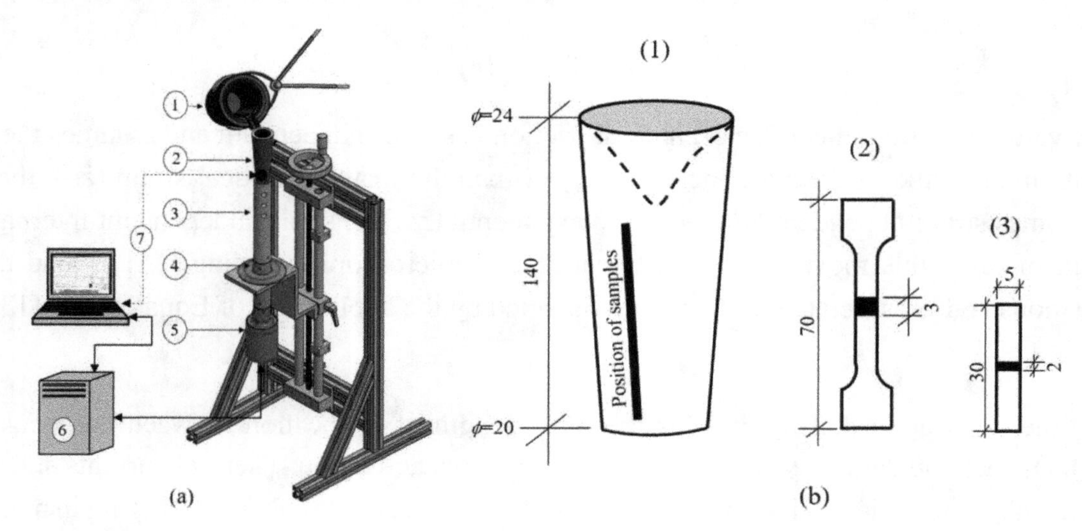


**Figure 1. (a)** Experimental setup: (1) crucible, (2) steel die, (3) waveguide, (4) booster and (5) 20 kHz Ttransducer, (6) MMM generator, (7) thermocouple sensor; **(b)** casted specimens: (1) casting, (2) tensile sample and (3) DMA sample.

The AZ91D alloy (whose chemical composition is presented in Table 1) was melted and held inside the crucible at 700 °C for 15 min for homogenization while protected by $CO_2$ + 0.5% $SF_6$ atmosphere to prevent the oxidation. The molten alloy was then allowed to cool to 680 °C and poured in the steel die (pre-heated to 250 °C). For every experimental condition, US vibration regulated for 400 W and 20.1 ± 0.25 kHz frequency was continuously applied to the bottom of the die until solidus ±10 °C and then stopped, allowing the melt to cool to room temperature. For the sake of comparison the alloy was also poured without ultrasonic vibration.

Specimens for microstructure characterization were taken from cast samples (Figure 1b(1)), by sectioning them perpendicularly to its longitudinal axis, at a distance of 60 mm to the waveguide/mold interface. The samples were etched in a solution containing 1 mL glacial acetic acid, 50 mL distilled water, and 150 mL anhydrous ethyl alcohol. The porosity was determined by the use of an image editing software (ImageJ) where a contrast tool was applied to reveal the fraction of area covered by pores in five image fields by processing conditions, with a magnification of 100×.

**Table 1.** Composition of the AZ91D Mg alloy used in this work, obtained by optical emission spectrometry.

| Element | wt. % |
|---|---|
| Al | 9.100 |
| Zn | 0.850 |
| Mn | 0.150 |
| Fe | 0.005 |
| Cu | 0.003 |
| Ni | 0.002 |
| Si | 0.050 |
| Other | 0.030 |
| Mg | Bal. |

To characterize the static mechanical behavior for both experimental conditions, specimens (10 of each processing conditions) were machined from the as-cast samples with a gauge length $L_0$ of 40 mm and a rectangular cross section of 5 × 3 mm$^2$ (Figure 1b(2)). Tensile tests were carried out at room temperature with 0.02 s$^{-1}$ strain rate. Vickers hardness measurements were performed on the as-cast US treated and non-treated samples. Ten measurements were performed for each processing, using a 5 kgf load and 20 s dwell.

To determine the dynamic thermomechanical behavior, the 30 × 5 × 2 mm$^3$ rectangular samples (Figure 1b(3)) were tested using Dynamic Mechanical Analysis in a single-cantilever configuration. This type of test applies a sinusoidal load to the sample and monitors the resultant deformation, evaluating the energy accumulated in the material ($E'$, storage modulus) and the energy that is lost in that process ($E''$, loss modulus). Consequently, the damping capacity of the material can be evaluated determining its internal friction ($Q^{-1}$), according to Equation 6 [25]. The samples were tested using a 4 × 10$^{-4}$ mm/mm strain amplitude, at 0.5, 10, and 20 Hz frequencies while heated at 10 °C/min from room temperature to 300 °C in nitrogen atmosphere. The strain amplitude value was selected to obtain damping values related to the strain independent region above the value of critical strain [26] and the adopted frequency values have been shown to reveal the viscous manner in grain boundaries at elevated temperatures [14].

## 3. Results and Discussion

Figure 2 shows the microstructure of as-cast AZ91D alloy for both non-treated and US treated specimens. The specimens treated by US (Figure 2b,d) reveal a more refined grain structure and homogeneous α-Mg matrix, when compared with the non-treated (Figure 2a,c) specimens. According to these microstructures, the supply of acoustic energy to the melt during the solidification has a high potential to change the morphology of the β-$Mg_{17}Al_{12}$ intermetallic phase (the predominant intermetallic phase present in the alloy grain boundaries [26]).

US treatment promotes a uniform dispersion in the matrix along the grain boundaries, as already demonstrated in previous works [27]. The efficiency of this refinement, caused by the effect of US, can be attributed to cavitation by two distinct mechanisms: (i) heterogeneous nucleation at high (above liquidus) temperature, and (ii) intermetallic compound fragmentation due to the streaming acoustic effect. The co-existence of two mechanisms, during the stage of solidification under effect of ultrasonic vibration, suggests an enhanced refinement of dendrite cell size, thinning, and dispersion of β-$Mg_{17}Al_{12}$ intermetallic phase in the α-Mg matrix. The β-$Mg_{17}Al_{12}$ intermetallic is confirmed by the SEM/EDS analysis presented in Figures 3 and 4.

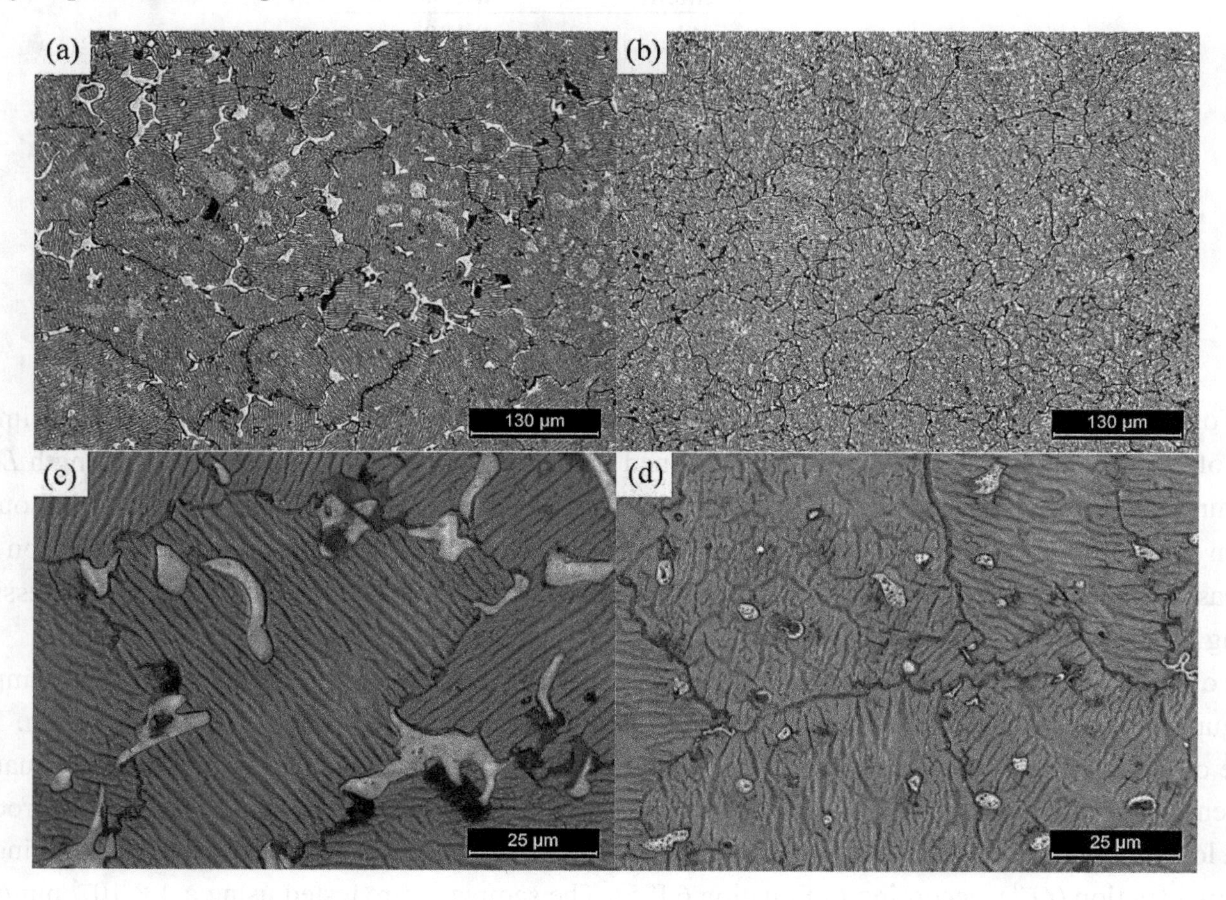


**Figure 2.** As-cast microstructures of AZ91D: (**a**,**c**) non-treated samples; (**b**,**d**) US treated samples.

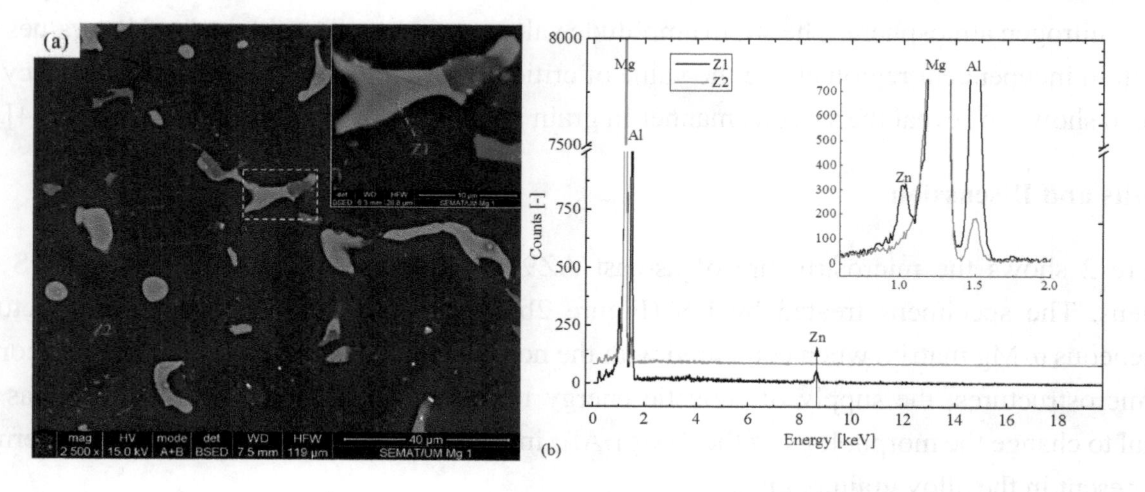


**Figure 3.** (**a**) Morphology and distribution of the β-$Mg_{17}Al_{12}$ intermetallic phase in non-treated samples; and (**b**) EDS spectrum of the β-$Mg_{17}Al_{12}$ intermetallic phase.

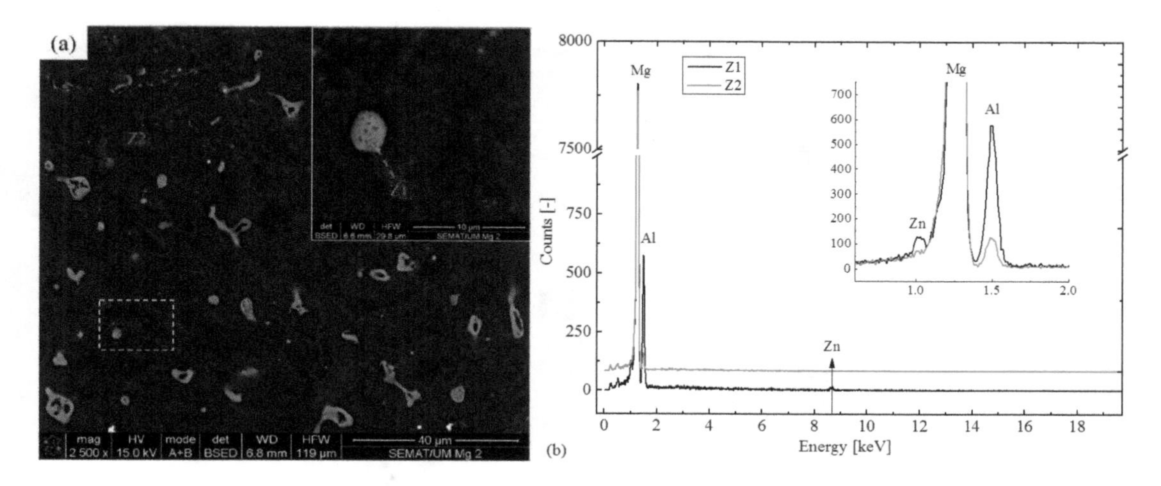


**Figure 4.** (**a**) Morphology and distribution of the β-$Mg_{17}Al_{12}$ intermetallic phase in US treated samples; and (**b**) EDS spectrum of the β-$Mg_{17}Al_{12}$ intermetallic phase.

Additionally to the microstructure refinement, US treatment also has a great impact on the porosity of Mg alloys [28]. Observing the microstructure shown in Figure 2a and according to Table 2, it is evident that the level of porosity is lower in US treated samples. The reduction of porosity in samples processed by US can be attributed to the effect of acoustic cavitation developed in the molten metal during the first stage of solidification.

**Table 2.** Characteristics of the microstructures.

| Samples | Grain Size Ø (μm) | Porosity (Area %) | β-$Mg_{17}Al_{12}$ | |
|---|---|---|---|---|
| | | | Area % | Morphology |
| Non-treated | 120 ± 20 | 2.1 ± 0.3 | 5.45 ± 0.5 | Coarse with blocky shape |
| US treated | 64 ± 5 | 0.8 ± 0.1 | 5.67 ± 0.4 | Fine with globular shape |

Figure 5 represents the mean static mechanical properties of the as-cast AZ91D alloy for both non-treated and US treated specimens, by showing their stress-strain curve. From this figure it is evident that US treatment effectively enhanced the ultimate tensile strength and the elongation of the material. Non-treated samples are characterized by an average tensile strength of 160 MPa, being increased to 225 MPa after US treatment. Additionally, an increase in elongation was also verified, which allowed an enhancement in toughness of approximately four times when compared with non-treated samples.

It is well known that static mechanical properties of Mg alloys depend on several factors, with particular emphasis to the microstructure morphology [29], presence of intermetallic phases [30], and porosity [28,29]. Focusing on the β-$Mg_{17}Al_{12}$ intermetallic phase, Li *et al.* [10] pointed that the body-centered cubic structure can deteriorate the ultimate tensile strength and elongation. Moreover, according to Du *et al.* [29] the uniform distribution of fine β-$Mg_{17}Al_{12}$ intermetallic phase enhances mechanical properties. The same authors also refer that the main defects that lower mechanical properties are gas pores and microstructural non-uniformity. The reduction of these defects contributes for the reduction of the stress concentration in the bulk material, thus improving ductility by the prevention of early fracture. This enhancement in ductility promoted by US treatment can be confirmed by the superior hardness values of the treated samples (78 ± 4 $HV_5$) when compared with non-treated samples (65 ± 6 $HV_5$).

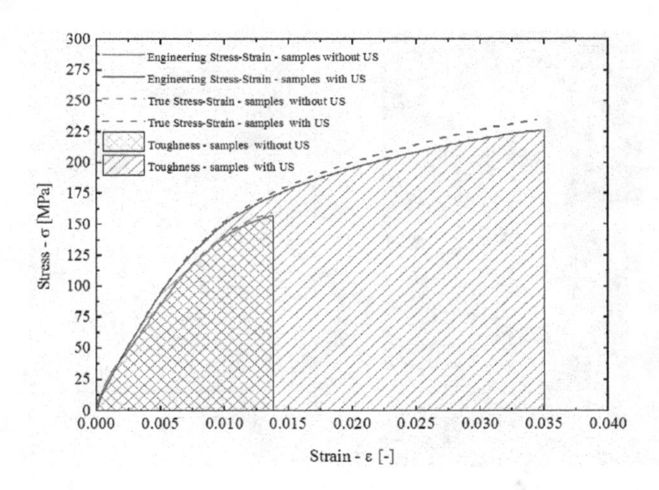


**Figure 5.** Mean static mechanical properties of as-cast AZ91 D alloy for both non-treated and US treated specimens.

According to Figure 2d and Table 2, it is suggested that the US treatment reduced the size of $\beta$-$Mg_{17}Al_{12}$ phase, promoted a globular shape and improved their uniformity and distribution in the $\alpha$-Mg matrix. It seems that the mechanism through which the US treatment affects the size and shape of the intermetallic phase depends on the solidification process under ultrasonic vibration. As the US treatment was performed continuously from the pouring temperature until solidus (±10 °C), the results can be explained as a conjugation of two mechanisms: cavitation-enhanced heterogeneous nucleation (of both grain and intermetallic phase) and the fragmentation of the $\beta$-$Mg_{17}Al_{12}$ phase.

During the first stage of solidification, the US treatment improves the grain refinement by heterogeneous nucleation, as well as the wettability of the $\beta$-$Mg_{17}Al_{12}$ by the metal further improving the refinement of the secondary phase. On a second stage, corresponding to the formation of the first solid metal, the cavitation can develop acoustic streaming caused by the collapse of the bubbles in the remaining liquid, promoting the fragmentation of grain dendrites and intermetallic clusters.

The coexistence of these two mechanisms seems to promote a high density of nuclei in the melt, thus, leading to the development of a large number of smaller grains [23]. These aspects are in agreement with this work, in which uniformity of microstructure, reduced porosity and refinement of $\beta$-$Mg_{17}Al_{12}$ intermetallic phase promoted the increase of static mechanical properties.

The internal friction of AZ91D is presented in Figure 6. It is shown that the frequency effect is not relevant at room temperature, however with increasing temperature the value internal friction tends to increase at a higher rate for low frequencies, as confirmed by the increase in the curves slope.

Observing the damping behavior shown in Figure 6, it can be seen that temperature increase generates an elevation in the internal friction in the samples, however, this effect is more pronounced for lower frequencies (0.5 Hz). At higher frequencies (10 and 20 Hz) it is suggested that there is no significant difference in terms of internal friction. This behavior was already confirmed previously in AZ91 alloys by Shu-wei Liu *et al.* [31]. Interpreting Figure 6, there seems to be no significant difference in terms of internal friction between US treated and non-treated samples.

It is known that the major source of internal friction in Mg alloys is the dislocation motion of basal planes in the hexagonal structure by stress induced inelasticity [32] and this movements are dependent to the grains size [33] and shape [34], presence of secondary phases [35], and porosity [10].

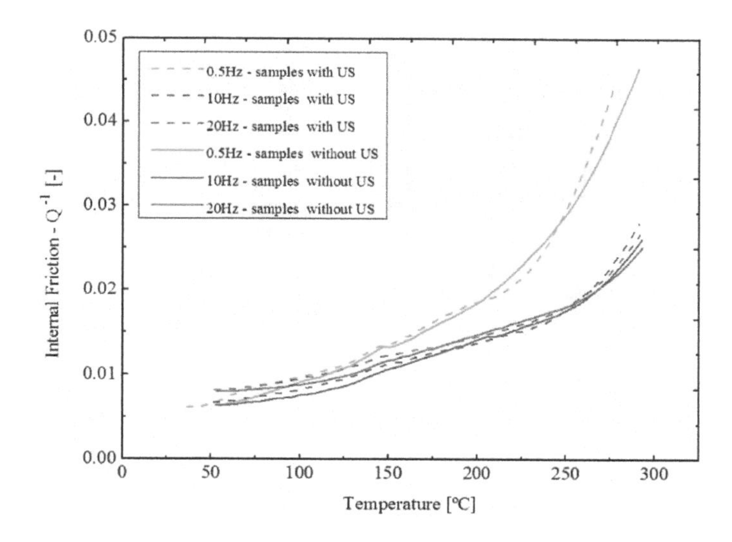


**Figure 6.** Internal friction ($Q^{-1}$) of as-cast AZ91D.

According to Figure 2b and Table 2, it can be observed that the grain size is reduced by US treatment and this is generally a route to enhance internal friction [13]. However, US treatment promotes the dispersion and homogenization of the secondary phase β-$Mg_{17}Al_{12}$ and a reduction of porosity in the bulk samples, and those are primary sources in internal friction by dislocation motion. The dispersion of β-$Mg_{17}Al_{12}$ and reduction of porosity implies the reduction of viscous flow and a loss in the dissipated energy under cyclic stress. Additionally, the homogenization of the grain shape by US treatment involves the reduction of dendrite grains, another source of internal friction due to high angular in the vicinity of boundaries. Since one of the fundamental damping mechanisms in polycrystals is originated by dislocation on grain boundaries and between the primary (α-Mg) and secondary (β-$Mg_{17}Al_{12}$) phases, rather than lattice dislocations in the grain interior [6], there are opposite damping transformations generated by US treatment.

Figure 7a presents the damping peaks extracted from the plots in Figure 6, originated by activation of thermoelastic effects [10]. Considering the values of peak damping, it may be observed that US treatment enhances the internal friction at lower frequencies. Additionally, it can be observed that with an increase in frequency promotes a narrowing of the damping peak in US treated samples, relative to the non-treated samples. It is visible that an increase in frequency generates an elevation in the peak temperature, however, it seems that US treatment has no relevant role in the change of peak temperature.

Figure 7b shows the Arrhenius plot of both US treated and non-treated samples. It can be observed by the slopes of the functions, that US treated samples shows an activation energy of 1.73 eV, while non-treated as-cast samples reveal an activation energy of 1.82 eV. Apparently, US-treatment has no relevant effect on the activation energy in AZ91D alloys.

According to the results, it is suggested that the enhancement of internal friction by the reduction of grain size is compensated by the dispersion of the intermetallic phase, grain homogenization and reduction of porosity generated by the US treatment. Thus, even though US treatment promotes a significant change in the material microstructure there is no relevant changes in the bulk dynamic mechanical properties of the AZ91D alloy.

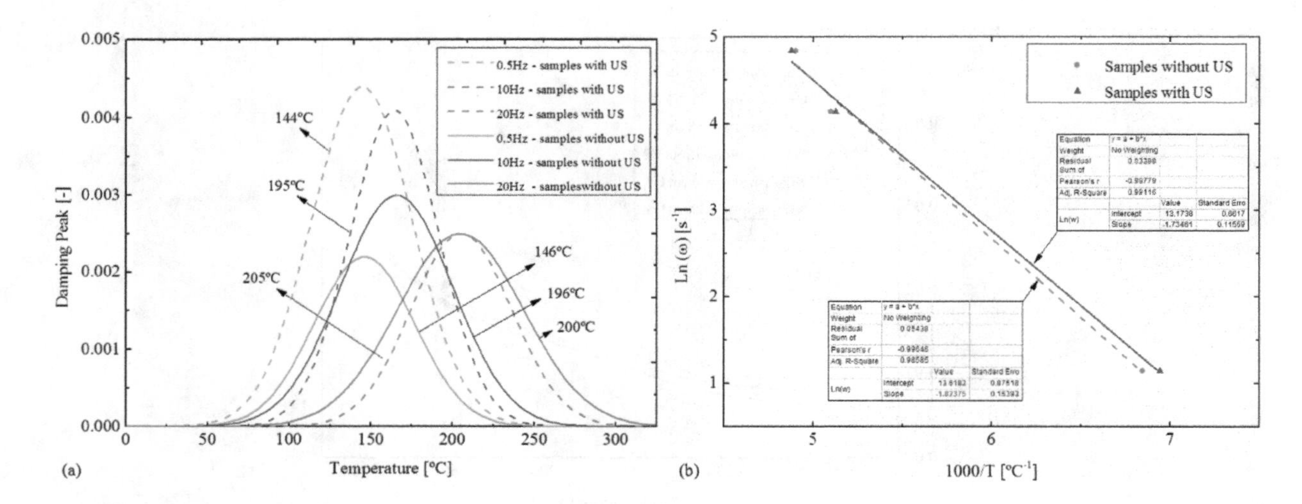


**Figure 7.** Damping peaks (**a**) and Arrhenius plot (**b**) in AZ91 alloy.

## 4. Conclusions

(1). The effect of US treatment during solidification refines the α-Mg matrix and β-$Mg_{17}Al_{12}$ intermetallic phase.

(2). The use of high intensity acoustic vibration promotes uniform dispersion of the β-$Mg_{17}Al_{12}$ intermetallic phase and reduces the level of porosity.

(3). Static tensile testing shows an increase of both tensile strength (40.7%) and strain (150%) when processing the alloys by US. Consequently, the overall toughness in US processed as-cast samples is increased by four times.

(4). US treated samples show an activation energy of 1.73 eV, while non-treated samples reveal an activation energy of 1.82 eV. It suggested that the internal friction, generated by the grain refinement by US treatment, is compensated by a loss of damping capacity due to the intermetallic phase dispersion and reduction porosity.

(5). US treatment in the AZ91D samples seems to improve static mechanical properties without compromising the dynamic mechanical properties of these alloys.

## Acknowledgments

This research was supported by FEDER/COMPETE funds and by national funds through FCT—Portuguese Foundation for Science and Technology and was developed on the aim of Post-Doctoral grant SFRH/BPD/76680/2011. Also, this work has been supported by the FCT in the scope of the project: UID/EEA/04436/2013.

## Author Contributions

H.P. and V.C. conceived and designed the experiments, wrote and edited the manuscript, and contributed in all activities. V.V. performed OM and SEM experiment. J.B. helped finishing casting experiments, analyzing the results and revising the manuscript.

## References


1. James, D.W. High damping metals for engineering applications. *Mater. Sci. Eng.* **1969**, *4*, 1–8.
2. Zhang, J.; Perez, R.J.; Lavernia, E.J. Documentation of damping capacity of metallic, ceramic and metal-matrix composite materials. *J. Mater. Sci.* **1993**, *28*, 2395–2404.
3. Ono, N.; Nowak, R.; Miura, S. Effect of deformation temperature on Hall-Petch relationship registered for polycrystalline magnesium. *Mater. Lett.* **2004**, *58*, 39–43.
4. Li, X.; Xiong, S.M.; Guo, Z. Correlation between Porosity and Fracture Mechanism in High Pressure Die Casting of AM60B Alloy. *J. Mater. Sci. Technol.*, in press, 2015.
5. Colakoglu, M. Factors effecting internal damping in aluminium. *J. Theor. Appl. Mech.* **2004**, *42*, 95–105.
6. Jiang, W.B.; Kong, Q.P.; Cui, P. Further evidence of grain boundary internal friction in bicrystals. *Mater. Sci. Eng. A* **2010**, *527*, 6028–6032.
7. Blanter, M.S; Golovin, I.S.; Neuhauser, H.; Sinning, H.-R. *Internal Friction in Metallic Materials: A Handbook*; Springer: Berlin, Germany, 2007.
8. Liu, C.; Pineda, E.; Crespo, D. Mechanical Relaxation of Metallic Glasses: An Overview of Experimental Data and Theoretical Models. *Metals* **2015**, *5*, 1073–1111.
9. Filmer, A.J.; Hutton, G.J.; Hutchison, T.S. Internal Friction in Aluminum at Low Temperatures. *J. Appl. Phys.* **1958**, doi:10.1063/1.1723055.
10. Zhang, J.; Gungor, M.N.; Lavernia, E.J. The effect of porosity on the microstructural damping response of 6061 aluminium alloy. *J. Mater. Sci.* **1993**, *28*, 1515–1524.
11. Hu, X.; Wang, X.; He, X.; Wu, K.; Zheng, M. Low frequency damping capacities of commercial pure magnesium. *Trans. Nonferrous Met. Soc. China* **2012**, *22*, 1907–1911.
12. Granato, A.; Lücke, K. Theory of Mechanical Damping Due to Dislocations. *J. Appl. Phys.* **1956**, doi:10.1063/1.1722436.
13. Liao, L.; Zhang, X.; Li, X.; Wang, H.; Ma, N. Effect of silicon on damping capacities of pure magnesium and magnesium alloys. *Mater. Lett.* **2007**, *61*, 231–234.
14. Watanabe, H.; Owashi, A.; Uesugi, T.; Takigawa, Y.; Higashi, K. Grain boundary relaxation in fine-grained magnesium solid solutions. *Philos. Mag.* **2011**, *91*, 4158–4171.
15. Fan, G.D.; Zheng, M.Y.; Hu, X.S.; Wu, K.; Gan, W.M.; Brokmeier, H.G. Internal friction and microplastic deformation behavior of pure magnesium processed by equal channel angular pressing. *Mater. Sci. Eng. A* **2013**, *561*, 100–108.
16. Randall, R.H.; Zener, C. Internal Friction of Aluminum. *Phys. Rev.* **1940**, *58*, 472–473.
17. Cao, X.; Huang, C. The annealing and aging effects of high temperature internal friction in pure aluminum. *Mater. Sci. Eng. A* **2004**, *383*, 341–346.
18. Chen, Y.; Wang, Q.; Lin, J.; Liu, M.; Hjelen, J.; Roven, H.J. Grain refinement of magnesium alloys processed by severe plastic deformation. *Trans. Nonferrous Met. Soc. China* **2014**, *24*, 3747–3754.
19. Lee, Y.C.; Dahle, A.K.; StJohn, D.H. The role of solute in grain refinement of magnesium. *Metall. Mater. Trans. A* **2000**, *31*, 2895–2906.
20. Liu, X.; Osawa, Y.; Takamori, S.; Mukai, T. Grain refinement of AZ91 alloy by introducing ultrasonic vibration during solidification. *Mater. Lett.* **2008**, *62*, 2872–2875.

21. Eskin, G.I.; Eskin, D.G. *Ultrasonic Treatment of Light Alloy Melts*, 2nd ed.; CRC Press: Boca Raton, FL, USA, 2014.
22. Ali, Y.; Qiu, D.; Jiang, B.; Pan, F.; Zhang, M.-X. Current research progress in grain refinement of cast magnesium alloys: A review article. *J. Alloys Compd.* **2015**, *619*, 639–651.
23. Puga, H.; Barbosa, J.; Costa, S.; Ribeiro, S.; Pinto, A.M.P.; Prokic, M. Influence of indirect ultrasonic vibration on the microstructure and mechanical behavior of Al–Si–Cu alloy. *Mater. Sci. Eng. A* **2013**, *560*, 589–595.
24. Ferguson, J.; Schultz, B.; Cho, K.; Rohatgi, P. Correlation *vs.* Causation: The Effects of Ultrasonic Melt Treatment on Cast Metal Grain Size. *Metals* **2014**, *4*, 477–489.
25. Anilchandra, A.R.; Surappa, M.K. Microstructure and damping behaviour of consolidated magnesium chips. *Mater. Sci. Eng. A* **2012**, *542*, 94–103.
26. Zhang, Z.; Zeng, X.; Ding, W. The influence of heat treatment on damping response of AZ91D magnesium alloy. *Mater. Sci. Eng. A* **2005**, *392*, 150–155.
27. Patel, B.; Chaudhari, G.P.; Bhingole, P.P. Microstructural evolution in ultrasonicated AS41 magnesium alloy. *Mater. Lett.* **2012**, *66*, 335–338.
28. Liu, X.; Zhang, Z.; Hu, W.; Le, Q.; Bao, L.; Cui, J.; Jiang, J. Study on hydrogen removal of AZ91 alloys using ultrasonic argon degassing process. *Ultrason. Sonochem.* **2015**, *26*, 73–80.
29. Du, X.; Zhang, E. Microstructure and mechanical behaviour of semi-solid die-casting AZ91D magnesium alloy. *Mater. Lett.* **2007**, *61*, 2333–2337.
30. Li, P.; Tang, B.; Kandalova, E.G. Microstructure and properties of AZ91D alloy with Ca additions. *Mater. Lett.* **2005**, *59*, 671–675.
31. Liu, S.; Jiang, H.; Li, X.; Rong, L. Effect of precipitation on internal friction of AZ91 magnesium alloy. *Trans. Nonferrous Met. Soc. China.* **2010**, *20*, s453–s457.
32. Schwaneke, A.; Nash, R. Effect of preferred orientation on the damping capacity of magnesium alloys. *Metall. Mater. Trans. B* **1971**, *2*, 3453–3457.
33. Wang, H.; Tian, X.-F.; Yin, C.; Huang, Z. The effect of heat treatment and grain size on magnetomechanical damping properties of Fe–13Cr–2Al–1Si alloy. *Mater. Sci. Eng. A* **2014**, *619*, 199–204.
34. Nowick, A.S. Dislocations and Crystal Boundaries. In *Anelastic Relaxation in Crystalline Solids*; Academic Press: New York, NY, USA, 1972; pp. 350–370.
35. Tanaka, M.; Iizuka, H. Effects of grain size and microstructures on the internal friction and Young's modulus of a high-strength steel HT-80. *J. Mater. Sci.* **1991**, *26*, 4389–4393.

# 6

# A Novel Bulk Acoustic Wave Resonator for Filters and Sensors Applications

**Zhixin Zhang, Ji Liang, Daihua Zhang, Wei Pang * and Hao Zhang**

State Key Laboratory of Precision Measuring Technology and Instruments, Tianjin University, Tianjin 300072, China; E-Mails: zxzhang10@tju.edu.cn (Z.Z.); liangjitju@tju.edu.cn (J.L.); dhzhang@tju.edu.cn (D.Z.); haozhang@tju.edu.cn (H.Z.)

* Author to whom correspondence should be addressed; E-Mail: weipang@tju.edu.cn

Academic Editor: Behraad Bahreyni

**Abstract:** Bulk acoustic wave (BAW) resonators are widely applied in filters and gravimetric sensors for physical or biochemical sensing. In this work, a new architecture of BAW resonator is demonstrated, which introduces a pair of reflection layers onto the top of a thin film bulk acoustic resonator (FBAR) device. The new device can be transformed between type I and type II dispersions by varying the thicknesses of the reflection layers. A computational modeling is developed to fully investigate the acoustic waves and the dispersion types of the device theoretically. The novel structure makes it feasible to fabricate both type resonators in one filter, which offers an effective alternative to improve the pass band flatness in the filter. Additionally, this new device exhibits a high quality factor ($Q$) in the liquid, which opens a possibility for real time measurement in solutions with a superior limitation of detection (LOD) in sensor applications.

**Keywords:** BAW resonator; dispersion type; pass band flatness; $Q$ factor in liquid

## 1. Introduction

Bulk acoustic wave (BAW) resonators based on aluminum nitride (AlN) are one of the success stories in modern microwave technologies. AlN based BAW resonators have two basic architectures: Thin Film Bulk Acoustic Wave Resonator (FBAR), which is suspended above an air cavity, and Solidly Mounted Resonator (SMR), which utilizes a series of high and low impedance reflectors to isolate the resonator from the substrate. Thanks to their miniature sizes, high quality factors ($Q$) in air, and compatibilities with integrated circuit processing, BAW resonators have been used in a widespread applications, such as filters and duplexers in communication electronics [1,2], oscillators for frequency control, gravimetric sensors for physical [3], chemical [4] and biological [5,6] sensing, and general physics applications [7].

For filter and duplexer applications, the spurious modes in resonators inevitably induce ripples in the pass band, which greatly degrade the flatness. Enlarging the resonator area in the series branch of the filter can subdue the ripples in the pass band, whereas it lowers the out-of-band rejection at a price. Though the flatness in the pass band can also be improved by introducing a frame or other structures into the device [8], the introduced structures need to be optimized in several processing cycles, which complicates the device fabrication. An alternative method to improve the pass band flatness of the filter is to deploy dispersion type I and type II resonators for series and shunt branches, respectively, in a filter. Since the locations of spurious modes in frequencies are different according to the different dispersion types of resonators, the ripples in the pass band can be moved to the edge, which contributes to the pass band flatness significantly. However, AlN FBARs exhibit fixed dispersion type II properties, failing to be applied for this method. Though the acoustic dispersion type of an SMR can be tailored by tuning the thickness of the reflection layers [9,10], it is difficult to fabricate both type devices on the same chip, since the reflectors are at the bottom of the device.

In this paper, we present a new BAW resonator device, in which a reflection stack with specific thicknesses is introduced onto the top of an FBAR. Though the dependence of dispersion types on the stack was investigated before [9,11], the quantitative relations between them are not clarified yet. In this paper, we theoretically analyze the device stack through the computational modeling. As a result, the dispersion type of a given stack can be predicted. By varying the thicknesses of the reflection layers, the acoustic dispersion type of the device can be tailored. Since the reflection layers are on the top of the device, the dispersion relations could be post-tuned when the fabrication is accomplished, and it is also possible in process to fabricate resonators with both type I and type II dispersion types on one chip.

In addition, for sensing applications, the quality factor ($Q$) is a key parameter to guarantee the limitation of detection (LOD) of sensors. Since most biochemical sensing experiments are conducted in liquids, and real time measurements of frequency shifts are desirable, it is in great demand for resonators with high $Q$ values in the liquid environment. Although the $Q$ factors of FBAR and SMR devices are extremely high in air, they decline sharply when immerging in a liquid [12,13]. Therefore, it is desirable to develop devices with high $Q$ factors in liquids. In this work, as the reflector layers are on the top of the device, they can serve as an isolator to attenuate the acoustic wave leakage when the device is immerged in the liquid, preserving a high $Q$ factor of the resonator.

## 2. Experimental Section

### 2.1. Device Architecture and Computational Modeling

The device is consisted of a conventional FBAR, which includes a piezoelectric layer sandwiched by top and bottom electrodes, and a reflection stack from the bottom up, as depicted in Figure 1. The reflection stack is composed of a silicon oxide ($SiO_2$) layer and a molybdenum (Mo) layer. In a BAW resonator, there exist two wave modes—thickness extensional (TE) mode and thickness shear (TS) mode.

If the longitudinal cutoff frequency ($f_{c,TE1}$) (or series resonant frequency $f_s$) is lower than the second shear wave cutoff frequency ($f_{c,TS2}$), the device is defined as a type II dispersion device. On the contrary, devices whose $f_{c,TE1}$ is higher than $f_{c,TS2}$ are type I dispersion devices [8].

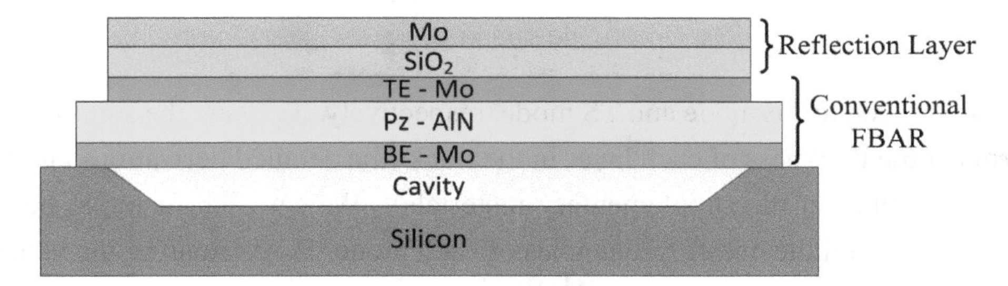


**Figure 1.** Illustration of the cross section of the new device: It is composed of a bottom electrode (BE), piezoelectric layer (Pz), top electrode (TE), and reflection layers.

The acoustic dispersion types are susceptible to the stack, and both $f_{c,TE1}$ and $f_{c,TS2}$ can be figured out through a Transfer Matrix method [14]. For each layer in the stack, the relations between the field quantities of interest on both surfaces can be described by matrix:

$$\begin{pmatrix} u \\ v \\ \sigma \\ \tau \end{pmatrix}_{bot} = M_i \cdot \begin{pmatrix} u \\ v \\ \sigma \\ \tau \end{pmatrix}_{top} \tag{1}$$

where $u$, $v$, $\sigma$, and $\tau$ are the normal displacement, the tangential displacement, the normal stress, and the tangential stress, respectively; the subscripts top and bot represent the top and the bottom surface, respectively; and $M_i$ is the Transfer Matrix of the $i$th layer. Take the bottom electrode as an example, its Transfer Matrix is:

$$\begin{pmatrix} \cos(\sqrt{\rho/c_{44}} \cdot t \cdot \omega) & 0 & 0 & \frac{\sin(\sqrt{\rho/c_{44}} \cdot t \cdot \omega)}{\omega \cdot \sqrt{\rho \cdot c_{44}}} \\ 0 & \cos(\sqrt{\rho/c_{11}} \cdot t \cdot \omega) & \frac{\sin(\sqrt{\rho/c_{11}} \cdot t \cdot \omega)}{\omega \cdot \sqrt{\rho \cdot c_{11}}} & 0 \\ 0 & -\omega \cdot \sqrt{\rho c_{11}} \cdot \sin(\sqrt{\rho/c_{11}} \cdot t \cdot \omega) & \cos(\sqrt{\rho/c_{11}} \cdot t \cdot \omega) & 0 \\ -\omega \cdot \sqrt{\rho c_{44}} \cdot \sin(\sqrt{\rho/c_{44}} \cdot t \cdot \omega) & 0 & 0 & \cos(\sqrt{\rho/c_{44}} \cdot t \cdot \omega) \end{pmatrix} \tag{2}$$

where $\omega$, $\rho$, $c$, and $t$ are the frequency, the density, the stiffness, and the thickness, respectively. According to the continuity conditions at the interfaces, the Transfer Matrix of the whole stack can be deduced as follows:

$$\begin{pmatrix} u \\ v \\ \sigma \\ \tau \end{pmatrix}_{last} = M_n M_{n-1} \cdots M_2 M_1 \cdot \begin{pmatrix} u \\ v \\ \sigma \\ \tau \end{pmatrix}_{first} = M \cdot \begin{pmatrix} u \\ v \\ \sigma \\ \tau \end{pmatrix}_{first} \tag{3}$$

where $M$ is a $4 \times 4$ Transfer Matrix, which condenses a multilayered system into a set of four equations correlating the boundary conditions at the first interface to that at the last interface.

Since both upper and lower surfaces of the device are exposed to air, the stresses on them are zero. Hence, the determination of the 2 × 2 bottom left sub-matrix of $M$ is zero. The determination can be rewritten in the form of a product of two factors, which are named as *ye* and *ys*, respectively:

$$ye \cdot ys = 0 \tag{4}$$

where *ye* and *ys* are related to TE mode and TS mode, respectively. They are the sums of a few sinusoidal functions in terms of the thickness of each layer in the stack (the detailed derivations and the expressions of *ye* and *ys* can be found in the supplementary materials). We set each factor to be zero and solve for the frequencies to obtain the cutoff frequencies of each mode. By comparing the values of $f_{c,TE1}$ and $f_{c,TS2}$, we could identify which type the device is. Since these two factors are too complex to be solved by analytic method, we plot *ye* and *ys versus* frequencies, as shown in Figure 2. Frequencies at zero crossing points are the cutoff frequencies we are looking for in each mode, and the values of cutoff frequencies can be read out on the axis directly. By comparing the values of $f_{c,TE1}$ and $f_{c,TS2}$, the dispersion type of the device can be identified clearly.

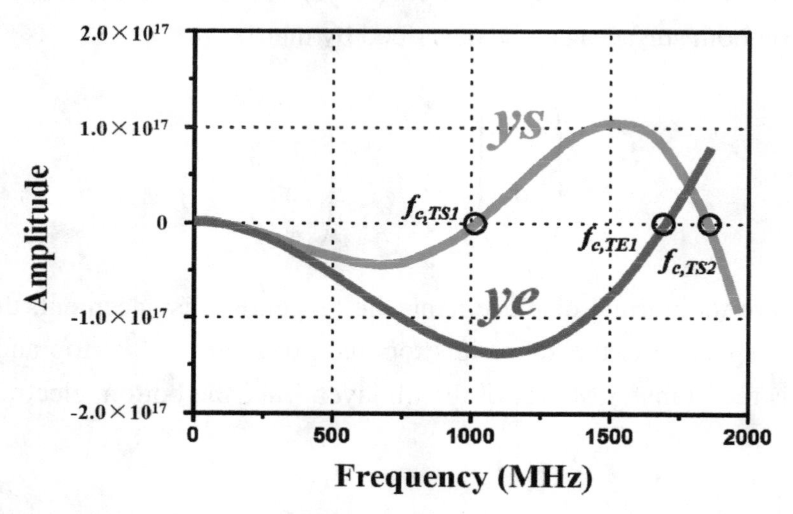


**Figure 2.** Plot of *ye* and *ys versus* frequencies. The frequencies at zero crossing points are the cutoff frequencies in each mode (*ye* and *ys* are unitless).

The relations between cutoff frequencies and stack layers are investigated. For computational simplicity, the thickness ratio of both the top and the bottom electrode to the piezoelectric layer is set to be $k$, and the ratio of the $SiO_2$ layer and the topmost Mo layer to the piezoelectric layer is set to be $k_{Ox}$ and $k_{Mo}$, respectively. Since the cutoff frequencies are functions of $k$, $k_{Ox}$, and $k_{Mo}$, the investigations are conducted first in the condition of a fixed $k$ value, and it is set to be 0.2. Either $k_{Ox}$ or $k_{Mo}$ is varied in a series of values while keeping the other as a constant, and both $f_{c,TE1}$ and $f_{c,TS2}$ are figured out and plotted in Figure 3a,b. It is known from the figures that the reflection layers have greater impact on $f_{c,TS2}$ than $f_{c,TE1}$. When the thickness of $SiO_2$ or Mo layer reaches a certain value, $f_{c,TE1}$ is equal to $f_{c,TS2}$, and this is defined as the critical condition for the dispersion type conversion. As the thickness of one of the reflection layers exceeds the critical value, $f_{c,TE1}$ surpasses $f_{c,TS2}$, which means that the device is transformed from type II to type I. On the other hand, if the thicknesses of the reflection layers are less than the threshold, the device preserves the characteristic of the type II.

The values of $k_{Ox}$ and $k_{Mo}$ in the critical conditions for FBAR type transfermation (*i.e.*, in the condition of $f_{c,TE1} = f_{c,TS2}$) with various $k$ values are calculated and plotted in Figure 3c. Each curve exhbits a negative correlation between $k_{Ox}$ and $k_{Mo}$; that is to say, in the critical conditions, the thickness increase of one reflection layer induces the thickness decrease of the other one. By comparing these curves, it is also observed that larger $k$ value corresponds to larger $k_{Ox}$ and $k_{Mo}$, suggesting that thicker reflection layers are needed for type conversion in the device with a smaller piezoelectric coupling coefficient ($k^2_{t,eff}$).

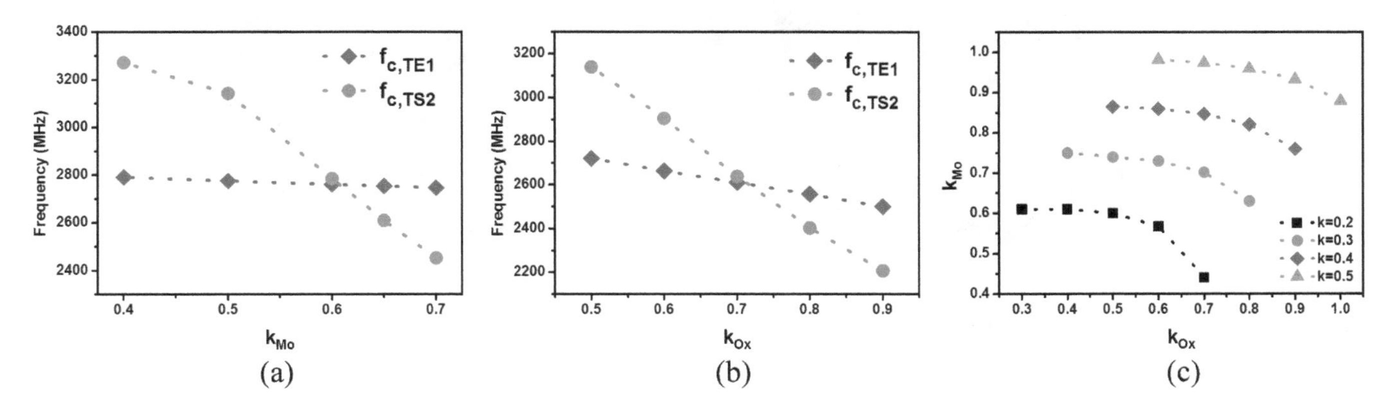


**Figure 3.** (**a**) Relations between cutoff frequencies and $k_{Mo}$; (**b**) Relations between cutoff frequencies and $k_{Ox}$; (**c**) Combinations of $k_{Mo}$ and $k_{Ox}$ in the critical conditions for type conversion.

### *2.2. Simulations and Fabrications*

A 2D finite element method (FEM) and solver software package (COMSOL MULTIPHYSISCS, Stockholm, Sweden) is adopted for simulations. Through the FEM simulations, a series of spatial amplitude distributions over different frequencies are available. Then, a fast Fourier transform (FFT) algorithm is applied to the amplitude dataset, yielding an amplitude distribution over the lateral propagation constant, thus the dispersion curves are obtained [15]. A type I device is designed following the theory, which is built by 2000 Å Mo, 5000 Å AlN, 2000 Å Mo, 5000 Å $SiO_2$, and 5000 Å Mo from the bottom up. The FEM simulation outcome is illustrated in Figure 4a. $f_{c,TE1}$ is greater than $f_{c,TS2}$, suggesting it is a type I resonator, which agrees with the theoretical calculations. The dispersion curves of TE1 mode has its real wave numbers higher in frequency than $f_{c,TE1}$, indicating that the spurious modes are in the frequency range higher than $f_s$ in the resonator. A type II device is also designed and simulated, whose stack is the same as that of the type I except that the top reflection layer of Mo is thinned to 1000 Å. The simulated dispersion relations are presented in Figure 4b. $f_{c,TE1}$ is less than $f_{c,TS2}$, exhibiting a characteristic of type II. Since real wave numbers are lower than $f_{c,TE1}$ in TE1 dispersion branch, the spurious modes locate in frequencies below $f_s$.

A CMOS-compatible AlN Micro-electromechanical Systems (MEMS) fabrication process is employed. The process starts by a RF sputtering deposition of 2000 Å Mo, and a plasma etching is used to pattern it as the bottom electrode. Then, 0.5 μm AlN, 2000 Å Mo, 5000 Å $SiO_2$, and 5000 Å Mo are deposited in turn onto the bottom electrode, respectively. Next, the top two reflection layers are patterned by plasma etching all at once. Wet etching is then used to pattern the top electrode. Subsequently, AlN is etched by a combination of plasma etching and potassium hydroxide (KOH) wet etching. After AlN etch, the bottom electrode can be accessed. Gold (Au) is then evaporated and patterned by lift-off process, serving as electrical connections and testing pads. Finally comes the release by trickling xenon difluoride ($XeF_2$) gas into the release holes to etch the silicon beneath the device, forming an air cavity to reflect acoustic waves.

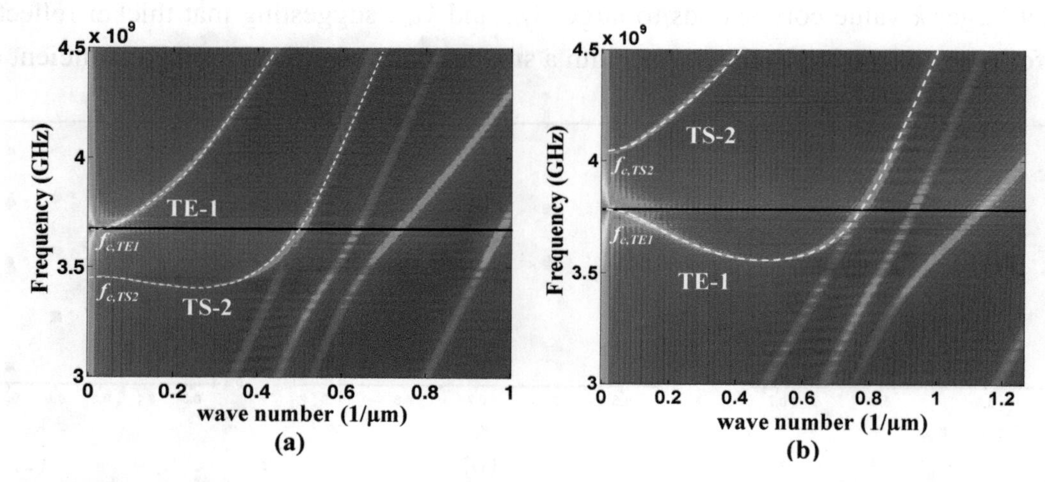


**Figure 4.** Dispersion Curves of the new device with the topmost Mo layer of (**a**) 5000 Å and (**b**) 1000 Å simulated by a 2D finite element method (FEM).

## 3. Results and Discussion

The scanning electron microscope (SEM) images of the top and the cross-sectional view of a fabricated resonator device are shown in Figure 5a,b. The scattering parameters of the resonator are measured with a ground-signal-ground (GSG) probe and a vector network analyzer on wafer level. One-port calibration with short, open, and load standards is applied with the reference plane at the probe tips. The electrical responses of the device are depicted in Figure 5c,d. From the figures, it is known that the spurious modes only appear in frequency range higher than $f_s$, presenting a typical characteristic of the type I dispersion, which coincides with the theoretical predictions and the FEM simulations. The measured $k^2_{t,eff}$ of the device is 4.7%, and the $Q$ factor at $f_s$ is 1500. Since the energy in reflection layers can not be converted into the electrical energy, which decreases the electromechanical conversion efficiencies, the $k^2_{t,eff}$ of the device is smaller than that of the FBAR. The $Q_s$ is slightly lower than that of the conventional FBAR [16], which is mainly due to the the energy leakages and the damping losses of the additional reflection layers.

Since reflection layers are on the top of the device, it is feasible to post tune the dispersion relations though the fabrication is accomplished. The thickness of the top reflection layer is thinned from 5000 Å to 1000 Å, and the measured input impedance is plotted in Figure 6. The spurious modes are in frequencies less than $f_s$, indicating that the dispersion relations are changed, and it has been transferred to a type II dispersion from a type I property. When the thicknesses of both reflection layers are decreased to zero, the device becomes a traditional FBAR. From this point, an FBAR can be recognized as one of the devices among them.

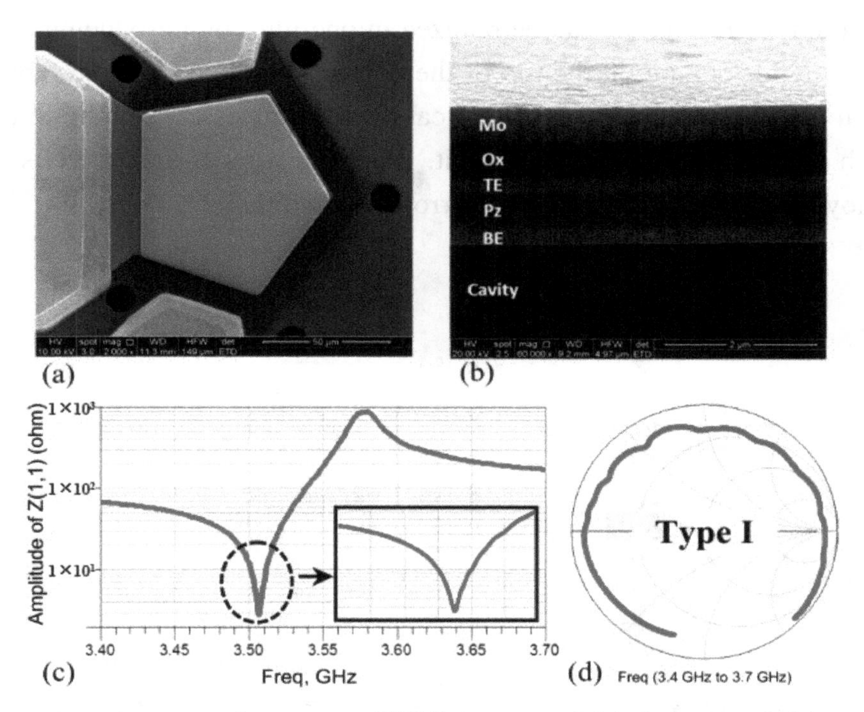


**Figure 5.** Scanning electron microscope (SEM) images of (**a**) the top and (**b**) cross-sectional view of the fabricated device; (**c**) The input impedance curve and (**d**) the Smith circle of the resonator device.

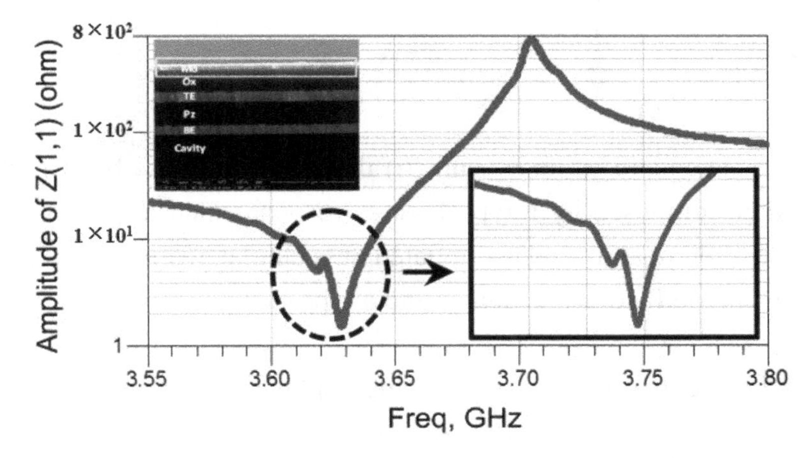


**Figure 6.** The measured impedance data of the device with a thinner top reflection layer. The inset is the SEM image of the stack.

In filter applications, Resonators are electrically connected in a ladder or a lattice topology. They are marked as series resonators or shunt resonators according to their positions in the topology. In conventional FBAR filters, spurious modes below $f_s$ in series resonators give rise to ripples in the pass band (Figure 7c,d). If type II resonators with the same area are replaced by type I resonators in the series branch, the ripples can be moved to the margin of the pass band, recovering the flatness in most of the pass band (Figure 7a,b). The out-of-band rejection level is influenced by the area ratio of the shunt resonators to the series ones. Since the resonator areas are not changed, the substitution between types can maintain the rejection level in the out-of-band, and meanwhile suppress ripples in the pass band effectively. In the realization of the filter building, as the thicknesses of the reflection layers dominate the dispersion types, both type resonators can be realized in one filter by introducing different thicknesses of reflectors. As the reflectors are on the top of the device, it is convenient to fabricate two different thicknesses of them in different resonators by typical MEMS processes, coming to the implementation of a filter with both dispersion type resonators in it. Since the $k^2_{t,eff}$ of the device is slightly lower, it is preferable to employ the device in filters with a narrow bandwidth.

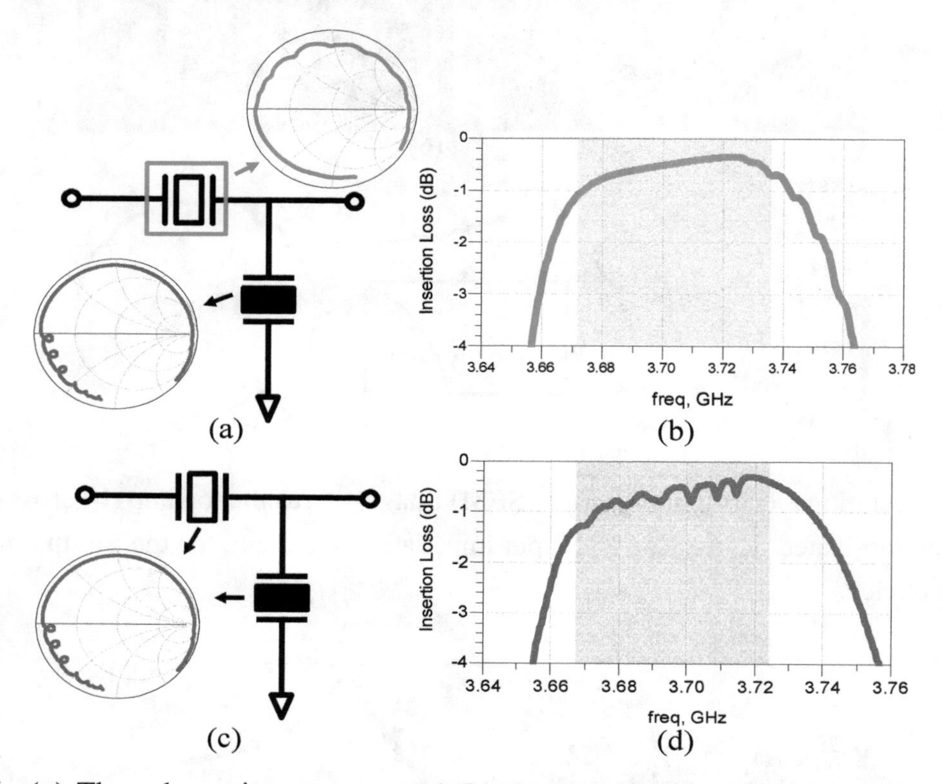


**Figure 7.** (**a**) The schematic structure and (**b**) the simulated pass band performance of a primary ladder filter composed by a type I series resonator and a type II shunt resonator; (**c**) The schematic structure and (**d**) the simulated pass band performance of a primary ladder filter composed by a type II series resonator and a type II shunt resonator.

In addition, the new device is also tested for sensing applications. It was reported that the minimum detectable mass (or LOD) is inversely proportional to the $Q$ value of the device [4], thus the high $Q$ of the device further contributes to minor LOD of the sensor in biochemical sensing applications. The novel device is measured in both air and water conditions, and Figure 8a shows the performances in a Smith chart. The $Q$ circle in water shrinks by a little scale compared to that in air. The $Q_s$ is 1500 in air, and 700 in water. The $Q$ value in water is almost half of that in air. For comparison, an FBAR device is also tested in both environments, and the electrical performances are compared in Figure 8b. The $Q_s$ in air is over 1000, whereas it declines to less than 10 in water. The new device can keep a high $Q$ in water, due to the two reflection layers on the top, which act as an isolator to prevent energy leakage into the water. Therefore, the novel device tends to have a better LOD.

Since the reflection layers act as a Bragg reflector, the top surface of the Mo reflection layer is insensitive to mass variations, leading to a mediocre LOD value. To solve this issue, we can adopt a localized sensing method [17]. If we punch a hole at the center of the device in the two reflection layers, the little center area of the top electrode (TE) can be accessed, which is highly sensitive to mass variations. The exposed TE area is used for mass detection, and the reflection layers around act as the energy protector to guarantee the high $Q$ of the device in the liquid environment.

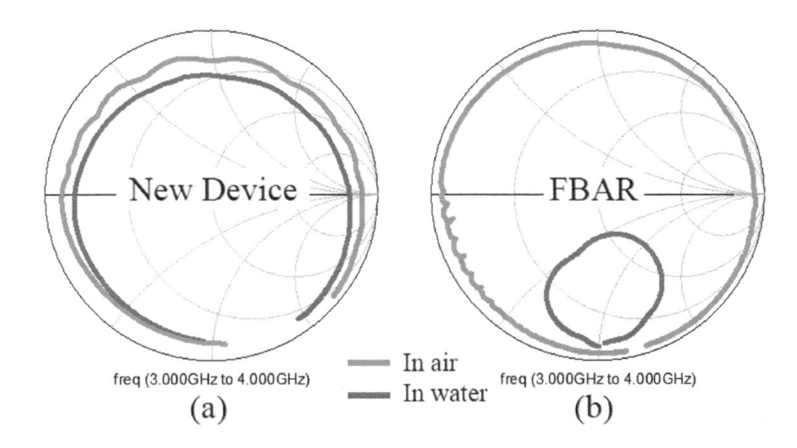


**Figure 8.** Electrical performances of (**a**) the new device and (**b**) the FBAR device measured in air and in water.

## 4. Conclusions

A new structure of BAW resonator is presented in this paper. A Transfer Matrix method is applied to develop the computational modeling of the acoustic waves in the device. The dispersion types are analyzed, and a set of design rules for resonators is extracted. The dispersion relations can be changed by varying the thicknesses of the reflection layers in the stack. The new structure makes it feasible to fabricate both type I and type II resonators on one chip, which provides a promising solution to improve the pass band flatness in filters. This novel device also presents a high $Q$ value in a liquid environment, which has great potential for real time measurements in liquids in biochemical sensing applications.

## Acknowledgments

This work was supported by Natural Science Foundation of China (NSFC No. 61176106) and the Program of Introducing Talents of Discipline to Universities (111 project No. B07014).

## Author Contributions

Zhixin Zhang, Hao Zhang and Wei Pang proposed the idea; Zhixin Zhang performed the computational modeling and the theoretical analysis; Zhixin Zhang and Wei Pang performed the FEM simulations; Zhixin Zhang, Ji Liang and Hao Zhang prepared the manuscript; and Daihua Zhang and Wei Pang contributed to the discussion and gave valuable suggestions on the manuscript revision according to the referee report.

**References**


1. Ruby, R.; Bradley, P.; Larson, J.; Oshmyansky, Y. PCS 1900 MHZ duplexer using thin film bulk acoustic resonators (FBARs). *Electron. Lett.* **1999**, *35*, 794–795. [CrossRef]
2. Ylilammi, M.; Ella, J.; Partanen, M.; Kaitila, J. Thin film bulk acoustic wave filter. *IEEE Trans. Ultrason. Ferroelectr. Freq. Control* **2002**, *49*, 535–539. [CrossRef] [PubMed]
3. Qiu, X.; Tang, R.; Zhu, J.; Oiler, J.; Yu, C.; Wang, Z.; Yu, H. The effects of temperature, relative humidity and reducing gases on the ultraviolet response of ZnO based film bulk acoustic-wave resonator. *Sens. Actuators B Chem.* **2011**, *151*, 360–364. [CrossRef]
4. Pang, W.; Zhao, H.; Kim, E.S.; Zhang, H.; Yu, H.; Hu, X. Piezoelectric microelectromechanical resonant sensors for chemical and biological detection. *Lab Chip* **2012**, *12*, 29–44. [CrossRef] [PubMed]
5. García-Gancedo, L.; Zhu, Z.; Iborra, E.; Clement, M.; Olivares, J.; Flewitt, A.; Milne, W.; Ashley, G.; Luo, J.; Zhao, X. Aln-based baw resonators with CNT electrodes for gravimetric biosensing. *Sens. Actuators B Chem.* **2011**, *160*, 1386–1393. [CrossRef]
6. Zhao, X.; Pan, F.; Ashley, G.M.; Garcia-Gancedo, L.; Luo, J.; Flewitt, A.J.; Milne, W.I.; Lu, J.R. Label-free detection of human prostate-specific antigen (HPSA) using film bulk acoustic resonators (FBARs). *Sens. Actuators B Chem.* **2014**, *190*, 946–953. [CrossRef]
7. O'Connell, A.D.; Hofheinz, M.; Ansmann, M.; Bialczak, R.C.; Lenander, M.; Lucero, E.; Neeley, M.; Sank, D.; Wang, H.; Weides, M.; *et al.* Quantum ground state and single-phonon control of a mechanical resonator. *Nature* **2010**, *464*, 697–703. [CrossRef] [PubMed]
8. Hashimoto, K.-Y. *RF Bulk Acoustic Wave Filters for Communications*; Artech House: Boston, MA, USA, 2009.
9. Jose, S.; Hueting, R.; Jansman, A. On the rule of thumb for flipping the dispersion relation in baw devices. In Proceedings of the 2011 IEEE International Ultrasonics Symposium (IUS), Orlando, FL, USA, 18–21 October 2011; pp. 1712–1715.
10. Zhou, C.; Zhang, D.; Pang, W.; Zhang, H. Spurious free and temperature stable Giga-hertz acoustic wave resonators with enhanced quality factor. In Proceedings of the Asia-Pacific Microwave Conference (APMC 2012), Kaohsiung, Taiwan, 4–7 December 2012; pp. 1286–1288.
11. Fattinger, G.; Marksteiner, S.; Kaitila, J.; Aigner, R. Optimization of acoustic dispersion for high performance thin film baw resonators. In Proceedings of the 2005 IEEE Ultrasonics Symposium, Rotterdam, The Netherlands, 18–21 September 2005; pp. 1175–1178.
12. Zhang, H.; Marma, M.S.; Kim, E.S.; McKenna, C.E.; Thompson, M.E. Implantable resonant mass sensor for liquid biochemical sensing. In Proceedings of the IEEE International Conference on Micro Electro Mechanical Systems, Maastricht, The Netherlands, 25–29 January 2004; pp. 347–350.
13. Zhang, H.; Kim, E.S. Micromachined acoustic resonant mass sensor. *J. Microelectromech. Syst.* **2005**, *14*, 699–706. [CrossRef]
14. Lowe, M.J. Matrix techniques for modeling ultrasonic waves in multilayered media. *IEEE Trans. Ultrason. Ferroelectr. Freq. Control* **1995**, *42*, 525–542. [CrossRef]

15. Fattinger, G.G.; Tikka, P.T. Laser measurements and simulations of fbar dispersion relation. In Proceedings of the 2001 IEEE MTT-S International Microwave Symposium Digest, Phoenix, AZ, USA, 20–24 May 2001; pp. 371–374.
16. Ruby, R. Review and comparison of bulk acoustic wave fbar, SMR technology. In Proceedings of the 2007 IEEE Ultrasonics Symposium, Vancouver, BC, Canada, 28–31 October 2007; pp. 1029–1040.
17. Campanella, H.; Esteve, J.; Montserrat, J.; Uranga, A.; Abadal, G.; Barniol, N.; Romano-Rodriguez, A. Localized and distributed mass detectors with high sensitivity based on thin-film bulk acoustic resonators. *Appl. Phys. Lett.* **2006**, *89*, 033507. [CrossRef]

7

# Optimization of Surface Acoustic Wave-Based Rate Sensors

**Fangqian Xu [1], Wen Wang [2,*], Xiuting Shao [2], Xinlu Liu [2] and Yong Liang [2]**

[1] Zhejiang University of Media and Communications, 998 Xueyuan Street Higher Education Zone Xia Sha, Zhejiang 310018, China; E-Mail: xufangqian2006@163.com

[2] State Key Laboratory of Acoustics, Institute of Acoustics, Chinese Academy of Sciences, No. 21, BeiSiHuan West Road, Beijing 100190, China; E-Mails: shaoxiuting10@mails.ucas.ac.cn (X.S.); liuxinlu1987@foxmail.com (X.L.); liangyong@mail.ioa.ac.cn (Y.L.)

* Author to whom correspondence should be addressed; E-Mail: wangwenwq@mail.ioa.ac.cn

Academic Editor: Vittorio M. N. Passaro

---

**Abstract:** The optimization of an surface acoustic wave (SAW)-based rate sensor incorporating metallic dot arrays was performed by using the approach of partial-wave analysis in layered media. The optimal sensor chip designs, including the material choice of piezoelectric crystals and metallic dots, dot thickness, and sensor operation frequency were determined theoretically. The theoretical predictions were confirmed experimentally by using the developed SAW sensor composed of differential delay line-oscillators and a metallic dot array deposited along the acoustic wave propagation path of the SAW delay lines. A significant improvement in sensor sensitivity was achieved in the case of 128° YX $LiNbO_3$, and a thicker Au dot array, and low operation frequency were used to structure the sensor.

**Keywords:** Coriolis force; delay line-oscillator; metallic dot array; partial-wave analysis; SAW rate sensor

---

## 1. Introduction

The surface acoustic wave (SAW)-based micro rate sensor has gained increasing attraction for inertial navigation applications because it exhibits many unique properties such as superior inherent shock robustness, a wide dynamic range, low cost, small size, and long working life compared to other current gyroscope types [1]. The typical working principle of the SAW rate sensor is so-called SAW gyroscopic effect [2,3], that is, as the Coriolis force induced by the applied rotation acts on the vibrating particles along the SAW propagation path, a pseudo running wave shifted by a quarter of a wavelength will arise, and it couples with the initial SAW generated by the interdigital transducers (IDTs) on the piezoelectric substrate, resulting in the change of trajectory of the wave particles and an acoustic wave velocity shift. Consequently, a frequency signal variation proportional to the applied rotation is expected. Referring to a certain differential oscillation structure, the SAW micro rate sensor-based gyroscopic effect can be implemented. Lee *et al.* first realized a prototype of a micro rate sensor based on SAW gyroscopic effect utilizing a temperature-compensated ST quartz substrate and a differential dual-delay-line oscillator configuration [4,5], but, the corresponding sensitivity was far too low, only 0.43 $Hz \cdot deg^{-1} \cdot s^{-1}$. To improve the sensor sensitivity, a X-112°Y $LiTaO_3$ substrate was suggested to form the SAW gyroscope because it exhibits a stronger gyroscopic effect, and a sensitivity of 1.332 $Hz \cdot deg^{-1} \cdot s^{-1}$ in a wide dynamic range (0~1000 $deg \cdot s^{-1}$) and good linearity are obtained [6]. Moreover, some other meaningful research works about SAW rate sensors were also reported [7,8], however, it is obvious that there is still no tangible improvement in the sensor performance because of its very weak Coriolis force.

To achieve a breakthrough in the performance of the SAW-based rate sensor, a creative idea was proposed whereby a metallic dot array strategically deposited on the SAW propagation path of SAW devices was considered to enhance the Coriolos force acting on the propagating SAW [9–11]. The scheme of the proposed SAW rate sensor incorporating a metallic dot array is depicted in Figure 1. The proposed sensor was composed of differential delay line-oscillators set in opposite directions, and metallic dot arrays deposited along the SAW propagation path of each SAW device. The centre distance of the dot element in the array is set to one wavelength in each direction, and also, the size of dots is a quarter-wave. When the sensor is subjected to an angular rotation, the Coriolis force acts on the vibrating metallic dots because of the Coriolis effect ($F_{coriolis} = 2m(v \times \Omega)$; $m$: mass of dot, $v$: velocity of the dot, $\Omega$: rotation rate). Moreover, the direction of the force is the same as the direction of wave propagation. Therefore, the amplitude and velocity of the wave are changed ($\Delta v_c$), and this change induces a shift in the oscillation frequency ($\Delta f_c$). Obviously, the enhanced Coriolis force will improve significantly the detection sensitivity. Exciting detection sensitivity results (16.7 $Hz \cdot deg^{-1} \cdot s^{-1}$) were achieved with a 80 MHz rate sensor on X-112°Y $LiTaO_3$ with a 900 nm Cu dot array distribution [9]. This provides a good start to break through the detection sensitivity bottleneck of SAW-based rate sensors, even though there is still a great gap between the obtained sensitivity and the demands for real applications.

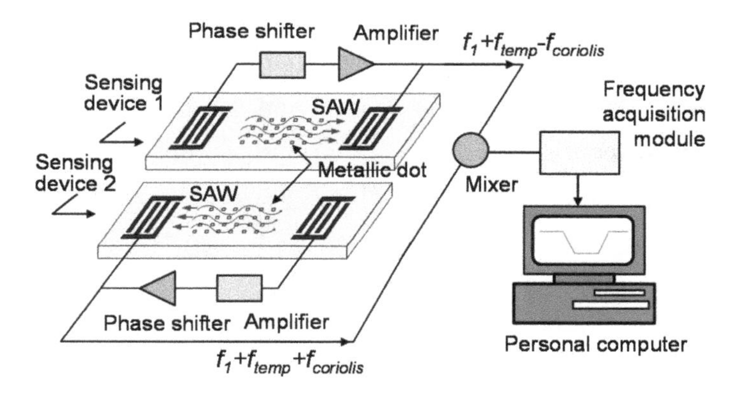


**Figure 1.** The scheme and working principle of the SAW micro rate sensor.

The main purpose of this work is to determine the optimal design parameters by analyzing the partial-wave in layered media utilizing the acoustic wave equation considering the contribution of the Coriolis force [12]. The materials for the piezoelectric crystal and metallic dots, dot geometry, and operation frequency were determined theoretically. The theoretical predictions were readily confirmed in rate sensing experiments by using the sensor scheme mentioned in Figure 1. Higher sensitivity and good linearity were achieved by using the 128°YX $LiNbO_3$, a thick Au dot array and lower operation frequency.

## 2. Theoretical Determination of Design Parameters

In this section, to simplify the theoretical analysis process, the metallic dot is considered as a semi-infinite surface, hence, the pre-rotated SAW propagation along the piezoelectric substrate with metallic dot distribution was analyzed by solving the partial-wave equations in layered structure described in our previous work [11], and the SAW propagates along the $x(x_1)$ axis on the $x$-$y(x_2)$-plane at $z(x_3) = 0$. The key factors which influence the rate sensor performance were studied theoretically. Hence, the optimal design parameters were extracted.

### *2.1. Theoretical Model*

Considering there is an anisotropic and piezoelectric medium occupying a half-space ($x_3 \leq 0$) with interdigital transducer (IDT) about the plane ($x_3 = 0$) and a metallic layer ($0 \leq x_3 \leq h$), as schematically illustrated in Figure 2, the dynamic wave equations considering the Coriolis force contribution of linear piezoelectricity in half-space piezoelectric substrate take the following forms in this coordinate system mentioned in Figure 2:

$$\begin{cases} C_{ijkl}{}^{I} u_{k,jl} + e_{kij}\varphi_{,jk} = \rho^{I}[\ddot{u}_i{}^{I} + 2\varepsilon_{ijk}\Omega_j \dot{u}_k{}^{I} - (\Omega_j^2 u_i{}^{I} - \Omega_i\Omega_j u_j{}^{I})] \\ e_{jkl}u_{k,jl}{}^{I} - \varepsilon_{jk}\varphi_{,jk} = 0 \end{cases} \tag{1}$$

where Einstein's summation rule is used, and the indices changed from 1 to 3. We denote by $u_i^I$ the mechanical displacements and by φ the electric potential. $c_{ijkl}^I$, $e_{kij}$, and $\varepsilon_{ij}^I$ stand for the elastic, piezoelectric and dielectric constants, and $\rho^I$ for the mass density of the piezoelectric substrate, respectively. $\varepsilon_{ijk}$ is the Levi-civita symbol. We assume a general solution of Equation (1), the particle displacement and electrical potential, are in the form:

$$\begin{cases} u_i{}^{I} = A_i{}^{I} \exp[-j(\omega t - \beta x_1 - \beta\alpha x_3)] \\ \varphi = A_4{}^{I} \exp[-j(\omega t - \beta x_1 - \beta\alpha x_3)] \end{cases} \tag{2}$$

where $\beta$ and ω are the wave numbers in the $x_1$ direction and the angular frequency, respectively. α is a decay constant along the $x_3$ direction. $A_i^I$ ($i$ = 1, 2, 3) and $A_4^I$ are wave amplitudes. Substitution of Equation (2) into Equation (1) leads to four linear algebraic equations (Cristoffel equation) for $A_i^I$ and $A_4^I$, that is:

$$\begin{bmatrix} \Gamma_{11} & \Gamma_{12} & \Gamma_{13} & \Gamma_{14} \\ \Gamma_{21} & \Gamma_{22} & \Gamma_{23} & \Gamma_{24} \\ \Gamma_{31} & \Gamma_{32} & \Gamma_{33} & \Gamma_{34} \\ \Gamma_{41} & \Gamma_{42} & \Gamma_{43} & \Gamma_{44} \end{bmatrix} \begin{bmatrix} A_1 \\ A_2 \\ A_3 \\ A_4 \end{bmatrix} = 0 \tag{3}$$

Then, for nontrivial solutions of $A_i^I$ and/or $A_4^I$, the determinant of the coefficient matrix of the linear algebraic equations must vanish, and this leads to a polynomial equation of degree eight for α. The coefficients of this polynomial equation are generally complex. To ensure the decrease in the displacement $u_i$ and the potential φ into the substrate, the generally complex constant α must have a negative imaginary part. Thus, we select four eigenvectors with negative imaginary part denoted by $\alpha_n$ ($n$ = 1, 2, 3, 4), and the corresponding eigenvectors by $A_i^{I(n)} = [A_1^{I(n)}\ A_2^{I(n)}\ A_3^{I(n)}\ A_4^{I(n)}]$, $n$ = 1, 2, 3, 4. Thus, the general wave solution to Equation (1) in the form of Equation (2) can be written as:

$$\begin{cases} u_i = \sum_{n=1}^{4} A_i^{(n)} C_n^I \exp\{-jk_s(x_1 + a_n x_3)\} \\ \varphi = \sum_{n=1}^{4} A_4^{(n)} C_n^I \exp\{-jk_s(x_1 + a_n x_3)\} \end{cases} ,i = 1,2,3 \tag{4}$$

where $C_n^I$ ($n$ = 1, 2, 3, 4) are the weight factors, and can be determined by the boundary condition.

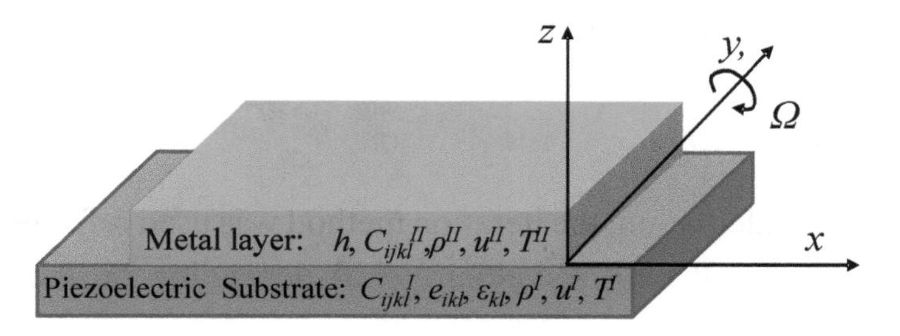


**Figure 2.** The coordinate system used in this study.

Then, the acoustic wave equation considering the contribution of Coriolis force in the isotropic metal layers is:

$$C_{ijkl}{}^{II} u_{k,jl} = \rho^{II} [\ddot{u}_i^{II} + 2\varepsilon_{ijk}\Omega_j \dot{u}_k^{II} - (\Omega_j^2 u_i^{II} - \Omega_i \Omega_j u_j^{II})],\ i,j,k,l{=}1,2,3 \tag{5}$$

Here, $\rho^{II}$ is the density of the metal layer, $u_i^{II}$ is the component of the acoustic wave displacement. The solution of Equation (3) is assumed as:

$$u_i^{II} = A_i^{II} \exp[-j(\omega t - \beta x_1 - \beta\eta x_3)] \tag{6}$$

Substituting the Equation (4) into Equation (2), the corresponding Cristoffel equation in metal layer can be written as:

$$\begin{bmatrix} \Gamma_{11} & \Gamma_{12} & \Gamma_{13} \\ \Gamma_{21} & \Gamma_{22} & \Gamma_{23} \\ \Gamma_{31} & \Gamma_{32} & \Gamma_{33} \end{bmatrix} \begin{bmatrix} A_1 \\ A_2 \\ A_3 \end{bmatrix} = 0 \tag{7}$$

Also, for nontrivial solutions of $A_i^{II}$, the determinant of the coefficient matrix of the linear algebraic equations must vanish, and there is an algebraic equation of the 6-th order in η. Then, substituting the four eigenvectors into Equation (3), the corresponding normalized amplitude $A_i^{II(n)}$ can be determined. Full solution of the wave equation was the linear combination of four basic groups:

$$u_i^{I} = \sum_{n=1}^{6} C_n^{I(n)} \exp\left[-j\left(\omega t - \beta x_1 - \beta\eta_n^{I} x_3\right)\right] \tag{8}$$

where the coefficient $C_n^I$ was determined by the boundary conditions.

Then, the solutions of the motion equations should satisfy both the mechanical boundary condition and the electrical boundary condition respectively. The mechanical boundary condition and electrical boundary condition at boundary between the piezoelectric substrate and metal layer ($x_3 = 0$) and boundary between metal layer and vacuum ($x_3 = h$) as schematically illustrated in Figure 2 are:

$$\left.\begin{cases} T_{i3}^{I} - T_{i3}^{II} = 0 \\ u_i^{I} - u_i^{II} = 0, i = 1,2,3 \\ \varphi^{II} = 0 \end{cases}\right\} x_3 = 0 \qquad (9)$$
$$\left\{T_{i3}^{II} = 0,\ i = 1,2,3\right\} x_3 = h$$

Then, substituting the solutions of the wave Equations (4) and (8) into the boundary conditions Equation (9), the following equation can be obtained:

$$H_m C_m = 0 \qquad (10)$$

The condition of nontrivial solution in Equation (10) was the determinant coefficient should be zero, that is:

$$|H_m| = 0 \qquad (11)$$

To simplify the theoretical calculation, the iteration method was used referring to the Matlab software. Based on the deduced formulas, the SAW velocity shift depending on the normalized rotation can be computed.

### *2.2. Numerical Results and Discussion*

The piezoelectric crystals used for the SAW sensors analyzed herein are YZ-$LiNbO_3$, X-112°Y $LiTaO_3$, ST-X quartz, and 128°YX $LiNbO_3$, and the metallic dot materials are assumed as copper (Cu) and gold (Au), respectively. The corresponding mechanical parameters are listed in Table 1. In the calculation, the piezoelectric crystals are assumed to be rotated around *y*-axis, and the SAW along the substrate propagates along the *x*-axis (Figure 2). Figure 3 illustrates the gyroscopic effect in above piezoelectric crystals without metallic dot array distribution. ST-X quartz and 128°YX $LiNbO_3$ exhibit larger gyroscopic effects than other materials in the relative rotation range of −0.02~0.02.

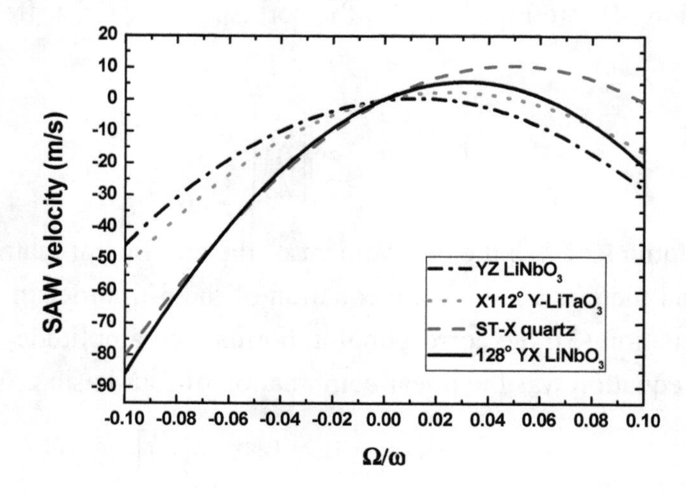


**Figure 3.** Gyroscopic effect in various piezoelectric substrates without metallic dots.

**Table 1.** Mechanical parameters for the piezoelectric substrate and metallic dots [13,14].

| Materials | Euler Angle | Stiffness Coefficients ($10^{10}$ N/m$^2$) | Piezoelectric Modules (C/m$^2$) | Permittivity Constants ($10^{-12}$ F/m) | Density (kg/m$^3$) |
|---|---|---|---|---|---|
| YZ $LiNbO_3$<br>128°YX $LiNbO_3$ | (0°, 90°, 90°)<br>(0°, 37.86°, 0°) | $C_{11}$: 23.3<br>$C_{33}$: 27.5<br>$C_{44}$: 9.4<br>$C_{12}$: 4.7<br>$C_{13}$: 8.0<br>$C_{14}$: −1.1 | $e_{15}$: 2.58<br>$e_{22}$: 1.59<br>$e_{31}$: −0.24<br>$e_{33}$: 1.44 | $\varepsilon_{11}$: 51 × $\varepsilon_0$<br>$\varepsilon_{33}$: 43 × $\varepsilon_0$<br>$\varepsilon_0$: 8.854 | 7450 |
| ST-X quartz | (0°, 132.75°, 0°) | $C_{11}$:8.674<br>$C_{12}$: 0.699<br>$C_{13}$:1.191<br>$C_{14}$: −1.791<br>$C_{33}$:10.72<br>$C_{44}$: 5.794 | $e_{x1}$: 30.171<br>$e_{x4}$: −0.0436<br>$e_{z6}$: 0.14 | $\varepsilon_{11}$: 4.5 × $\varepsilon_0$<br>$\varepsilon_{33}$: 4.6 × $\varepsilon_0$<br>$\varepsilon_0$: 8.854 | 2651 |
| X-112°Y $LiTaO_3$ | (90°, 90°, 112.2°) | $C_{11}$: 23.28<br>$C_{12}$: 4.65<br>$C_{13}$: 8.36<br>$C_{14}$: −1.05<br>$C_{33}$: 27.59<br>$C_{44}$: 9.49 | $e_{x5}$: 2.64<br>$e_{y2}$: 1.86<br>$e_{z1}$: −0.22<br>$e_{z3}$: 1.71 | $\varepsilon_{11}$:40.9 × $\varepsilon_0$<br>$\varepsilon_{33}$:42.5 × $\varepsilon_0$<br>$\varepsilon_0$: 8.854 | 7454 |
| Cu | | $C_{11}$: 17.69<br>$C_{33}$: 7.96 | | | 8900 |
| Au | | $C_{11}$:18.6<br>$C_{12}$: 15.7<br>$C_{44}$: 4.2 | | | 19,300 |

Additionally, to compare the gyroscopic effect in various piezoelectric substrates with metallic dot array distributions, the SAW velocity shift induced by the external rotation was calculated in the case of X-112°Y $LiTaO_3$ and 128°YX $LiNbO_3$ with 900 nm thick Cu dots, as shown in Figure 4a. It is obvious that the 128°YX $LiNbO_3$ shows a stronger gyroscopic effect compared to X-112°Y $LiTaO_3$. Also, the contributions from the various dot materials (900 nm Au dot and Cu dot) are studied in Figure 4b in case where 128°YX $LiNbO_3$ was applied, and it is clear that a heavy metallic dot induces a larger SAW gyroscopic effect; hence, high sensor sensitivity will be expected. Similarly, increasing the metallic dot thickness can also improve the gyroscopic effect, as depicted in Figure 4c, in the case where various Au thicknesses from 300 nm to 900 nm are applied on 128°YX $LiNbO_3$. Moreover, it is easy to make a conclusion in the above calculation that decreases in sensor operation frequency at a given rate angular will increase the SAW velocity shift, that is, the sensor sensitivity will be improved by decreasing the sensor operation frequency.

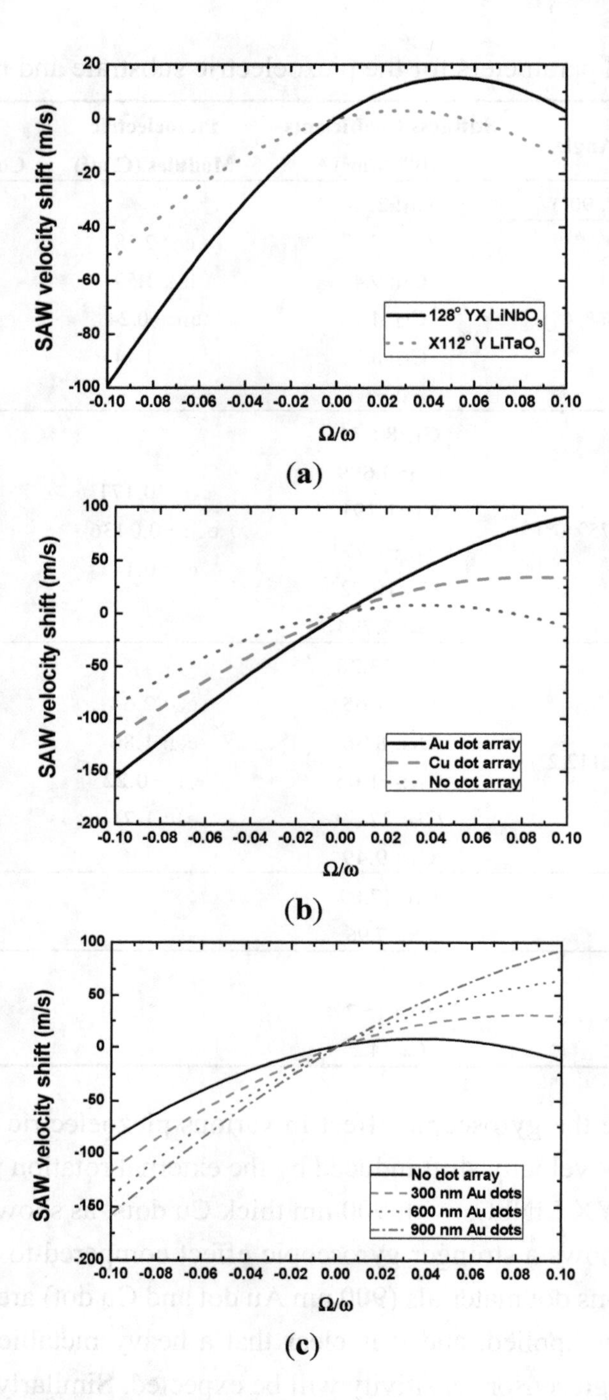


**Figure 4.** Calculated effects from the piezoelectric substrate (**a**), metallic dots (**b**), and geometry of dots (**c**).

## 3. Sensor Experiments

### *3.1. Physical Structure of the SAW Rate Sensor*

A series of SAW rate sensors utilizing various design parameters listed in Table 2 were constructed to confirm the theoretical predictions. The sensor scheme is depicted in Figure 1, in which, two parallel SAW delay lines with opposite directions and metallic (Cu and Au) dot array distribution were fabricated on a same piezoelectric substrate by a photolithography technique. Single phase unidirectional transducers (SPUDTs) and combed transducers were the structures used in the SAW delay lines to reduce the insertion loss and improve the frequency stability of the oscillator [15]. Using a HP 8753D network analyzer, the amplitude responses ($S_{21}$) of the developed SAW delay lines (Figure 5a) were measured under matched conditions. Figure 5 shows the typical $S_{21}$ plots from the SAW devices on 128°YX $LiNbO_3$ with 900 nm thick Au dots and operation frequency of 30 MHz, 80 MHz, and 95 MHz, respectively. It is worth noting that the effect of the metallic dot array distribution on device performance is insignificant and can be neglected (Figure 5b).

**Table 2.** Design parameters for the SAW sensor chips.

| Items | Design Parameters |
|---|---|
| Piezoelectric substrates | X-112°Y $LiTaO_3$, 128°YX $LiNbO_3$ |
| Operation frequency | 95 MHz, 80 MHz, 30 MHz |
| Metallic dot materials | Cu, Au |
| Metallic dot thickness | 300 nm, 600 nm, 900 nm |
| Metallic dot size | $1/4\lambda \times 1/4\lambda$ |

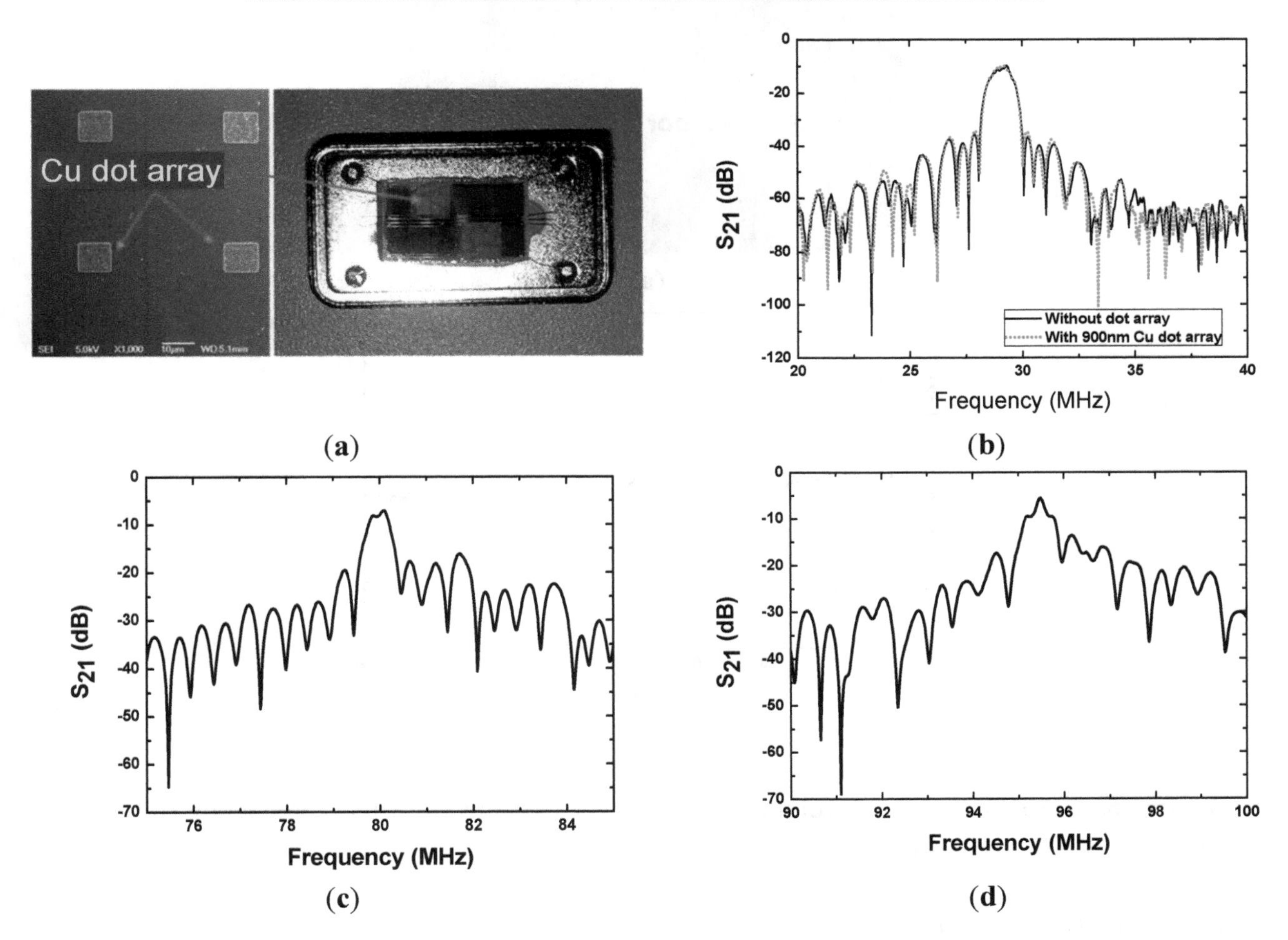


**Figure 5.** Developed SAW sensor chip (**a**), measured $S_{21}$ of 30 MHz SAW device (**b**), and 80 MHz SAW device (**c**), and 95 MHz SAW device (**d**).

Next, the fabricated SAW chips were loaded in a standard metal base (Figure 5a), and acted as the oscillation feedback elements. The launching and readout transducers of the SAW devices were connected by an oscillation circuit composed of a discrete elements (amplifier, phase shifter, mixer and LPF) on a printed circuit board (PCB). The outputs of the oscillators were mixed to obtain a differential frequency in kHz range. This technique allows doubling the detection sensitivity and a reduction of the influence of the thermal expansion of the substrate. The differential frequency signals was picked up by the frequency acquisition module (FAM) on the PCB and output to a PC, as shown in Figure 6a. To further improve the frequency stability of the oscillator, the oscillation was modulated at the frequency point with lowest insertion loss by a strategically phase modulation [16]. The typical short term frequency stability of the oscillator at room temperature (20 °C) is characterized as 0.8 Hz/s, as shown in Figure 6b.

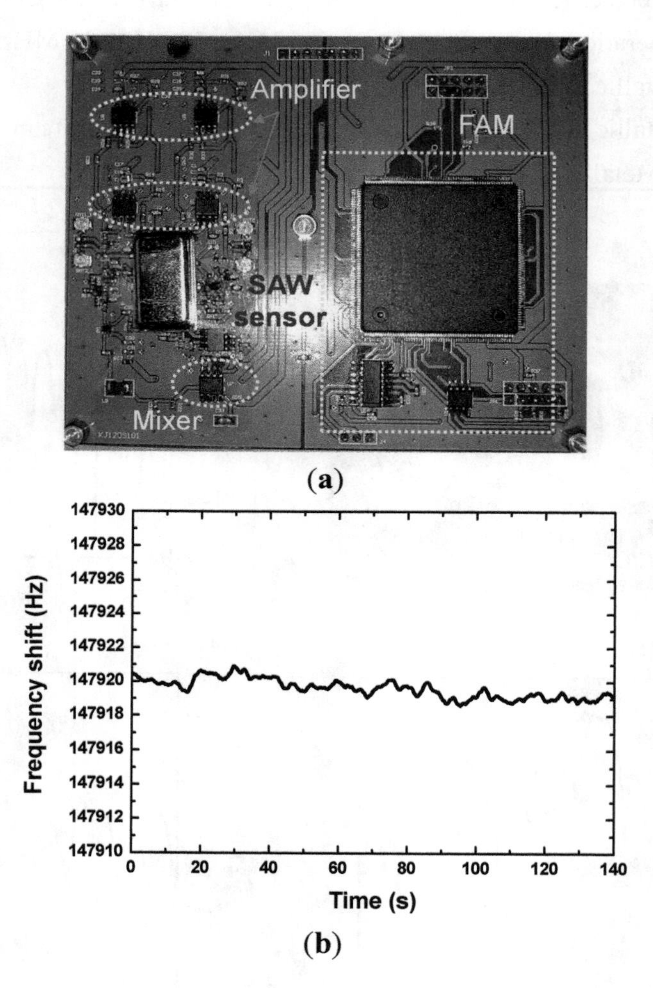


**Figure 6.** The PCB for the developed SAW sensor (**a**), and short-term frequency stability testing of SAW oscillator (**b**).

### *3.2. Sensor Experiments and Discussions*

Next, the sensor performance of the packaged SAW micro rate sensors were evaluated experimentally by using a precision temperature-controlled rate table. A self-made interface display program was used to record and plot the sensor responses in real time. The rate sensor is subjected to a rotation in the

$y$-axis. Figure 7 shows the continuous response of the stimulate sensor on X-112°Y $LiTaO_3$ with 300 nm thick Cu dot array. The data sampling time of the FAM is 18 ms, which means the FAM collects experimental data every 18 ms, so that one point in Figure 7 corresponds to a 18 ms interval. Next, the response of 95 MHz rate sensors on 128°YX $LiNbO_3$ and X-112°Y $LiTaO_3$ in case of 300 nm Cu dot array were tested as shown in Figure 8a. It is clear that the 128°YX $LiNbO_3$ displays a larger gyroscopic effect, which agrees well with the theoretical calculation in Figure 5a. Thus, the 128°YX $LiNbO_3$ substrate is adopted in the following experiments. Figure 8b illustrates the effect from the metallic dot materials on sensor response; obviously, the Au dot will provide the strongest sensor response. Also, thicker dots will obtain larger sensor responses, as shown in Figure 8c, where Au dots with thicknesses of 300 nm, 600 nm, and 900 nm are used. Moreover, the effect from the operation frequency on sensor response is also analyzed experimentally, as described in Figure 9. With the decrease of the sensor operation frequency, the sensor sensitivity increases, and the highest sensitivity was observed from the rate sensor on 128°YX $LiNbO_3$ substrate with 900 nm Au dots and operation frequency of 30 MHz. All the measured results indicate the validity of the theoretical predictions. The measured detection frequency is evaluated as ~43 $Hz \cdot deg^{-1} \cdot s^{-1}$, and good linearity was observed in the dynamic range of 0~500°/s. The measured sensitivity is over 2.7 times larger than that of reported similar rate sensors [9].

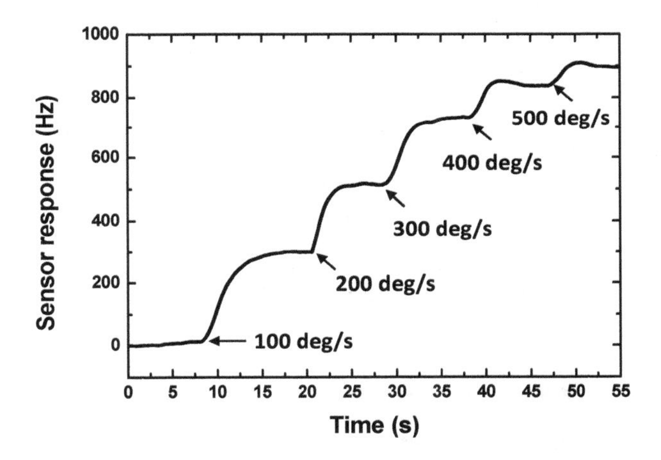


**Figure 7.** The continuous response of the stimulate 95 MHz sensor on X-112°Y $LiTaO_3$ with 300 nm thick Cu dot array.

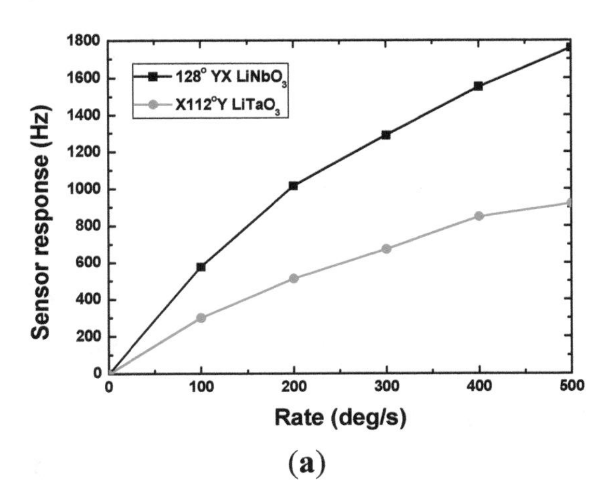


(a)

**Figure 8.** *Cont.*

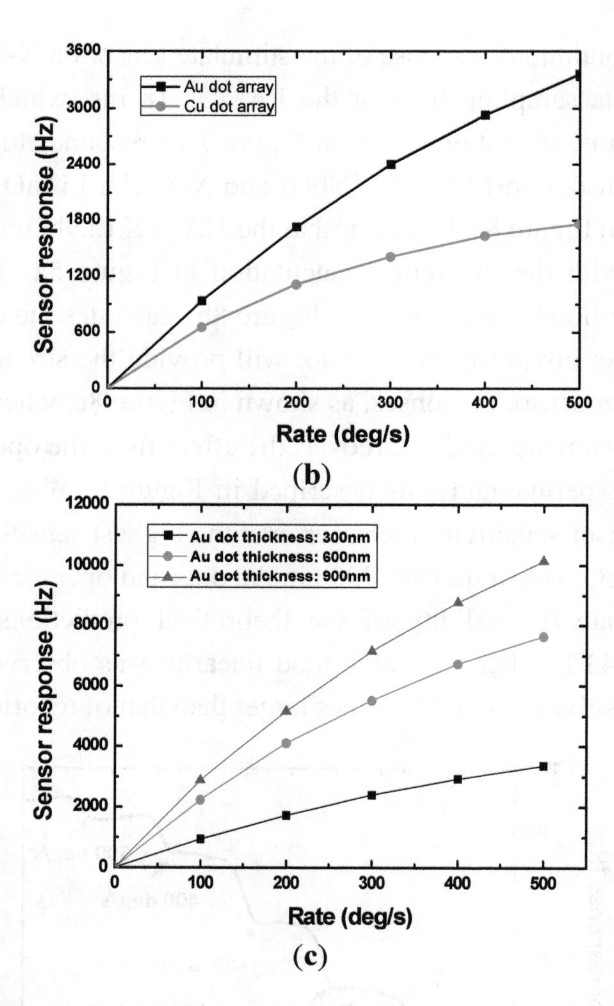


**Figure 8.** Gyroscopic effect comparison among various piezoelectric substrate (**a**), metallic dot material (**b**), and dot thickness (**c**), sensor operation frequency: 95 MHz.

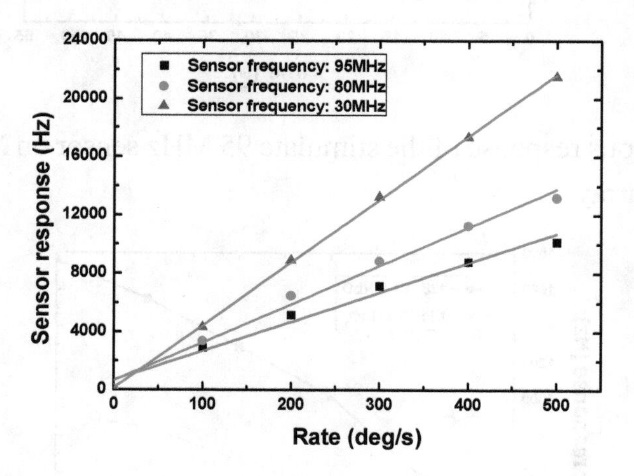


**Figure 9.** The experimental sensor response depending on various operation frequencies, piezoelectric substrate: 128°YX $LiNbO_3$, Au dot thickness: 900 nm.

Additionally, due to the differential oscillation structure, the temperature effect is compensated well as shown in Figure 10. The changes in the detection sensitivity at various temperatures were less than 5% (the detection sensitivities at temperature of 15 °C, 25 °C, 35 °C and 45 °C are 43.17 Hz·deg$^{-1}$·s$^{-1}$,

43.83 $Hz{\cdot}deg^{-1}{\cdot}s^{-1}$, 43.96 $Hz{\cdot}deg^{-1}{\cdot}s^{-1}$, and 44.26 $Hz{\cdot}deg^{-1}{\cdot}s^{-1}$, respectively). This implies that the temperature effect was effectively removed by using the differential oscillation structure.

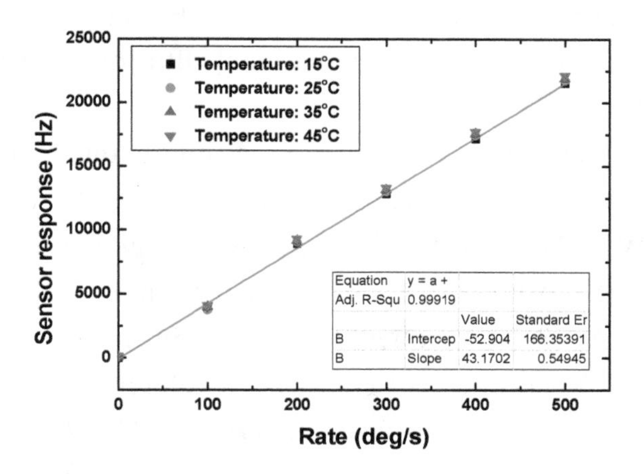


**Figure 10.** Testing of temperature effect on sensor response.

## 4. Conclusions

An optimization of a SAW rate sensor was performed by using the method of partial-wave analyses in layered media. The optimal design parameters were determined theoretically. The theoretical predictions were confirmed well in the subsequent rate sensing experiments. Significant improvements of the sensor performance were observed with the optimized SAW rate sensor. Higher sensitivity of ~43 $Hz{\cdot}deg^{-1}{\cdot}s^{-1}$ and good linearity in larger dynamic range of 0~500 Hz were achieved with the developed SAW rate sensor adopting 128°YX $LiNbO_3$, 900 nm thick Au dots and an operation frequency of 30 MHz.

## Acknowledgments

The author gratefully acknowledges the support of the "The Hundred Talents Program" of Chinese Academy of Sciences, and National Natural Science Foundation of China: 11274340 and 11374254.

## Author Contributions

All authors participated in the work presented here. Fangqian Xu and Wen Wang defined the research topic, Xiuting Shao contributed the theoretical analysis, Xinlu Liu and Yong Liang helped with the sensor design and experiments. All authors read and approved the manuscript.

## References

1. Lukyanov, D.P.; Filatov, Y.V.; Shevchenko, S.Y.; Shevelko, M.M.; Peregudov, A.N.; Kukaev, A.S.; Safronov, D.V. State of the art and prospects for the development of SAW-based soli-state gyros. *Gyroscopy Navig.* **2011**, *4*, 214–221.

2. Kurosaws, M.; Fukula, Y.; Takasaki, M.; Higuchi, T. A surface-acoustic-wave gyro sensor. *Sens. Actuators A Phys.* **1998**, *66*, 33–39.
3. Biryukov, S.V.; Chmidt, H.; Weihnacht, M. Gyroscopic Effect for SAW in Common Piezoelectric Crystals. In Proceedings of the 2009 IEEE International Ultrasonics Symposium (IUS), Rome, Italy, 20–23 September 2009; pp. 2133–2136.
4. Lee, S.W.; Rhim, J.W.; Park, S.W.; Yang, S.S. A micro rate gyroscope based on the SAW gyroscopic effect. *J. Micromech. Microeng.* **2007**, *17*, 2272–2279.
5. Lee, S.W.; Rhim, J.W.; Park, S.W.; Yang, S.K. A Novel Micro Rate Sensor using a Surface-Acoustic-Wave (SAW) Delay-Line Oscillator. In Proceedings of the 2007 IEEE on Sensors, Atlanta, GA, USA, 28–31 October 2007; pp. 1156–1159.
6. Wang, W.; Liu, J.L.; Xie, X.; Liu, M.H.; He, S.T. Development of a new surface acoustic wave based gyroscope on a X-112°Y $LiTaO_3$ substrate. *Sensors* **2011**, *11*, 10894–10906.
7. Yan, Q.; Wei, Y.; Shen, M.; Zhu, J.; Li, Y. Theoretical and Experimental Study of Surface Acoustic Wave Gyroscopic Effect. In Proceedings of the International Conference on Mechatronics and Automation, Harbin, China, 5–8 August 2007; pp. 3812–3816.
8. Wang, W.; Xu, F.Q.; Li, S.Z.; He, S.T.; Li, S.Z.; Lee, K. A new micro-rate sensor based on shear horizontal SAW gyroscopic effect. *Jpn. J. Appl. Phys.* **2010**, *49*, doi:10.1143/JJAP.49.096602.
9. Oh, H.; Lee, K.; Yang, S.S.; Wang, W. Enhanced sensitivity of a surface acoustic wave gyroscope using a progressive wave. *J. Micromech. Microeng.* **2011**, *21*, doi:10.1088/0960-1317/21/7/075015.
10. Wang, W.; Shao, X.T.; Liu, X.L.; Liu, J.L.; He, S.T. Enhanced Sensitivity of Surface Acoustic Wave-Based Rate Sensors Incorporating Metallic Dot Arrays. *Sensors* **2014**, *14*, 3908–3920.
11. Wang, W.; Shao, X.T.; Liu, J.L.; He, S.T. Theoretical analysis on SAW gyroscopic effect combining with metallic dot array. In Proceedings of the 2012 IEEE International Ultrasonics Symposium (IUS), Dresden, Germany, 7–10 October 2012; pp. 2412–2415.
12. Wang, W.; Li, S.Z.; Liu, M.H. Theoretical Analysis on Gyroscopic Effect in Surface Acoustic Waves. In Proceedings of the 2011 16th International Conference on Solid-State Sensors, Actuators and Microsystems, Beijing, China, 5–9 June 2011; pp. 1042–1045.
13. Auld, B. *Acoustic Fields and Waves in Solids*; Wiley: New York, NY, USA, 1973; Volume 1.
14. Polatoglou, H.M.; Bleris, G.L. Comparison of the constrained and unconstrained Monte-Carlo method: The case of $Cu_3Au$. *Solid State Commun.* **1994**, *90*, 425–430.
15. Wang, W.; He, S.T.; Li, S.Z.; Liu, M.H.; Pan, Y. Enhanced Sensitivity of SAW Gas Sensor Coated Molecularly Imprinted Polymer Incorporating High Frequency Stability Oscillator. *Sens. Actuators B Chem.* **2007**, *125*, 422–427.
16. Wang, W.; He, S.T.; Li, S.Z.; Liu, M.H.; Pan, Y. Advances in SXFA-Coated SAW Chemical Sensors for Organophosphorous Compound Detection. *Sensors* **2011**, *11*, 1526–1541.

# 8

# Study of the Tensile Damage of High-Strength Aluminum Alloy by Acoustic Emission

**Chang Sun [1], Weidong Zhang [1,†,*], Yibo Ai [1,†] and Hongbo Que [2]**

[1] National Center for Materials Service Safety, University of Science and Technology Beijing, Beijing 100083, China; E-Mails: schsky2009@163.com (C.S.); ybai@ustb.edu.cn (Y.A.)

[2] Qishuyan Institute Co. Ltd., China South Locomotive & Rolling Stock Corporation Limited, Changzhou 213011, China; E-Mail: quehongbo@csrqsyri.com.cn

† These authors contributed equally to this work.

* Author to whom correspondence should be addressed; E-Mail: zwd@ustb.edu.cn

Academic Editor: Hugo F. Lopez

---

**Abstract:** The key material of high-speed train gearbox shells is high-strength aluminum alloy. Material damage is inevitable in the process of servicing. It is of great importance to study material damage for in-service gearboxes of high-speed train. Structural health monitoring methods have been widely used to study material damage in recent years. This study focuses on the application of an acoustic emission (AE) method to quantify tensile damage evolution of high-strength aluminum alloy. First, a characteristic parameter was developed to connect AE signals with tensile damage. Second, a tensile damage quantification model was presented based on the relationship between AE counts and tensile behavior to study elastic deformation of tensile damage. Then tensile tests with AE monitoring were employed to collect AE signals and tensile damage data of nine samples. The experimental data were used to quantify tensile damage of high-strength aluminum alloy A356 to demonstrate the effectiveness of the proposed method.

**Keywords:** acoustic emission; aluminum alloy; tensile damage; quantification model

---

## 1. Introduction

The gearbox is one of the important parts of a high-speed train, which is subjected to damage in its service life. For example, the gearbox bracket area suffers static and dynamic loads during service. The static load is mainly caused by the weight and load of the train, while the dynamic load is related to a lot of factors including the natural wind, shock and vibration generated by operation of the train, the centrifugal force generated by curve movement of the train and so on. Static and dynamic loads bring both tensile damage and fatigue damage to gearbox bracket. Thus, it is of great importance to ensure the service safety of the gearbox. In this paper, high-strength aluminum alloy A356 is the key material of high-speed train gearbox shells. The performance analysis of the service process for this material has not yet been done [1,2]. It is necessary to develop a structural health monitoring method to study the tensile performance of this material.

In recent years, many studies have focused on developing methods to study in-service material performance by structural health monitoring techniques. Acoustic emission (AE), one of the efficient structural health monitoring techniques, which is defined as a phenomenon of rapid release of energy and generation of transient elastic wave from a localized source of the material [3]. Most of the material damage is related to microscopic processes involving some stress relaxation. Monitoring AE signals from a damaging material can give significant information about the microscopic mechanisms involved, because the magnitude of this stress relaxation is dependent upon the particular process [4].

The AE technique is widely used to detect the occurrence and growth of damage and quantify damage of in-service materials and is not limited by the materials. It has been identified that AE monitoring is capable of detecting material damage over the past three decades [5–7]. Previous studies have also proved that AE technique is sensitive and reliable in the detection of material damage for in-service structures [8,9]. The key of studying material tensile damage by using the AE technique is to relate AE parameters to tensile processes. Several attempts have been made to find the relationship between AE parameters and material damage during tensile processes. Haneef *et al.* [4] studied the tensile behavior of AISI type 316 stainless steel using AE and infrared thermography techniques. They discovered that AE root mean square voltage increased with an increase in strain rate due to the increase in source activation. The dominant frequency of the AE signals generated during different regions of tensile deformation has also been used to compare the results for different strain rates. They found that the dominant frequency of AE signals increased from elastic region to around 580–590 kHz during work hardening and 710–730 kHz around ultimate tensile strength for different strain rates, but they did not give an exact model to quantify tensile damage of the material. Lugo *et al.* [10] used AE to quantify the microstructural damage evolution under tensile loading for a 7075 aluminum alloy. They proved that the AE activity is related directly with damage progression in this alloy by building a model correlating AE counts to number density, but they focused on capturing ductile material failure. The difference between the elastic stage and plastic stage of tensile processes was ignored. Patrik *et al.* [11] monitored and analyzed the AE signal during tensile tests of pure Mg and Mg alloys of the AZ series in order to study the influence of alloy composition on plastic deformation. The Kaiser effect was used to determine the stability of the microstructure. They found the post-relaxation effect was sensitive to alloy composition and the strain at which the stress relaxation was performed, but they aimed to study the deformation mechanisms of plastic stage. The deformation mechanisms of the elastic stage were ignored. Godin *et al.* [12] used the AE technique to discriminate between the different types of damage occurring in a constrained composite. Two main types of signals were identified, originating from the two expected damage mechanisms *i.e.*, matrix cracking and decohesion. They used the k-means algorithm to split the AE data into two classes by counts, duration, and average frequency. Then k-nearest neighbors (KNN) method was used to classify the AE data. However, in their paper, AE signals were from composite materials, the AE signal of aluminum alloys is different. Cousland *et al.* [13] recorded AE signals during the unidirectional tensile deformation of aluminum alloys 2024 and 2124 to identify the sources of the emission. They concluded that the fracture of brittle inclusions in the primary source of the AE detected during the tensile testing of the alloys in the temper condition T351, but their main work was to find the source of AE signals during tensile fracture. They did not pay attention to quantify elastic deformation. Wen and Morris [14] investigated the effect of different thermal treatment temperatures (from 472 to 783 K) on the characteristics of serrated yielding of three commercial aluminum alloys, AA5052, AA5754, and AA5182 by using the AE technique. They discovered that the acoustic emission appears to be related to the number of Mg atoms actually participating in the dynamic strain aging process, but they concentrate on the yield stage of tensile process.

The AE technique is mainly applied to study plastic deformation during tensile processes. Bohlen *et al.* [15] observed and analyzed AE signals during plastic deformation of an AZ31 sheet in an H24 original condition, as well as after a heat treatment at elevated temperature. The AE count rates show a well-known correlation with the stress-strain curves. Máthis *et al.* [16] investigated mechanisms of plastic deformation of a commercial AM60 magnesium alloy by using AE measurements, TEM, and light microscopy. They found that the deformation behavior of the AM60 alloy exhibits three significant stages. Cakir *et al.* [17] measured the AE response of an implant-quality 205L stainless steel during slow strain rate tensile testing at a constant strain rate of $7.35 \times 10^{-6}$. They believed the attenuation of the AE activity beyond necking is attributed to the localization of plastic deformation. Vinogradov *et al.* [18] performed AE measurements during room temperature tensile deformation of high-alloyed cast model steels with different austenite stability to get a better understanding of the kinetics of TRIP/TWIP-assisted plastic deformation. They identified four different microstructure-related major mechanisms of plastic deformation as AE sources. Kocich *et al.* [19] discovered the special character of AE signal during plastic deformation, which can be called white noise with low energy. They used UFG materials to identify the limit of detectability. Therefore, the AE technique is available and effective for studying tensile deformation and quantifying tensile damage of materials. Presently, articles focused on quantifying elastic deformation during tensile process by the AE technique are relatively few. There is a need to develop an AE method to quantify elastic deformation of tensile damage.

In this paper, a method of tensile damage quantification by the AE technique was presented. The proposed method was based on the relationship between AE counts and tensile damage. This research focused on quantifying the elastic deformation of tensile damage. The method presented in this paper was a prognostic method only if data obtained from tensile tests was applied. The application of the developed method in high-strength aluminum alloy A356 of high-speed train gearbox shells was given to demonstrate the effectiveness of the proposed method.

The paper is organized as follows. Experimental procedures are described in Section 2. Section 3 then introduces the theoretical method for tensile damage quantification. The application of the proposed method is also introduced in this section. Conclusions will be discussed and summarized in the last section.

## 2. Experimental Procedures

An AE instrument was used to detect and record AE signals during tensile tests. Tensile tests with AE monitoring contained two parts of data, which are tensile damage data and AE monitoring data. The purpose of this study is to analyze the relationship between tensile damage and AE signals to quantify tensile damage of the high-strength aluminum alloy A356 by using AE signals.

### *2.1. Tensile Test*

The material used in this study is high-strength aluminum alloy A356, which has reached excellent properties of rigidity, strength, ductility, fatigue, casting, and shrinkage tendency [20]. Its nominal chemical composition is presented in Table 1.

**Table 1.** Chemical compositions (wt. %) of the improved aluminum alloy type A356.

| Composition | Si | Mg | Ti | Sr |
|---|---|---|---|---|
| wt. % | 6.5~7.5 | 0.20~0.35 | 0.08~0.2 | 0.005~0.015 |

An 810 Material Test System (MTS) was used to carry out tensile tests. The strain rate was $10^{-4}$/s, and test environment was 27 °C and 40% RH. There are a total of nine tensile samples, which were numbered S1 to S9. Figure 1 shows the geometry of the sample [21]. During the test, increased longitudinal tensile force was applied to the sample. Samples were broken in the end.

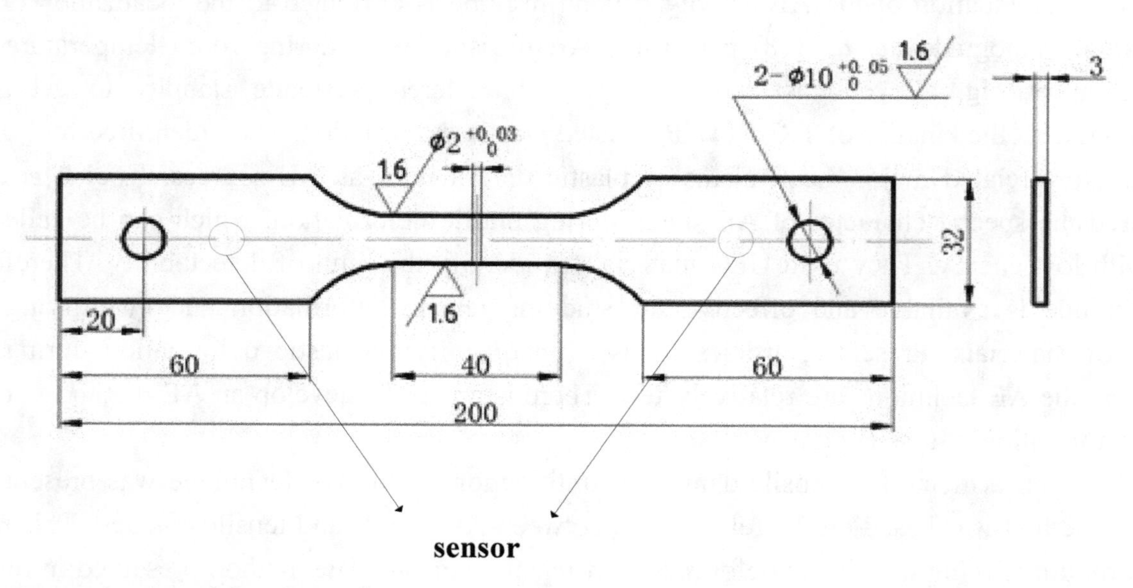


**Figure 1.** Geometry of tensile sample and arrangement of AE sensors, dimensions in mm [21].

### *2.2. Acoustic Emission Technology*

AE is defined as "a phenomenon of rapid release of energy and generation of transient elastic wave from a localized source of the material" [3]. AE signals are generated from the sudden release of strain energy at the damage sources, which are plastic deformation, crack propagation, wear, friction, and so on [12,22]. AE registration is an effective methodology, which allows "hearing" and registering damage during loading of samples [12].

An AE instrument PCI-2 supplied by American PCA company was performed to record and process AE signals during tensile tests. AE signals were recorded from tensile tests. During elastic deformation of tensile process the AE signals are mainly of continuous type signals. Typical parameters of AE signal are count, energy, amplitude, duration, rise time, and so on. The most commonly used AE parameter is count, which is defined as the number of times that the AE signal amplitude exceeds a predefined subjective threshold value [23]. Figure 2 shows some typical parameters in a burst-type waveform of AE signal.

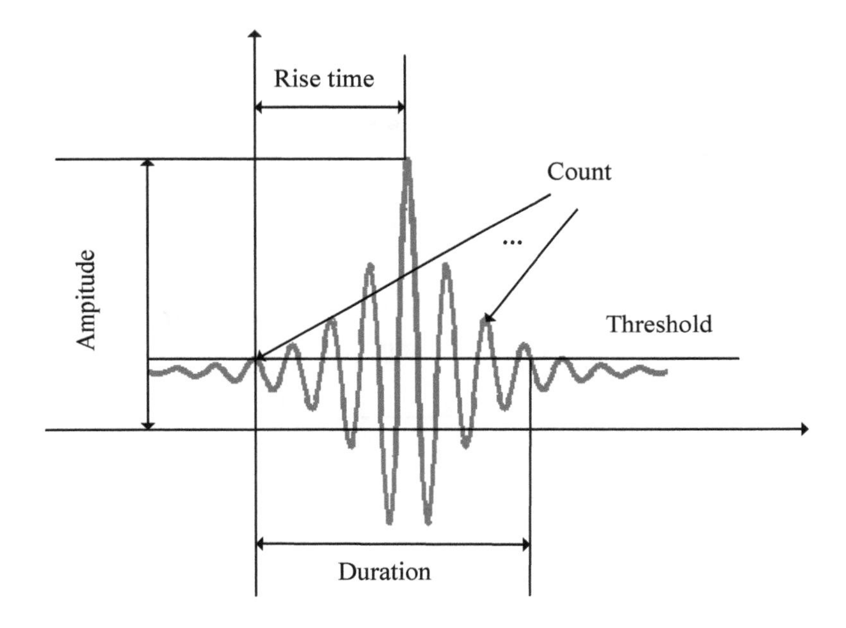


**Figure 2.** Typical parameters of AE signal.

The sensor used in tensile test was standard resonant sensor of type R15A, a 40dB pre-amplification was employed to amplify signals. For all the nine samples Vaseline was used to mount the sensor to the sample surface to ensure the extreme sensitivity of the sensor. Figure 1 shows the arrangement of sensors. Threshold values were set to remove noise, the fixed threshold (trigger level) value was 45 dB in all tests; the energy threshold value was 1. The sampling frequency is 1 MHz; peak identification time (PDT) was 300 μs; impact identification time (HDT) was 600 μs; hit lockout time (HLT) was 1000 μs; and the crash file length was 2kB.

## 3. Results and Discussion

In this section, a characteristic parameter was developed to connect AE signals with tensile processes. Then, a tensile damage quantification model was presented based on the relationship between AE counts and tensile behavior to quantify tensile damage evolution of high-strength aluminum alloy A356. This paper focused on quantifying the elastic deformation of tensile damage.

### *3.1. Material Tensile Damage*

The tensile process of metal can be divided into four stages; elastic stage, yield stage, plastic stage, and fracture stage [20]. Figure 3 displays the four stages of tensile process identified on the axial load-axial elongation curve. The four stages can be divided according to different characteristics. At the elastic stage, the material deformation is elastic, the force and elongation is proportional. The elongation will disappear with the force gone, and no residual elongation will occur. When the in-service material is in the elastic stage, it can be considered safe [21], but when the material comes into the yield stage, unrecoverable deformation will be produced. The reliability of the material will be reduced, and the final fracture will come quickly. Thus, the in-service material can be considered unsafe. In this paper, the transition point from elastic stage to yield stage is the failure point of the tensile process [21].

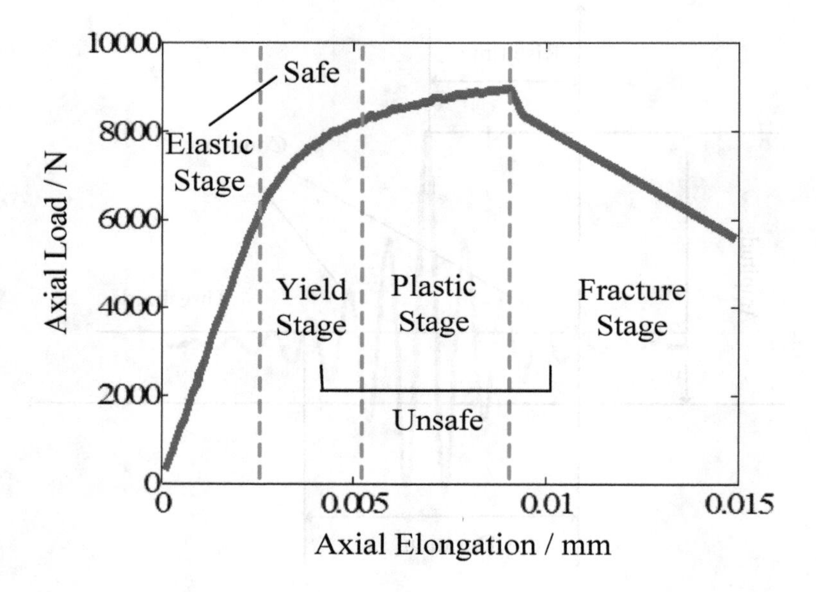


**Figure 3.** The axial load-axial elongation curve of the gearbox shell material.

If in-service materials reach the failure point of the tensile process, catastrophic failure may occur. It is of great importance to study tensile damage of in-service materials, especially the elastic stage of the tensile process. This paper aims to quantify the elastic deformation of material by the AE technique.

### *3.2. Characteristic of AE Signal for the Tensile Damage*

Figure 4 displays the relationship between AE counts and axial elongation in the same tensile process. It is indicated that AE counts increased significantly at the transition point from elastic stage to yield stage [21]. AE counts can be used to characterize the tensile damage, but the AE counts of each sample at the transition point are different, and the values have a great range. Thus, it is necessary to develop a characteristic parameter involving AE counts to find the transition point between the elastic stage and yield stage of the tensile process. The characteristic parameter is the key to relate tensile damage with AE signals.

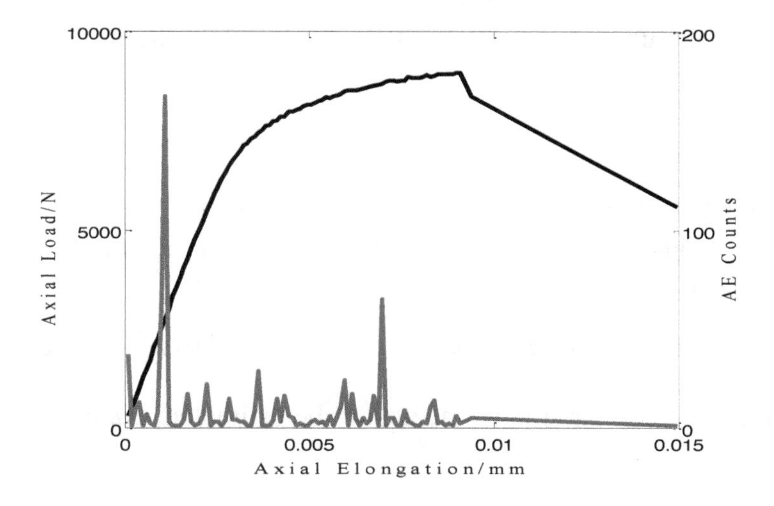


**Figure 4.** AE counts and axial elongation curve.

In this paper, the characteristic parameter is named $AC(t)$. It can be calculated by Equations (1) and (2):

$$M_j(t) = \text{median}_m\left(\frac{t_{j+1} - t_j}{c_j} + \frac{t_{j+2} - t_{j+1}}{c_{j+1}} \cdots \frac{t_{j+m} - t_{j+m-1}}{c_{j+m-1}}\right) \qquad j = 1, 2, \cdots, n \tag{1}$$

$$AC_j(t) = M_{j+1}(t) - M_j(t) \tag{2}$$

where $c_j$ is the AE count at $t_j$ time; $m$ is the quantity of values used to calculate the median of them; $j$ is the AE count number. Once the value of $AC(t)$ comes into the preset threshold interval, it means that the tensile process reaches the transition point from elastic stage to yield stage. In other words, the duration of the elastic stage $t_f$ can be obtained.

AE signals were collected and recorded from tensile tests with the AE instrument. The characteristic parameter $AC(t)$ was calculated based on Equations (1) and (2). Herein, $m$ = 10 and the threshold interval was $[7.65 \times 10^{-3}, 7.85 \times 10^{-3}]$, Once the value of $AC(t)$ reached the preset threshold interval, the duration of the elastic stage $t_f$ can be obtained. Table 2 shows the characteristic parameter $AC_f(t)$ at the transition point and duration of the elastic stage $t_f$ of all nine samples.

**Table 2.** The results of characteristic parameter $AC(t)$ and duration of the elastic stage $t_f$.

| Sample | $t_f$/s | $AC_f(t)$ |
|---|---|---|
| S1 | 88.6 | 0.000780 |
| S2 | 73.0 | 0.000769 |
| S3 | 88.9 | 0.000779 |
| S4 | 95.8 | 0.000782 |
| S5 | 122.0 | 0.000773 |
| S6 | 129.0 | 0.000781 |
| S7 | 111.8 | 0.000777 |
| S8 | 159.0 | 0.000775 |
| S9 | 92.6 | 0.000776 |

### *3.3. Tensile Damage Quantification Model*

A tensile damage quantification model was presented to quantify the remaining time of the elastic stage. The actual remaining time of the elastic stage in tensile process $TR_j(t)$ can be described by Equation (3):

$$TR_j(t) = t_f - T_j(t) \qquad j = 1, 2, \cdots, n_1 \tag{3}$$

where $T_j(t)$ is the tensile test running time; $j$ is the AE count number.

The remaining time of elastic stage in tensile process calculated from AE signals $TPF_j(t)$ can be described by the following function:

$$TPF_j(t) = TPM_j + TDM_j = am \cdot \exp(bm \cdot cn_j) + fm \cdot cn_j^2 + gm \cdot cn_j + hm \tag{4}$$

where $am$, $bm$, $fm$, $gm$ and $hm$ are parameters needed to be estimated; $cn_j$ is the normalized cumulative AE counts.

### *3.4. Model Parameters Estimation*

AE signals were collected and recorded from tensile tests with AE instrument. There were nine samples numbered from S1 to S9. Eight samples were selected randomly as training data, while the remaining sample remained as testing data. Training data were used to estimate parameters of the elastic stage remaining time quantification model. Testing data were used to verify the effectiveness of the model. If the testing sample was S9, then the following steps should be applied to get model parameters by using training samples S1–S8.

Step 1. There are a total of eight samples in training data. According to the relationship between cumulative counts and the remaining time, the remaining time $TP_{i,j}(t)$ Can be described as Equation (5):

$$TP_{i,j} = a_i \cdot \exp(b_i \cdot cn_{i,j}) \qquad i = 1, 2, \cdots, 8 \qquad j = 1, 2, \cdots, n_2 \tag{5}$$

where $cn_{i,j}$ is the normalized cumulative counts; $i$ is the sample number; $j$ is the AE count number; $a_i$ and $b_i$ are parameters obtained by performing a linear least squares regression, herein, $TP_{i,j} = TR_{i,j}$. Therefore, eight groups of $(a_i, b_i)$ can be obtained.

Step 2. Set $cn = [1,2,\ldots,M]$, herein, $M = 1000$. Calculate $TP_{i,j}'$ based on Equation (5), $cn$, $a_i$ and $b_i$. Each element in $cn$ corresponds to 8 groups of $TP_{i,j}'$. Calculate the mean value of eight groups of $TP_{i,j}'$. The quantity of mean values is $M$. Then get the values of parameters $am_9$ and $bm_9$ based on calculated mean values and $cn$ by performing a linear least squares regression. Parameters $am_9$ and $bm_9$ are taken as the parameter of sample S9. Thus, parameters $am_9$ and $bm_9$ in $TPM_{9,j}(t)$ can be obtained:

$$TPM_{9,j} = am_9 \cdot \exp(bm_9 \cdot cn_{9,j}) \qquad j = 1, 2, \cdots, n_3 \tag{6}$$

Step 3. Samples S1–S8 are training samples. Select samples S1–S7 to calculate $a_k$ and $b_k$ according to Equation (5) by performing a linear least squares regression, herein, $k = 1,2,\ldots,7$, $TP_{k,j} = TR_{k,j}$. Calculate $TP_{k,j}'$ based on Equation (5), $cn$, $a_k$ and $b_k$. Each element in $cn$ corresponds to 7 groups of $TP_{k,j}'$. Calculate the mean value of 7 groups of $TP_{k,j}'$. The quantity of mean values is $M$. Then get the values of parameters $ad_8$ and $bd_8$ based on calculated mean values and $cn$ by performing a linear least squares regression. So parameters $ad_8$ and $bd_8$ in can $TPD_{8,j}(t)$ be obtained:

$$TPD_{8,j} = ad_8 \cdot \exp(bd_8 \cdot cn_{8,j}) \qquad j = 1,2,\cdots,n_4 \tag{7}$$

Step 4. The variance between $TPD_{8,j}(t)$ and $TR_{8,j}(t)$ can be given by Equation (8). According to the relationship between cumulative counts and $TD_{8,j}(t)$, $TD_{8,j}(t)$ can be described as Equation (9):

$$TD_{8,j} = TR_{8,j}(t) - TPD_{8,j}(t) \qquad j = 1,2,\cdots,n_4 \tag{8}$$

$$TD_{8,j} = f_8 \cdot cn_{8,j}^2 + g_8 \cdot cn_{8,j} + h_8 \qquad j = 1,2,\cdots,n_4 \tag{9}$$

where $cn_{8,j}$ is the normalized cumulative counts; $f_8$, $g_8$ and $h_8$ are parameters obtained by performing a linear least squares regression. Repeat Step 3 and Step 4 seven times to get ($f_i$, $g_i$, $h_i$); herein, $i = 1,2,\ldots,7$. Hence, eight groups of ($f_i$, $g_i$, $h_i$) can be obtained.

Step 5. There are eight groups of ($f_i$, $g_i$, $h_i$). Calculate $TD_{i,j}'$ based on Equation (9), $cn$, $f_i$, $g_i$ and $h_i$. Each element in $cn$ corresponds to eight groups of $TD_{i,j}'$. Calculate the mean value of eight groups of $TD_{i,j}'$. The quantity of mean values is $M$. Then get the values of parameters $fm_9$, $gm_9$, and $hm_9$ based on calculated mean values and $cn$ by performing a linear least squares regression. Thus, parameters $fm_9$, $gm_9$, and $hm_9$ in $TDM_{9,j}(t)$ can be obtained:

$$TDM_{9,j} = fm_9 \cdot cn_{9,j}^2 + gm_9 \cdot cn_{9,j} + hm_9 \qquad j = 1,2,\cdots,n_3 \tag{10}$$

Similarly, model parameters of samples S1–S8 can be calculated according to step1–step5. Figure 5 shows the values of parameter $am_l$ and $bm_l$, $l = 1,2,\ldots,9$. Figure 6 shows the values of parameter $fm_l$, $gm_l$, and $hm_l$, $l = 1,2,\ldots,9$.

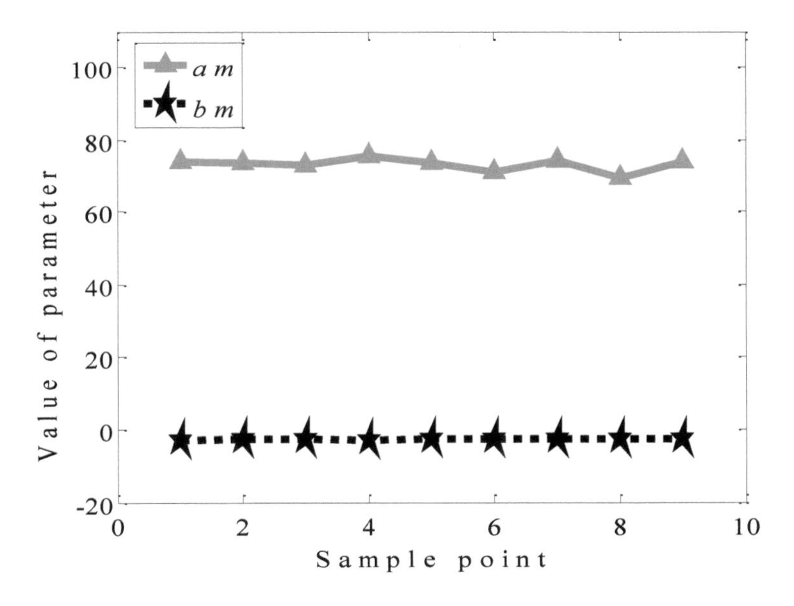


**Figure 5.** Values of parameter *am* and *bm*.

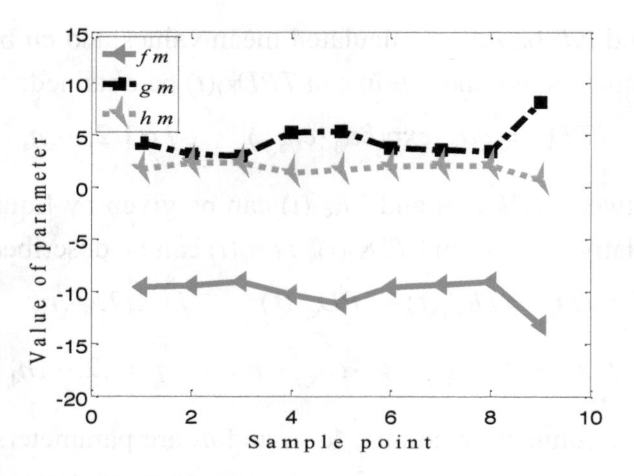


**Figure 6.** Values of parameter *fm*, *gm*, and *hm*.

### *3.5. Results and Verification of the Model*

Once the values of parameter $am_l$, $bm_l$, $fm_l$, $gm_l$, and $hm_l$ were obtained, the remaining time of samples can be calculated based on Equation (4). Figure 7 displays the comparison between the remaining time of the elastic stage calculated from AE data and actual remaining time of the elastic stage in all samples.

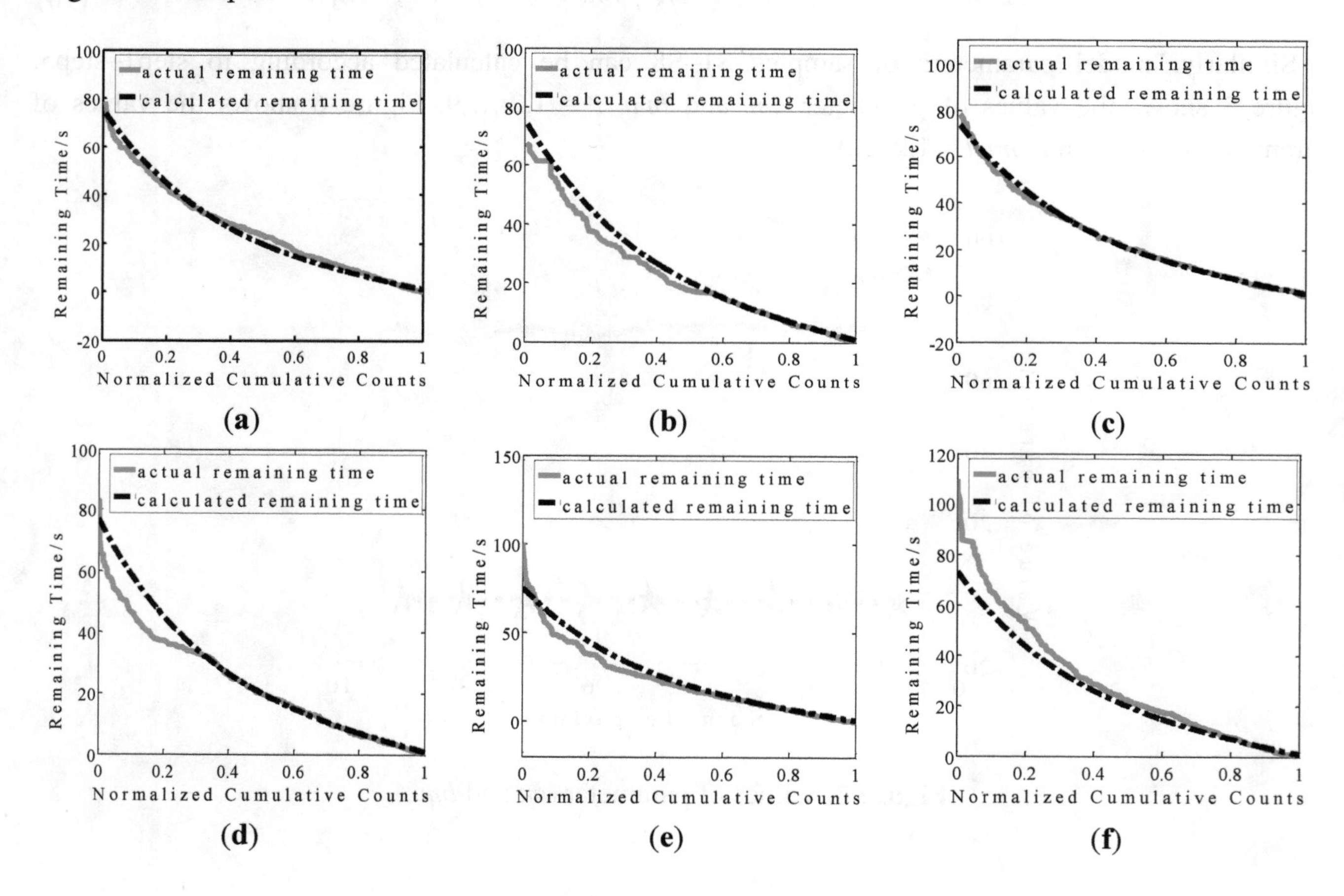


**Figure 7.** *Cont.*

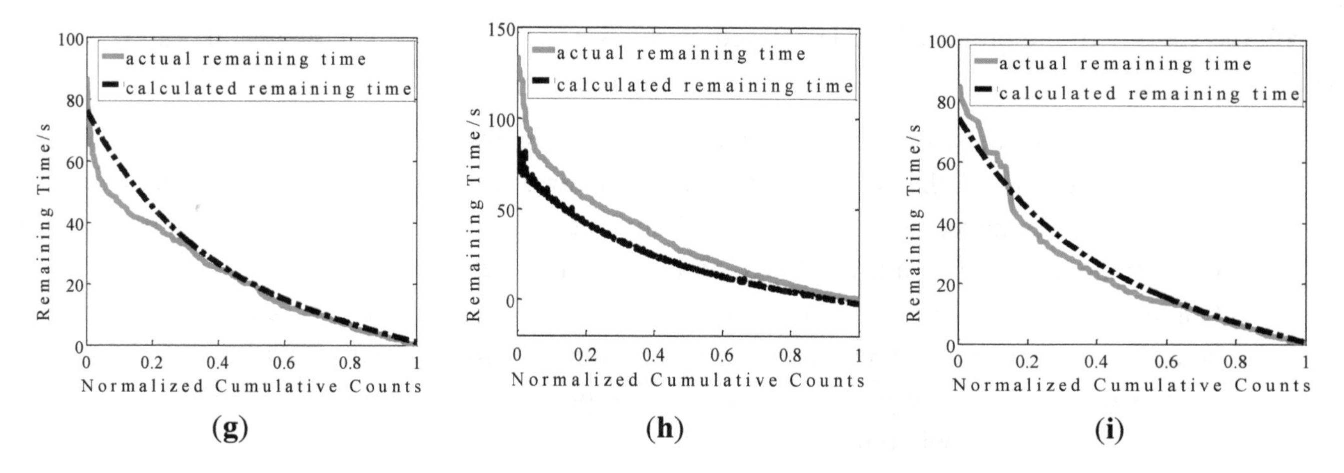


**Figure 7.** Comparison between the remaining time of elastic stage calculated from AE data and actual remaining time of the elastic stage in S1–S9. (**a**) S1; (**b**) S2; (**c**) S3; (**d**) S4; (**e**) S5; (**f**) S6; (**g**) S7; (**h**) S8; and (**i**) S9.

It is indicated in Figure 7 that the proposed model can be used to quantify remaining time of the elastic stage of high-strength aluminum alloy A356. The error between the calculated remaining time of the elastic stage and the actual remaining time is getting smaller when the material is close to the failure point.

Table 3 shows the variance between remaining time of elastic stage calculated from AE data and actual remaining time. It is indicated in Table 3 that the maximum absolute error is 59.9 s, the maximum average absolute error is 11.0 s for all samples. The average error in Table 3 refers to the mean of variance between the remaining time calculated from AE data and the actual remaining time of all the data points in each sample. Average errors of all nine tensile samples are acceptable. In other words, the proposed tensile damage quantification model is effective to quantify material tensile damage evolution of high-strength aluminum alloy A356.

**Table 3.** Variance between the predicted and actual remaining life.

| Sample | Maximum Error/s | Minimum Error/s | Average Error/s |
|---|---|---|---|
| **S1** | 5.7 | 0.001 | 2.0 |
| **S2** | 7.9 | 0.003 | 2.6 |
| **S3** | 4.6 | 0.003 | 1.0 |
| **S4** | 14.5 | 0.001 | 3.3 |
| **S5** | 25.0 | 0.144 | 3.4 |
| **S6** | 36.5 | 0.001 | 4.1 |
| **S7** | 14.8 | 0.002 | 3.8 |
| **S8** | 59.9 | 2.046 | 11.0 |
| **S9** | 10.8 | 0.224 | 3.2 |

## 4. Conclusions

In this paper, a tensile damage quantification model based on the relationship between AE counts and tensile behavior was presented to quantify material tensile damage evolution of high-strength aluminum alloy A356. Specialized tensile tests with AE monitoring were developed and carried out to verify the validity of the proposed model. Based on this study, the following conclusions can be drawn as follows:

(1) The correlation between tensile damage and AE signals was established by characteristic parameter *AC*(*t*), which can be used to monitor material elastic deformation of tensile damage.
(2) The proposed model is effective to quantify elastic deformation of tensile damage of high-strength aluminum alloy A356 of high-speed train gearbox shells.
(3) Cumulative counts, as one of the most commonly-used AE parameters, can be performed in combination with the proposed model to provide warning signs for gearbox of high-speed trains when tensile damage comes to the failure point, where the final fracture will be attained quickly, and catastrophic failure may occur.
(4) The method presented in this paper was a prognostic method only if data obtained from tensile tests is applied. In other words, the proposed elastic stage remaining time quantification model in this paper is offline. Hence, building an online elastic stage remaining time prediction model is work that needs to be done in the future.

## Acknowledgments

The authors would like to acknowledge the financial support provided by the National Natural Science Foundation of China (Grant No. 61273205), the Fundamental Research Funds for the Central Universities of China (Grant No. FRF-SD-12-028A), the 111 Project (Grant No. B12012).

## Author Contributions

Chang Sun and Weidong Zhang conceived and designed the study. Chang Sun and Yibo Ai performed the experiments. Hongbo Que provided the basic performance analysis of the material. Chang Sun made the data analysis. Chang Sun and Weidong Zhang wrote the paper. All authors read and approved the manuscript.

## References

1. Wang, B.M. *Overall System and Bogie of High-Speed Motor Train Units*; Southwest Jiaotong University Press: Chengdu, China, 2008.
2. Li, X.J. CRH bullet trains overview. *Railw. Tech. Superv.* **2007**, *9*, 26–28.
3. Yang, M.W. *Acoustic Emission Testing*; Machinery Industry Press: Beijing, China, 2005.
4. Haneef, T.; Lahiri, B.B.; Bagavathiappan, S.; Mukhopadhyay, C.K.; Philip, J.; Rao, B.P.C.; Jayakumar, T. Study of the tensile behavior of AISI type 316 stainless steel using acoustic emission andinfrared thermography techniques. *J. Mater. Res. Technol.* **2015**, *137*, 1–13.
5. Jalaj, K.; Sony, P.; Mukhopadhyay, C.K.; Jayakumar, T.; Vikas, K. Acoustic emission during tensile deformation of smooth and notched specimens of near alpha titanium alloy. *Res. Nondestruct. Eval.* **2012**, *23*, 17–31.

6. Luo, X.; Haya, H.; Inaba, T.; Shiotani, T.; Nakanishi, Y. Damage evaluation of railway structures by using train-induced AE. *Constr. Build Mater.* **2004**, *18*, 215–223.
7. Re, V.D. Acoustic emission applications for defect detection in steels and GFRP. *Int. J. Mater. Prod. Technol.* **1988**, *3*, 38–53.
8. Chen, H.L.; Choi, J.H. Acoustic emission study of fatigue cracks in materials used for AVLB. *J. Nondestruct. Eval.* **2004**, *23*, 133–151.
9. Johnson, M. Waveform based clustering and classification of AE transients in composite laminates using principal component analysis. *NDT E Int.* **2002**, *35*, 367–376.
10. Lugo, M.; Jordon, J.B.; Horstemeyer, M.F.; Tschopp, M.A.; Harris, J.; Gokhale, A.M. Quantification of damage evolution in a 7075 aluminum alloy using an acoustic emission technique. *Mater. Sci. Eng. A* **2011**, *528*, 6708–6714.
11. Patrik, D.; Jan, B.; Frantisek, C.; Pavel, L.; Dietmar, L.; Karl, U.K. Acoustic emission during stress relaxation of pure magnesium and AZ magnesium alloys. *Mater. Sci. Eng. A* **2007**, *462*, 307–310.
12. Godin, N.; Huguet, S.; Gaertner, R.; Salmo, L. Clustering of acoustic emission signals collected during tensile tests on unidirectional glass/polyester composite using supervised and unsupervised classifiers. *NDT E Int*. **2004**, *37*, 253–264.
13. Cousland, S.M.; Scala, C.M. Acoustic emission during the plastic deformation of aluminum alloys 2024 and 2124. *Mater. Sci. Eng.* **1983**, *57*, 23–29.
14. Wen, W.; Morris, J.G. An investigation of serrated yielding in 5000 series aluminum alloys. *Mater. Sci. Eng. A* **2003**, *354*, 279–285.
15. Bohlen, J.; Chmelík, F.; Dobrŏn, P.; Kaiser, F; Letzig, D.; Lukáč, P.; Kainer, K.U. Orientation effects on acoustic emission during tensile deformation of hot rolled magnesium alloy AZ31. *J. Alloys Compd.* **2004**, *378*, 207–213.
16. Máthis, K.; Chmelík, F.; Janeček, M.; Hadzima, B.; Trojanová, Z.; Luká, P. Investigating deformation processes in AM60 magnesium alloy using the acoustic emission technique. *Acta Mater.* **2006**, *54*, 5361–5366.
17. Cakir, A.; Tuncell, S.; Aydin, A. AE response of 205L SS during SSR test under potentiostatic control. *Corros. Sci.* **1999**, *41*, 1175–1183.
18. Vinogradov, A.; Lazarev, A.; Linderov, M.; Weidner, A.; Biermann, H. Kinetics of deformation processes in high-alloyed cast transformation-induced plasticity/twinning-induced plasticity steels determined by acoustic emission and scanning electron microscopy: Influence of austenite stability on deformation mechanisms. *Acta Mater.* **2013**, *61*, 2434–2449.
19. Kocich, R.; Cagala, M.; Crha, J.; Kozelsky, P. Character of acoustic emission signal generated during plastic deformation. In Proceedings of the 30th European Conference on Acoustic Emission Testing & 7th International Conference on Acoustic Emission, Granada, Spain, 12–15 September 2012; 1–8.
20. Zhang, W.D.; Zhang, X.W.; Yang, B.; Ai, Y.B. Damage characterization and recognition of aluminum alloys based on acoustic emission signal. *J. Univ. Sci. Technol. Beijing* **2013**, *35*, 626–633.
21. Ai, Y.B.; Sun, C.; Que, H.B.; Zhang, W.D. Investigation of material performance degradation for high-strength aluminum alloy using acoustic emission. *Metals* **2015**, *5*, 228–238.

22. Okafor, A.C.; Natarajan, S. Acoustic emission monitoring of tensile testing of corroded and un-corroded clad aluminum 2024-T3 and characterization of effects of corrosion on AE source events and material tensile properties. *AIP Conf. Proc.* **2014**, *1581*, 492–500.
23. Keshtgar, A.; Modarres, M. Acoustic emission-based fatigue crack growth prediction. In Proceedings of the Reliability and Maintainability Symposium: Product Quality & Integrity, Orlando, FL, USA, 28–31 January 2013; pp. 1–5.

# 9

# Managing Measurement Uncertainty in Building Acoustics

**Chiara Scrosati * and Fabio Scamoni**

Construction Technologies Institute, National Research Council of Italy, via Lombardia 49, 20098 San Giuliano Milanese (MI), Italy

* Author to whom correspondence should be addressed; E-Mail: c.scrosati@itc.cnr.it

Academic Editor: Umberto Berardi

---

**Abstract:** In general, uncertainties should preferably be determined following the principles laid down in ISO/IEC Guide 98-3, the Guide to the expression of uncertainty in measurement (GUM:1995). According to current knowledge, it seems impossible to formulate these models for the different quantities in building acoustics. Therefore, the concepts of repeatability and reproducibility are necessary to determine the uncertainty of building acoustics measurements. This study shows the uncertainty of field measurements of a lightweight wall, a heavyweight floor, a façade with a single glazing window and a façade with double glazing window that were analyzed by a Round Robin Test (RRT), conducted in a full-scale experimental building at ITC-CNR (Construction Technologies Institute of the National Research Council of Italy). The single number quantities and their uncertainties were evaluated in both narrow and enlarged range and it was shown that including or excluding the low frequencies leads to very significant differences, except in the case of the sound insulation of façades with single glazing window. The results obtained in these RRTs were compared with other results from literature, which confirm the increase of the uncertainty of single number quantities due to the low frequencies extension. Having stated the measurement uncertainty for a single measurement, in building acoustics, it is also very important to deal with sampling for the purposes of classification of buildings or building units. Therefore, this study also shows an application of the sampling included in the Italian Standard on the acoustic classification of building units on a serial type building consisting of 47 building units. It was found that the greatest variability is observed in the façade and it depends on both the great variability of window's typologies and on workmanship. Finally, it is suggested how to manage the uncertainty in building acoustics, both for one single measurement and a campaign of measurements to determine the acoustic classification of buildings or building units.

**Keywords:** measurement uncertainty; building acoustics; Round Robin Test (RRT); sampling; acoustic classification

## 1. Introduction

This paper is a revised and expanded version of the paper "Uncertainty in Building Acoustics" [1] presented at the 22nd International Congress on Sound and Vibration ICSV22.

When reporting the result of the measurement of a physical quantity, it is compulsory that some quantitative indications of the quality of the result be given so that those who use it can assess its reliability. Without such indications, measurement results cannot be compared, either with one another or with reference values given in a specification or standard. It is therefore necessary, in order to characterize the quality of the result of a measurement, to evaluate and to express its uncertainty. Generally, it is widely recognized that, when all of the known or suspected components of error have been evaluated and the appropriate corrections have been applied, an uncertainty about the correctness of the stated result still remains; that is, a doubt about how well the result of the measurement represents the value of the quantity being measured.

The word "uncertainty" means doubt, and thus in its broadest sense "uncertainty of measurement" means doubt about the validity of the result of a measurement. The formal definition of the term "uncertainty of measurement" developed in the Guide to the expression of uncertainty in measurement (GUM) [2] is as follows. *Uncertainty (of measurement): parameter, associated with the result of a measurement, that characterizes the dispersion of the values that could reasonably be attributed to the measurand.*

This definition of uncertainty of measurement is an operational definition that focuses on the measurement result and its evaluated uncertainty. However, it is not inconsistent with other concepts of uncertainty of measurement, such as a measure of the possible error in the estimated value of the measurand as provided by the result of a measurement; or an estimate characterizing the range of values within which the true value of a measurand lies. Although these two traditional concepts are ideally valid, they focus on unknowable quantities: the "error" of the result of a measurement and the "true value" of the measurand (in contrast to its estimated value), respectively.

## 2. The Uncertainty in Terms of Repeatability, Reproducibility and *in Situ* Standard Deviation

Tests performed on samples made of materials presumed to be the same, in identical conditions, generally do not give the same results. This condition is due to inevitable errors (systematic and random) in test procedures, caused by the difficulties in controlling the several factors that influence the test. To determine the accuracy of a measurement method, both accuracy and precision should be considered; in particular, the latter indicates the correlation between the test results.

Precision is a general term for the variability between repeated tests. Two measures of precision, termed repeatability and reproducibility, have proved necessary and, for many practical cases, sufficient for describing the variability of a test method. Repeatability refers to tests performed on the same test object with the same method under conditions that are as constant as possible, with the tests performed during a short interval of time, in one laboratory by one operator using the same equipment. On the other hand, reproducibility refers to tests performed on identical test items with the same method, in widely varying conditions, in different laboratories with different operators and different equipment. Thus, repeatability and reproducibility are two extremes, the first measuring the minimum and the latter the maximum variability in results.

The building acoustic quantities include airborne sound insulation of internal partitions, airborne sound insulation of façades, impact sound insulation of floors and sound pressure level from service equipment in buildings. The quantities that have to be measured and their measurement methods, for all aspect involved, are described in the international standard series EN ISO 10140 [3] for laboratory measurements and in the international standard series ISO 16283 [4] for field measurements. The accuracy of these measurement method depends on several factors that influence the test, such as acoustic instrumentation, acoustic method (microphones and sources position), context (regular rooms or semi-open space, of any size), constructive details of the building (that could have effect on acoustic measures) and workmanship, and, concerning sound levels, influence of instrumentation working conditions (repeat configuration). Detailed information for each of these factors is hardly available. Both random and systematic errors affect the acoustic measurements results. The random effects can be determined by repeated independent measurements in essentially identical conditions. The systematic effects, however, are not easy to determine, but, as a general rule, they can be determined thanks to comparative measurements to be executed in different test facilities (for laboratory measurements) or carried out by different laboratories (for field measurements), and the knowledge of the random errors in those conditions. Therefore, it is necessary to refer to the concepts of repeatability and reproducibility, which provide a simple means for the expression of the precision of a test method and of the measurements performed according to the test method.

The best methodology to study the repeatability and reproducibility of building acoustic measurements is to carry out an Inter-Laboratory Test (ILT), or a Round Robin Test (RRT), tests consisting of independent measurements executed several times by different operators. Due to the particular nature of the sample in building acoustics, in addition to repeatability and reproducibility standard deviations, another standard deviation is defined, the *in situ* standard deviation (defined, for the first time, in ISO 12999-1 [5]), which could be useful to estimate. The *in situ* standard deviation is a particular kind of reproducibility standard deviation that is measured in the same location on the same object. In fact, in the case of RRT field measurements, when different operators, with their own equipment, perform measurements on a particular building element, both the location and the object under test are the same. Therefore, location is the only difference between reproducibility and *in situ* standard deviation: for the *in situ* standard deviation, the location is exactly the same as is the test object, while in the case of reproducibility standard deviation the locations are different and the test object can be either the same test object or identical test objects tested in the different locations. The *in situ* standard deviation, therefore, corresponds to a reproducibility standard deviation of the same object in the same location.

### *2.1. Round Robin Test*

Generally, cooperative tests (ILT or RRT) assess the uncertainty of measurement methods using a reference value. One of the main aspects of these tests is the determination of this reference and its uncertainty. A reliable, low-uncertainty reference value is required in order to minimize the uncertainty of a cooperative test. Due to the typology of the sample test in acoustic measurements, a reference value does not exist; therefore an estimated value is used. The best measuring reference is the mean value. A RRT of sound insulation field measurements of building elements was carried out as part of a research sponsored by the Lombardy Region [6–8]; this study was based on the cooperation of three different bodies: a research body, ITC-CNR (Construction Technologies Institute of the National Research Council of Italy); a university laboratory, DISAT (Department of Earth and Environmental Sciences of the University of Milano-Bicocca); and a control organization, ARPA-Lombardy (Regional Agency for environmental protection) and it was coordinated by ITC-CNR. In the first approach to the problem [6], the analysis was centered on the single number values of the Italian regulation [9] and on the narrow frequency range (from 100 to 3150 Hz). In later studies [7,8], the analysis considered all the possible descriptors of the different European national legislations and was extended to the enlarged frequencies range (from 50 to 5000 Hz). Another study on the uncertainty of façade sound insulation [10] was carried out at the initiative of the Building Acoustics Group (GAE) of the Italian Acoustic Association (AIA). This study was focused on the low frequencies (from 50 to 80 Hz), in particular on the comparison between the procedure stated in ISO 140-5 [11] and the new low frequency procedure stated in ISO 16283 [4]. The main results of these studies are summarized in the following section.

#### 2.1.1. Airborne Sound Insulation

Notwithstanding the importance of the uncertainty of the measurement method in building acoustics, the uncertainty of field measurements was not comprehensively investigated. There are only few examples in the literature [12,13] compared to those of laboratory tests [14–18]. The studies regarding laboratory tests conclude that the main influences are caused by the laboratory geometry and materials, the flanking transmissions, the type of border material, and the different test opening dimensions [15,16].

Nine teams coordinated by ITC-CNR were involved in the study about the uncertainty of airborne sound insulation [7]; each of them has replicated the tests five times, including the reverberation time.

No deviations occurred from the test procedure laid down in ISO 140-4 [19] but, repeating the measurements several times, the parameters left open in the measurement procedure were represented as best as possible. In particular, the set of microphone positions and source positions were selected anew, more or less randomly, for each repeated measurement. The measurands were a floor without floating floor (surface mass of 550 kg/m$^2$ and surface of about 19 m$^2$) and a lightweight wooden partition wall (surface mass of 30 kg/m$^2$ and surface of about 8.5 m$^2$). Considering the goal of European harmonization of acoustic parameters [20], the differences between the various descriptors (R', $D_n$ and $D_{nT}$) were analyzed in terms of average, maximum and minimum values, and in terms of standard deviation of repeatability and reproducibility (*in situ* standard deviation, referring to ISO 12999-1 [5], where the reproducibility standard deviation of the same element is measured in the same location).

Figure 1 shows the standard deviations of repeatability $s_r$ and *in situ* reproducibility standard deviation $s_{situ}$ of all analyzed quantities. The descriptors extension at low frequencies (from 50 to 80 Hz) (LF) was also analyzed. From the graphs of Figure 1, it is evident that the uncertainty at LF is much greater than the uncertainty in the narrow frequencies range from 100 to 5000 Hz. From the comparison of the RRT $s_{situ}$ values with the values of the ISO 12999-1 [5] for situations A ($s_R$) and B ($s_{situ}$) (see Figure 1), it was found that the values of situation B underestimate the uncertainty of *in situ* measurements in particular at low-medium frequencies. Moreover, the values of $s_{situ}$ [7] obtained are higher also than the $s_R$ values, in particular for the floor at low-medium frequencies from 80 to 200 Hz, and for the wall from 160 to 250 Hz.

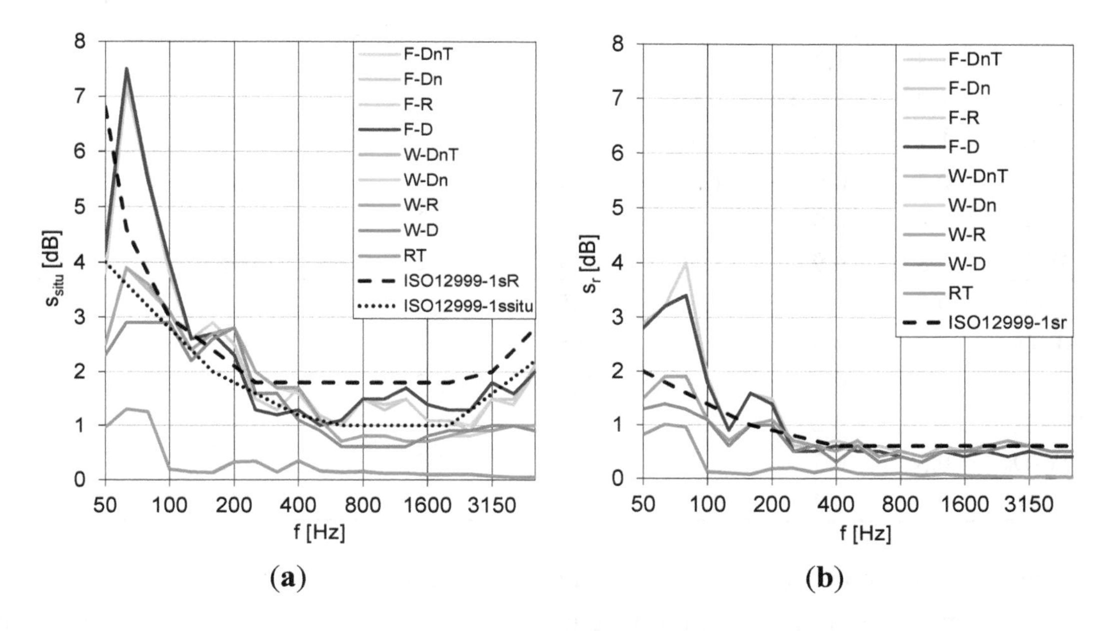


**Figure 1.** $s_{situ}$ (**a**) and $s_r$ (**b**) of floor (F) and wall (W) of R’, $D_n$, $D_{nT}$, D and RT [7], with the comparison with the reproducibility, *in situ* (a) and repeatability (b) standard deviation of ISO 12999-1 [5].

The results of SNQ calculations are shown in Table 1. Two different ways to determine the SNQs have been considered for the above-mentioned study [7]. The former is to determine SNQ according to ISO 717-1 [21] by shifting the reference curve (value in the range from 100 to 3150) in steps of 1 dB toward the measured curve, until the mean unfavorable deviation is as large as possible but not more than 32 dB; all the laboratories involved in the RRT have followed this procedure. The latter is to determine SNQ plus the spectrum adaptation terms C and $C_{tr}$ according to ISO 717-1 [21] both in the narrow frequency range from 100 to 3150 Hz, and in the enlarged frequency range from 50 to 5000 Hz; in both cases rounded to integer and with 1 decimal place (subscript 01), using Equation (1) [21]. The SNQs plus the spectrum adaptation terms were determined using a 0.1 dB resolution, following from the work of Wittstock [22], to obtain more accurate data for the analysis of standard deviation than the 1 dB resolution.

$$X_{Aj} = -10\lg\sum_i 10^{(L_{ij}-X_i)/10} = X_w + C_j[dB] \quad (1)$$

where *j* is the index of the spectrum No. 1 to calculate C or No. 2 to calculate $C_{tr}$ according to ISO 717-1 [21]; *i* is the index of frequencies; $L_{ij}$ is the level indicated in ISO 717-1 [21] at frequency *i* for spectrum *j*; $X_i$ is one of the quantities considered, $R_i$, $D_{ni}$ or $D_{nTi}$; at frequency *i* for the spectrum *j*; $X_w$ is the single number; and $C_j$ is the spectrum adaptation term C or $C_{tr}$ if calculated with spectrum No. 1 or No. 2, respectively.

**Table 1.** $s_r$ and $s_{situ}$ of SNQs of floor (F) and wall (W) in narrow (100–3150 Hz) and enlarged (50–5000 Hz) range [7].

| | | Narrow Range 100–3150 Hz | | | | | Enlarged Range 50–5000 Hz | |
|---|---|---|---|---|---|---|---|---|
| | | **X** | **X + C** | **X + $C_{tr}$** | **$X_{01}$ + C** | **$X_{01}$ + $C_{tr}$** | **$X_{01}$ + C** | **$X_{01}$ + $C_{tr}$** |
| $s_{situ}$ | F-$D_{nT}$ | 1.3 | 1.3 | 1.5 | 1.3 | 1.5 | 1.4 | 2.8 |
| | F-$D_n$ | 1.2 | 1.2 | 1.5 | 1.3 | 1.4 | 1.4 | 2.8 |
| | F-R’ | 1.2 | 1.2 | 1.5 | 1.3 | 1.5 | 1.4 | 2.7 |
| | W-$D_{nT}$ | 0.7 | 0.9 | 1.2 | 0.9 | 1.2 | 0.8 | 1.4 |
| | W-$D_n$ | 0.9 | 0.9 | 1.3 | 0.8 | 1.2 | 0.8 | 1.4 |
| | W-R’ | 0.8 | 0.9 | 1.3 | 0.9 | 1.2 | 0.8 | 1.4 |
| $s_r$ | F-$D_{nT}$ | 0.7 | 0.6 | 0.6 | 0.5 | 0.7 | 0.6 | 1.3 |
| | F-$D_n$ | 0.5 | 0.5 | 0.7 | 0.5 | 0.7 | 0.6 | 1.3 |
| | F-R’ | 0.5 | 0.6 | 0.9 | 0.5 | 0.7 | 0.6 | 1.3 |
| | W-$D_{nT}$ | 0.2 | 0.2 | 0.3 | 0.2 | 0.2 | 0.2 | 0.3 |
| | W-$D_n$ | 0.3 | 0.3 | 0.3 | 0.2 | 0.2 | 0.2 | 0.4 |
| | W-R’ | 0.2 | 0.2 | 0.4 | 0.2 | 0.2 | 0.2 | 0.4 |

The internal partitions considered in this RRT were a lightweight wall and a heavy floor. It was demonstrated that the uncertainties of lightweight samples are lower than the uncertainties of heavy types of construction; therefore it will be important for datasets of different constructions to be considered separately. A similar difference between the uncertainty of heavy and lightweight test samples was shown by Dijckmans and Vermeir [23] who made a numerical investigation of the repeatability and reproducibility of laboratory sound insulation measurements by investigating both the pressure method and the intensity method. Dijckmans and Vermeir [23] found that for large, heavy test elements, like concrete walls, the reproducibility in the lowest frequency bands is not improved by using the intensity method, while, for double plasterboard walls, the theoretical uncertainty is decreased by 1 dB by using the intensity method.

The results of Table 1 show that the one-third-octave band uncertainty at LF slightly affects the SNQs in the enlarged range plus C spectrum adaptation term but greatly affects (almost double than the narrow range standard deviation) the SNQs in the enlarged range plus $C_{tr}$ spectrum adaptation term. This is mainly due to the fact that the spectrum adaptation term $C_{tr}$ considers predominantly the low-medium frequencies noise components.

In their recent study on the correlations and implications of SNQ for rating airborne sound insulation in the frequency range 50 Hz to 5 kHz, Garg and Maij [24] showed that $R_{traffic}$ (as defined in ISO CD 16717-1 [25] and corresponding to $R_w$ + $C_{tr50–5000}$) is highly sensitive to low frequency sound insulation as compared to the current SNQ and $R_{living}$ (as defined in ISO CD 16717-1 [25] and corresponding to $R_w$ + $C_{50–5000}$). Finally, the measurement uncertainty in the low frequency range (due to the presence of the normal modes of vibration, that imply that at the first three one-third-octave bands the measured levels can be strongly influenced by the measurement position) is too high to justify the decision to perform field measurements down to low frequencies, and therefore the scientific evidence for including the low frequency range should be significantly improved. Moreover, the fact that the higher uncertainty at LF is not well represented in the SNQs uncertainty confirms that further studies are needed to better understand all the implications of the inclusions of LF in the SNQs, from both a physical point of view and from a legislation point of view. Garg and Maij [24] found interconversion equations applicable for sandwich gypsum constructions and roof constructions. They stressed the fact that testing of sound transmission loss characteristics in the extended frequency range of 50 Hz to 5 kHz also implies the need to reformulate the sound regulation requirements in buildings including the low frequency spectrum adaptation terms.

Some recent studies [26–29] on the uncertainty of SNQs extended to the low frequencies range show an increase in the SNQs uncertainty due to the LF extension, confirming the results found in this RRT. Mahn and Pearse [26] studied the effect on uncertainty of expanding the frequency range included in the calculation of the single number ratings, using laboratory measurements of 200 lightweight walls as data. They found that the uncertainty of the single number ratings is highly dependent on the shape of the sound reduction index curve. The uncertainty obtained for $R_{living}$ ($R_w$ + C in the enlarged frequency range) was greater than that of the traditional weighted sound reduction index for 98% of the 200 lightweight building elements included in the evaluation.

Hongisto *et al.* [27] focused their study on the two most important SNQs proposed by ISO CD 16717-1 [25]; that is, $R_{traffic}$ ($R_w + C_{tr}$ in the enlarged frequency range) and $R_{living}$ ($R_w$ + C in the enlarged frequency range), and how their reproducibility values differ from the reproducibility values of their counterparts $R_w + C_{tr}$ and $R_w$. They found that the reproducibility values of the proposed single-number quantities (50–5000 Hz; $R_{living}$, $R_{traffic}$) are larger than the reproducibility values of the present SNQs (100–3150 Hz; $R_w$, $R_w + C_{tr}$) with sound insulation measurements made with the pressure method; with the sound intensity method, the reproducibility values increased very little.

Machimbarrena *et al.* [28] presented an alternative procedure, aiming at evaluating the need of performing individual uncertainty calculations and the effect of extending the frequency range used to calculate sound insulation single number quantities. For this purpose they performed calculation in a set of 2081 field airborne sound insulation measurements on 22 different types of separating walls partitions of *in situ* airborne sound insulation measurements. The results of Machimbarrena *et al.* [26] show that the frequency range used for the evaluation affects the uncertainty of the single number quantity. In almost all the cases shown in their paper, the uncertainty is increased when the frequency range is extended.

António and Mateus [29] studied the influence of low frequency bands on airborne and impact sound insulation single numbers for typical Portuguese buildings. They found that the uncertainty is higher for the $D_{nT,w} + C_{tr}$ descriptor than for $D_{nT,w}$ + C, confirming what was found in this RRT. They also found that when the low frequency bands are included in the calculation, the uncertainty of the descriptor increases on average and this increase is more evident when the adaptation term is for a spectrum of traffic noise.

### 2.1.2. Façade Sound Insulation

The uncertainty of field measurements, in particular façade sound insulation, has not been comprehensively investigated. There is only one example in the literature of a Round Robin Test conducted on a window of a façade [12].

In the study about the uncertainty of façade sound insulation [8], the measurand was a prefabricated concrete façade with a 4 mm single glazing wood-aluminum frame window with a MDF (Medium Density Fiberboard) shutter box. The façade is situated at first floor level. Nine teams coordinated by ITC-CNR were involved in this study; each of them has replicated the tests five times, including the reverberation time. One laboratory showed a significant presence of stragglers and outliers. After a statistical examination of this result, the laboratory was excluded. In fact, it turned out that the random effect estimated for laboratory was, in absolute value, the highest value [8]: the Grubbs test [30,31] for one outlier identified the laboratory as the first outlier. Therefore here are the eight reported laboratories results.

In this study, the highest values of $s_r$ and $s_{situ}$ were found at the frequencies of 50, 63 and 80 Hz. That paper [8] also underlined that the uncertainties in $D_{ls,2m,nT}$ are heavily contaminated by the inappropriateness of the reverberation time correction at low-frequencies and a comparison between the uncertainties of the standardized level difference $D_{ls,2m,nT}$ and the level difference $D_{ls,2m}$ shows the magnitude of the reverberation time at low frequencies (see Figure 2). This influence is noticeable in particular at 63 Hz and at 80 Hz, while at 50 Hz the uncertainties of $D_{ls,2m,nT}$ and $D_{ls,2m}$ are coincident.

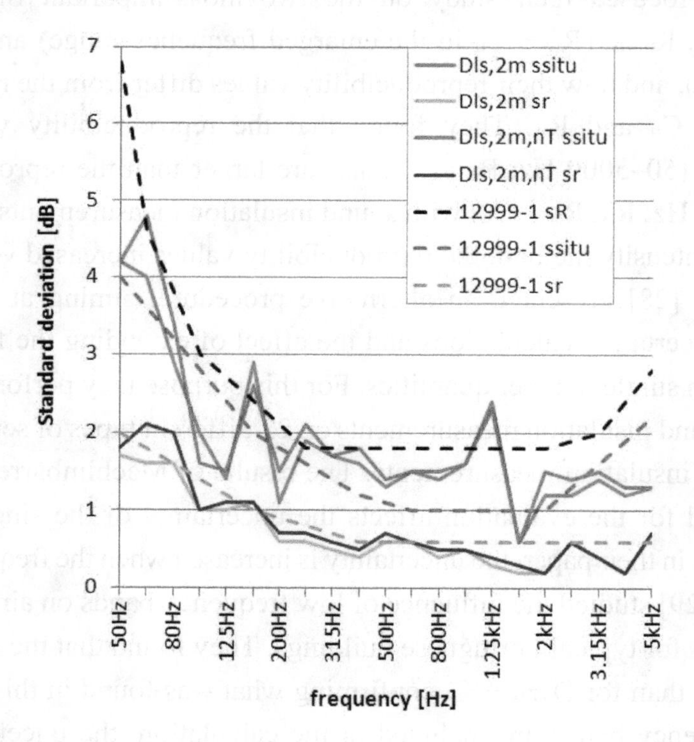


**Figure 2.** Comparison between the *in situ* and repeatability standard deviation of $D_{ls,2m,nT}$ and $D_{ls,2m}$ [8] and the reproducibility, *in situ* and repeatability standard deviation of ISO 12999-1 [5].

The variations between laboratories at low frequencies are still very high even if the reverberation time correction is not included in the calculation (*i.e.*, just considering $D_{ls,2m}$), which implies that for the sound pressure level measurements the low frequencies also have a high uncertainty. The $s_{situ}$ and $s_r$ behavior of $D_{ls,2m}$ is similar to the behavior of the uncertainties of ISO12999-1 [5], which increase steadily and rapidly below 100 Hz. Thus the trend of the standard deviation curve at low frequencies of *in situ* reproducibility and repeatability standard deviation calculated from the RRT study is attributable to the reverberation time measurements.

In Table 2 are shown the SNQs uncertainties, in terms of repeatability and *in situ* standard deviations. The SNQs were determined according to ISO 717-1 [21] shifting the reference curve both in steps of 1 dB and 0.1 dB (subscript 01), toward the measured curve, until the mean unfavorable deviation is as large as possible, but not more than 32 dB; all the laboratories involved in the RRT have followed this procedure. The shift in increments of 0.1 dB was evaluated because the 2013 update of the ISO 717-1 [21] provides for increments of 0.1 dB for the expression of uncertainty. The SNQs plus spectrum adaptation terms C and $C_{tr}$ according to ISO 717-1 [21] in the extended range (from 50 to 5000 Hz), both at integer and with one decimal place (subscript 01) were calculated using Equation (1).

**Table 2.** $s_{situ}$ and $s_r$ of SNQs, calculated as one of the levels *j* of RRT [8].

| Frequency Range | SNQs | $s_{situ}$ | $s_r$ |
|---|---|---|---|
| narrow range 100–3150 Hz | $D_{ls,2m,nT,w}$ | 0.8 | 0.3 |
| | $D_{ls,2m,nT,w} + C$ | 1.0 | 0.4 |
| | $D_{ls,2m,nT,w} + C_{tr}$ | 1.1 | 0.3 |
| | $D_{ls,2m,nT,w01}$ | 0.9 | 0.3 |
| | $D_{ls,2m,nT,w01} + C$ | 1.0 | 0.2 |
| | $D_{ls,2m,nT,w01} + C_{tr}$ | 1.1 | 0.3 |
| enlarged range 50–5000 Hz | $D_{ls,2m,nT,w01} + C$ | 0.9 | 0.2 |
| | $D_{ls,2m,nT,w01} + C_{tr}$ | 1.1 | 0.3 |

In the study about the airborne sound insulation [7], it was found that the extension at low frequencies range increases the uncertainty of the SNQs. In the case of the façade, calculating the SNQs uncertainty handling the SNQs values as a level of the RRT itself (see Table 2), no significant differences are observed whether including or excluding the low frequencies. In this case, the low frequency uncertainty is not well reflected in the SNQs uncertainty. Considering the extension to low frequencies, the suitability of the reference spectra for rating airborne sound insulation should be validated.

On this topic, Masovic *et al.* [32] made a study on the suitability of ISO CD 16717-1 [25] reference spectra for rating airborne sound insulation. The ISO CD 16717-1 [25] spectra living and traffic correspond to the reference spectra C (50–5000 Hz) and $C_{tr}$ (50–5000 Hz) of ISO 717-1 [21], respectively. Masovic *et al.* [32] demonstrated, with an extensive noise monitoring in a number of dwellings recordings of 38 potentially disturbing activities, that the reference spectrum for living noise ($L_{living}$), should be redefined to better match the typical spectrum of noise in dwellings because it seems to be rather high at lower frequencies, especially below 100 Hz. Moreover, in the case of noise generated by sources of music with strong bass content the reference spectrum for traffic noise ($L_{traffic}$) seems to be more appropriate above 100 Hz than $L_{living}$. This could suggest one of the reasons why the low frequencies uncertainty is not adequately reflected by the SNQs uncertainty extended to low frequencies and should be considered deeper before deciding to perform measurements down to LF range.

Therefore, considering this kind of façade (prefabricated concrete façade with a single glazing window and with a shutter box) including the low frequencies range in the façade sound insulation measurements, brings no obvious advantage, but rather the disadvantage of complicating and lengthening the measurement. In literature, there are some studies (e.g., Rindel [33] and Park and Bradley [34]) on the annoyance of noise from neighborhood at low frequencies that stress the importance of investigating the LF noise; nevertheless, at present time, effective protection systems against low frequency noise are still an open challenge both for researchers and components manufacturers, as underlined by Prato and Schiavi [35]. Hongisto *et al.* [27] suggested that scientifically valid socio-acoustic evidence for the need to include the frequency range 50–80 Hz should be significantly improved before deciding that the low frequency measurements are included in the calculation of the SNQs. Last but not least, if LF measurements are aimed at the protection against LF noise, the fact that the high uncertainty of the one-third octave LF band affects the reliability of the performance of the test element implies that the potential effectiveness of the protection system against low frequency noise is not quantifiable.

A prefabricated concrete façade with a PVC frame with double glazing 4/12/4 window was tested in the further RRT study concerning façade sound insulation uncertainty [10], focused on the new low frequencies measurement procedure stated in ISO/DIS 16283-3 [36], that will soon replace the standard ISO 140-5 [11]. Ten teams, coordinated by ITC-CNR were involved in this RRT, each of them operating with its own equipment and replicates the tests 5 times, including the new low frequencies procedure (explained below) and the reverberation time measurements. All teams performed measurements following the global loudspeaker method, which yields the level difference of a façade in a given place with respect to a position 2 m in front of the façade. All teams positioned the outside microphone 2 m in front of the façade, and the loudspeaker on the ground, with the angle of sound incidence equal to 45° ± 5°; as positioned directly in front of the façade by some teams, and in a lateral position by other teams. The statistical analysis of the data provides a three-step procedure for the identification of stragglers and outliers. Following this procedure, two teams were identified as outliers and excluded because they showed a significant presence of stragglers and outliers starting from 500 Hz to 3150 Hz [10]. The comparison of standard deviation values, repeatability and *in situ* standard deviation, from RRT (calculated for both $D_{ls,2m,nT}$ and $D_{ls,2m}$) and from ISO 12999-1 [5] are plotted in Figure 3.

Regarding the low frequency range (from 50 to 80 Hz), the reasons for the high values of $s_r$ and $s_{situ}$ can be sought in the presence of the normal modes of vibration, in fact at the first three one-third- octave bands (50, 63 and 80 Hz), the measured levels can be strongly influenced by the measurement position.

At low frequencies, the $s_{situ}$ and $s_r$ behavior of both $D_{ls,2m,nT}$ and $D_{ls,2m}$ is not similar to the behavior of the uncertainties of ISO 12999-1 [5], in terms of reproducibility $s_R$ and *in situ* standard deviation, which increase steadily and rapidly below 100 Hz, as it can be seen in graphs of Figure 3. Contrary to what was found in the previous RRT [8], this difference is not attributable to the reverberation time measurements. This different behavior could be attributable to the differences of the façade test samples: the façade of the previous RRT [8] is a prefabricated concrete façade with a 4 mm single glazing wood-aluminum frame window with a MDF shutter box; the façade of the second study is a prefabricated concrete façade with a PVC frame with double glazing 4/12/4 window. Also the loudspeaker position could be relevant and its influence is under investigation.

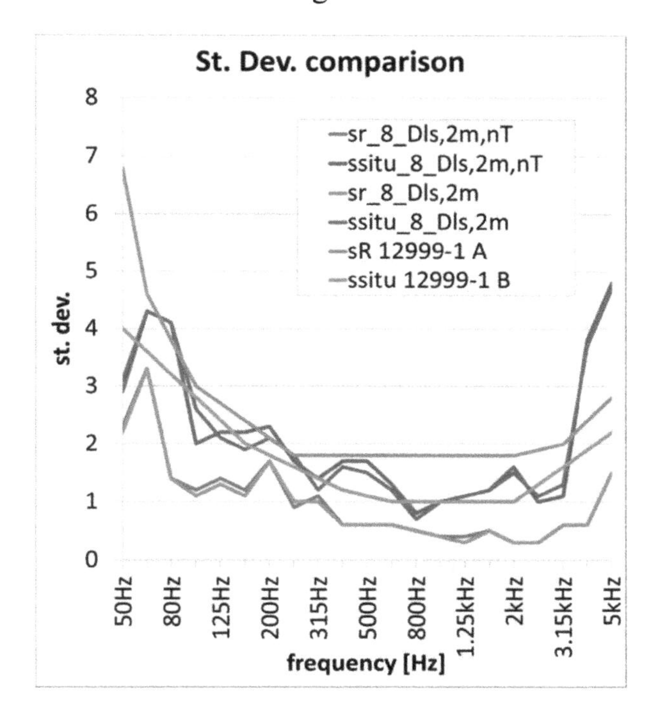


**Figure 3.** Comparison of standard deviation values from RRT (calculated for both $D_{ls,2m,nT}$ and $D_{ls,2m}$) and from ISO 12999-1 [10].

With respect to the high frequency range, in particular at 4000 and 5000 Hz, the RRT and ISO 12999-1 [5] standard deviations values show the same behavior, *i.e.*, an increase with frequency, but the RRT $s_{situ}$ values are higher than the ISO 12999-1 [5] values. Moreover the RRT $s_{situ}$ values are higher than the low frequency $s_{situ}$ values of both RRT and ISO 12999-1 [5]. This is probably due to the different positions of the loudspeaker with respect to the façade [10] and it is still under investigation. In the previous RRT [8], where all the teams involved placed the loudspeaker in the same position (directly in front of the façade), the high frequency uncertainty was lower, in particular lower than ISO 12999-1 [5] values and much lower than the low frequencies uncertainty.

In the first RRT on façade sound insulation [8] a behavior similar to the behavior found by Lang [12] in the Austrian RRT was observed, where the RRT values exceed the values of the ISO 140-2 [37] (the standard on acoustics measurement uncertainty available at the time of Lang's RRT) in the range of mass-spring-mass resonance frequency and in the range of the coincidence frequency of the double glazing. Lang suggests that such behavior may be caused by the difficulty of arranging the loudspeaker at an angle of incidence of 45°.

The first RRT [8] faced no difficulty with the arrangement of the loudspeaker at an angle of incidence of 45°. Such behavior is thus exclusively attributable to the nature (*i.e.*, critical frequencies) of the measurand itself. However, the uncertainty dependence from the loudspeaker position could be found at high frequencies as shown in the second RRT [10] and, as already said, it must be more deeply investigated. Berardi *et al.* [38] and Berardi [39] considered the position of the loudspeaker as a variable, but its influence on the high frequencies was not comprehensively evaluated.

In this RRT [10] all the participating laboratories repeated the measurements with the low-frequency procedure included in the upcoming standard ISO 16283-3 (ISO/DIS 16283-3 [36]). In his recent paper Hopkins [40] gives the background to the revision of the ISO 140 standards relating to field measurement of airborne, impact and façade sound insulation that form the new ISO 16283 series. The low-frequency procedure was first studied and proposed by Hopkins and Turner [41] in a work about the airborne sound insulation between rooms. For each of the 50, 63 and 80 Hz bands, they proposed that the average low frequency sound pressure level in the room, $L_{LF}$, be calculated from $L_{ISO140-4}$ (the average sound pressure level in a room measured according to the normative guidance in ISO 140-4) and $L_{corner}$ (the corner sound pressure level measured according to the normative guidance in ISO 16283-1) according to:

$$L_{LF} = 10\lg\left[\frac{2\left(10^{0.1L_{ISO140-4}}\right)+10^{0.1L_{corner}}}{3}\right][dB] \tag{2}$$

The low-frequency (LF) procedure is mandatory in case of room volume lower than 25 m$^3$. As the volume of the receiving room in this RRT is 40 m$^3$, it was possible to compare the results of the two procedures: the LF procedure and the default procedure. The results of this comparison, for the LF range are shown in Table 3. The results refer both to 8 and to 10 teams, as the two outlier teams that are excluded from the calculation of standard deviation for the all frequencies considered (from 50 to 5000 Hz), can be included in the evaluation of the LF standard deviation because these teams showed a significant presence of stragglers and outliers starting from 500 Hz to 3150 Hz.

**Table 3.** Low frequency $s_r$ and $s_{situ}$ values for the two measurement methods (default and LF) for both 8 and 10 teams [10].

| Standard | 50 Hz | | 63 Hz | | 80 Hz | |
|---|---|---|---|---|---|---|
| Deviations | Default | LF | Default | LF | Default | LF |
| $s_r$_10 | 2.7 dB | 2.5 dB | 3.1 dB | 4.5 dB | 1.4 dB | 2.3 dB |
| $s_{situ}$_10 | 3.1 dB | 3.1 dB | 4.8 dB | 5.5 dB | 4.0 dB | 4.1 dB |
| $s_r$_8 | 2.3 dB | 2.3 dB | 3.3 dB | 5.0 dB | 1.4 dB | 2.5 dB |
| $s_{situ}$_8 | 2.9 dB | 3.2 dB | 4.3 dB | 5.2 dB | 4.1 dB | 4.2 dB |

With the low-frequency procedure there is an increase of the uncertainty, particularly noticeable at 63 Hz: the repeatability standard deviation increases by about 1.5 dB while the *in situ* standard deviation increases by about 1 dB. The results shown in Table 3 indicate that the low-frequency measurement procedure does increase the uncertainty. This cannot be attributed to the operators whose experience is well proven; this aspect is still under investigation.

To deal with the measurement issue in the low frequency domain, Prato and Schiavi [35] and Prato *et al.* [42] suggest the modal approach. At frequencies below 100 Hz, the acoustic field is non-diffuse, as it is characterized by large fluctuations of sound pressure levels in space and frequency domains. Because of the inhomogeneity of the acoustic field, Prato *et al.* [42] suggest to move from a statistical approach typical of diffuse sound field (average sound energy) to a discrete one, focused at highest noise and annoyance points, *i.e.*, the points of highest sound pressure level in space (corners) and frequency (resonance modes): the so-called modal approach.

In this RRT [10], it was found that the differences between including and excluding low frequencies are a little higher for SNQ plus $C_{tr}$ when using standard measurement procedure and are very high for SNQ plus $C_{tr}$ when using the LF measurement procedure, as shown by comparing Tables 4 (SNQs without LF) and 5 (SNQs with LF), contrary to what was found in the previous RRT [8] that showed that the differences between including or not the low frequencies were practically negligible.

**Table 4.** Standard uncertainties of SNQs without low frequencies for the 8 teams [10].

| Descriptor (SNQs) | $s_r$ (dB) | $s_{situ}$ (dB) |
|---|---|---|
| $D_{ls,2m,nT,w}$ | 0.4 | 0.7 |
| $D_{ls,2m,nT,w} + C_{(100–3150)}$ | 0.6 | 0.8 |
| $D_{ls,2m,nT,w} + C_{tr(100–3150)}$ | 0.8 | 1.0 |
| $D_{ls,2m,nT,w01}$ | 0.3 | 0.7 |
| $D_{ls,2m,nT,w01} + C_{(100–3150)}$ | 0.5 | 0.8 |
| $D_{ls,2m,nT,w01} + C_{tr(100–3150)}$ | 0.7 | 1.0 |
| $D_{ls,2m,nT,w01} + C_{(100–5000)}$ | 0.6 | 1.2 |
| $D_{ls,2m,nT,w01} + C_{tr(100–5000)}$ | 0.7 | 1.0 |

**Table 5.** Standard uncertainties of SNQs with low frequencies for the 8 teams [10].

| Descriptor (SNQs) | $s_r$ (dB) | | $s_{situ}$ (dB) | |
|---|---|---|---|---|
| | Default | LF | Default | LF |
| $D_{ls,2m,nT,w01} + C_{(50–3150)}$ | 0.5 | 0.6 | 0.8 | 1.0 |
| $D_{ls,2m,nT,w01} + C_{tr(50–3150)}$ | 0.8 | 1.9 | 1.0 | 2.1 |
| $D_{ls,2m,nT,w01} + C_{(50–5000)}$ | 0.6 | 0.6 | 1.2 | 1.3 |
| $D_{ls,2m,nT,w01} + C_{tr(50–5000)}$ | 0.8 | 1.9 | 1.0 | 2.1 |

This different behavior could be attributable to the differences of the façade test samples: the façade of the previous RRT [8] is a prefabricated concrete façade with a 4 mm single glazing wood-aluminum frame window with a MDF shutter box; the façade of the second study is a prefabricated concrete façade with a PVC frame with double glazing 4/12/4 window.

In fact, from the experience derived from many measurements of façade sound insulation [43,44], the lower the insulation of a window, the lower the spectrum adaptation term $C_{tr}$ and *vice versa*, the higher the window insulation, the higher $C_{tr}$. For this reason, in the case of the previous RRT [8] (a façade with low insulation window) the difference between $D_{ls,2m,nT,w}$ and $D_{ls,2m,nT,w} + C_{tr}$ averages, was not a large one, only 1.5 dB, while in the case of the present study (a façade with higher insulation window), the difference between the average values of $D_{ls,2m,nT,w}$ and of $D_{ls,2m,nT,w} + C_{tr,50–5000}$ is 5.3 dB for default measurements and 6.8 dB for the low-frequency method.

## 3. How to Manage the Cooperative Tests Uncertainty

As stated in the introduction, current knowledge in building acoustics suggests that the best methodology to study the measurements uncertainty is to carry out an Inter-Laboratory Test or a Round Robin Test. Therefore the results of ILTs and RRTs are very important to know the uncertainty magnitude that is reasonably expected for a measurement result. However, even if an ILT or RRT gives the uncertainty of a measurement method, the uncertainty magnitude depends also on the measurand.

An example of the dependence on the method can be drawn from the results of uncertainty of façade sound insulation measurements discussed in Section 2.1.2, where the high frequency uncertainty depends on the loudspeaker position (which is still under study). On the other hand, the uncertainty magnitude also depends on the test sample, as showed in Section 2.1.1 concerning the sound insulation of internal partitions where it was found that the uncertainties of lightweight samples are lower than the uncertainties of heavy types of construction. The dependence on the measurand, in particular for including or not the LF in SNQs, was also found in the case of façade sound insulation uncertainty (see Section 2.1.2) where the comparison of the two RRTs results highlighted that that the differences are attributable to the windows, on which the $C_{tr}$ coefficient depends: a single glazing window and a double 4/12/4 glazing window.

ISO 12999-1 [5] gives the medium uncertainty on all the ILTs and RRTs considered (and available at the time when the standard draft was being written) in that standard, for airborne sound insulation, without distinction of the type of measurand. At the current level of knowledge and due to the number of cooperative tests available, this seems to be the only way to give an idea of the uncertainty magnitude. The fact that the values of ISO 12999-1 [5] are the best estimates for the uncertainty of sound insulation measurements that can be obtained today, was also underlined by Wittstock [45] in his paper that describes how the average uncertainty values standardized in ISO 12999-1 [5] were derived. Therefore, it is important to keep that standard constantly updated in order to increase the number of available data on which the average uncertainty values could be calculated. This specific standard is inaccurate as far as the façade sound insulation is concerned, because its uncertainty is considered equal to the airborne sound insulation uncertainty; indeed, the façade sound insulation measurement method is extremely different from the airborne sound insulation measurement method for party walls and floors. *A priori*, the reproducibility standard deviation is higher than the *in situ* standard deviation because of, as far as reproducibility is concerned, the geometry of the rooms and wall can change, while this is not the case for the *in situ* standard deviation as defined in Section 2. Because the geometry (*i.e.*, modal behavior) has a large influence at low frequencies, $s_R$ is larger than $s_{situ}$ (*cf.* Table 6). The use of $s_{situ}$ is thus only appropriate when the geometry is the same. In the case of façade sound insulation, however, there are no literature data that referred to RRT of the same object in different situations and it will be appropriate in the future that ISO 12999-1 [5] include this difference (*i.e.*, reproducibility and *in situ* standard deviation for façade sound insulation), considering the following: the measurement method of façade sound insulation is extremely different from the laboratory measurement of airborne sound insulation; the uncertainty at high frequencies (which exceed, in the case of the second façade RRT [10], the $s_R$ values of ISO 12999-1 [5] as shown in Figure 3) is mainly dependent on the loudspeaker position (as supposed in the case of the second RRT of Façade [10], as said before), and the RRT [12] values exceed the values of the ISO 140-2 [37] in the range of mass-spring-mass resonance frequency and in the range of the coincidence frequency of the double glazing. At the present state of knowledge, the reproducibility standard deviation values included in ISO 12999-1 [5] seem to be the only available uncertainty that could be used also in the case of façade sound insulation, keeping in mind that the façade sound insulation measurement method is very different.

**Table 6.** Standard uncertainties for single-number values in accordance with ISO 717-1, as per ISO 12999-1 [5].

| Descriptor | $s_R$ dB | $s_{situ}$ dB | $s_r$ dB |
|---|---|---|---|
| $R_w$, $R'_w$, $D_{nw}$, $D_{nT,w}$ | 1.2 | 0.9 | 0.4 |
| ($R_w$, $R'_w$, $D_{nw}$, $D_{nT,w}$) + $C_{100-3150}$ | 1.3 | 0.9 | 0.5 |
| ($R_w$, $R'_w$, Dnw, $D_{nT,w}$) + $C_{100-5000}$ | 1.3 | 0.9 | 0.5 |
| ($R_w$, $R'_w$, $D_{nw}$, $D_{nT,w}$) + $C_{50-3150}$ | 1.3 | 1.0 | 0.7 |
| ($R_w$, $R'_w$, Dnw, $D_{nT,w}$) + $C_{50-5000}$ | 1.3 | 1.1 | 0.7 |
| ($R_w$, $R'_w$, $D_{nw}$, $D_{nT,w}$) + $C_{tr,100-3150}$ | 1.5 | 1.1 | 0.7 |
| ($R_w$, $R'_w$, $D_{nw}$, $D_{nT,w}$) + $C_{tr,100-5000}$ | 1.5 | 1.1 | 0.7 |
| ($R_w$, $R'_w$, $D_{nw}$, $D_{nT,w}$) + $C_{tr,50-3150}$ | 1.5 | 1.3 | 1.0 |
| ($R_w$, $R'_w$, $D_{nw}$, $D_{nT,w}$) + $C_{tr,50-5000}$ | 1.5 | 1.0 | 1.0 |

Therefore, in the case of a single measurement, the uncertainty that should be associated to this measurement is the reproducibility standard deviation given in ISO 12999-1 [5] multiplied by the appropriate coverage factor to obtain the expanded uncertainty. Now, considering what was stated in the introduction, when reporting the result of the measurement of a physical quantity, it is compulsory that some quantitative indications of the quality of the result be given. Such an indication should be independent on the final use of the results (verification of a requirement or determination of predicted values), and shall be stated as follows, as provided by GUM [2] and ISO 12999-1 [5]:

$$Y = y \pm U \tag{3}$$

where $Y$ is the measurand; $y$ is the best estimate (obtained through the measurement) of the value attributable to the measurand; and $U$ is the expanded uncertainty, calculated for a given confidence level for the two-sided test, defined as the product of the measurement uncertainty $u$ (which is the reproducibility standard deviation $s_R$) with a coverage factor $k$.

Therefore, for example, for a single measurement of the airborne sound insulation of a partition floor $R'_w$ ($C$;$C_{tr}$) = 53 (−1;−4), considering the values given in ISO 12999-1 (see Table 6), the airborne sound insulation of this partition wall shall be given to one decimal place ($R'_w$ = 52.6; C = −1.0; $C_{tr}$ = −4.1) to state also its uncertainty and should be designated as [5]:

$$R'_w = (52.6 \pm 2.4) dB (k = 1.96, two - sided) \tag{4}$$

$$R'_w + C = (51.6 \pm 2.6) dB (k = 1.96, two - sided) \tag{5}$$

$$R'_w + C_{tr} = (48.5 \pm 2.9) dB (k = 1.96, two - sided) \tag{6}$$

where $k$ = 1.96 corresponds to a confidence level of 95% for a two-sided test.

On the other hand, when a measurement is made in order to verify a requirement, the expanded uncertainty that should be given with the result, should be calculated using a coverage factor for one-sided test, as laid down in ISO 12999-1 [5]. Then the expanded uncertainty should be added to or subtracted from the measurement result to check whether that measurement result is smaller or larger than the requirement, respectively.

The Italian standard on the acoustic classification of building units UNI 11367 [46] first considers the measurement uncertainty from RRTs as a basis for the expanded uncertainty $U$. When a national regulation has to be met, the choice of the confidence level is very important. The Italian standard on the acoustic classification [46] has faced for the first time the problem related to the confidence level. In the case of measurement uncertainty, the standard recommends to use a coverage factor $k$ for one-sided test equal to 1, which corresponds to an 84% probability; for buildings performances, in fact, in order to meet the limit, it is not realistic to use a 95% or 90% confidence level, which is normally used in other contexts. As the update of the ISO 717-1 [21] allows applying the weighting procedure by 0.1 dB steps for the expression of measurement uncertainty, it could now possible also be to use, in building acoustics, a coverage factor $k$ for one-sided test equal to 1.65 corresponding to a 95% confidence level.

Generally, when measurements are made to verify the acoustic requirements of buildings, one single measurement might not be enough to this end, and therefore more measurements and more results for the same requirement are necessary. In this case, the measurement uncertainty is combined in a certain way with the uncertainty due to the number of tests performed.

## 4. Sampling

There are two different types of surveys that can be used to analyze the acoustic requirements of building units, or buildings: a census (the entire population is taken into account) or a sample survey (only a part of the elements that make up the population are considered). For building acoustics, a sample survey is the best solution in terms of cost and time. To make meaningful comparisons with both national regulations and acoustic classification, it is therefore necessary to determine the type and amount of the measurements. In order to make any sample survey on certain features (acoustical) of a finite population, it is essential to formulate a strategy of selection, which is closely connected with the purposes, the cost and the execution time of the survey. In addition, the sample obtained from it, is the only valuable information that could be used for the interpretation of the results.

Among the different sampling strategies currently available, the two main ones used in building acoustics, for the time being, are the following: the stratified sampling as adopted by UNI 11367 [46] (see next section) and a sampling procedure taking into account a certain percentage associated with a selection criterion as adopted by UNI 11444 [47] and proposed by ISO/WD 19488 [48]. Only the former strategy (stratified sampling) includes the sampling uncertainty. The strategy of UNI 11444 [47] consists in the selection of a minimum number of Building Units (BUs): not less than 10% of the total amount of BUs composing the building system and not less than 2 BUs, if the total amount is 4, and not less than 3 BUs for building systems up to 30 BUs. These BUs must be the most critical BUs from an acoustic point of view. The selection of the most critical BUs must take into account all the critical acoustic features of the building elements of the BU. The selection criteria for each type of acoustic performance (façade sound insulation, sound insulation of horizontal and vertical partitions, impact sound insulation and equipment noise level) are stated in standard UNI 11444 [47]. This standard does not include the sampling uncertainty but, for each measurement, it includes the measurement uncertainty as stated in UNI 11367 [46].

The standard proposal ISO/WD 19488 [48] considers, as a general principle, that, when verifying the acoustic class of a unit, a sufficient number of measurements of each relevant acoustic characteristic must be performed in order for the result to represent the unit. It also suggests that care should be taken to include the critical site/rooms, e.g., partitions with critical flanking constructions. At the current stage, the proposal includes neither the sampling uncertainty nor the measurement uncertainty, but it considers that compliance is granted if the average results comply with the class limits and no individual result deviates unfavorably by more than 2 dB. Moreover, if classification for different dwellings, rooms or acoustic characteristics varies, the classification assigned is the minimum class obtained.

Considering the pros and cons of these two sampling strategies, the first thing to keep in mind is the scope of the measurements; *i.e.*, to determine a class within the acoustic classification or to verify the legal requirements. In the former case, it is obvious that a value as close as possible to the value of all the elements is suitable. In the latter case, the scope is to identify the worst acoustic performances and to verify if also the critical site/rooms is/are in compliance with the legal requirements.

The stratified sampling strategy allows increasing the efficiency of a sampling plan, without increasing the sample size. With this strategy it is possible to obtain the best representative value of a class to be attributed to the entire building system, as if the entire population were taken into account. Another pro is the stratified sampling uncertainty related to the final result that gives a confidence level, which is important both for the owners and the builders. The con of this strategy is that it requires a large number of measurements (a minimum of three measurements for each homogeneous group).

A strategy that takes into account a certain percentage of the population, including all the critical site/rooms, could not be representative of all situations but would give the worst results and therefore, if this result complies with the legal requirements, the whole building complies with them. On the other hand, not all the critical site/rooms may have been taken into account and therefore the confidence level and the sampling uncertainty to be associated with the results is not known. Moreover, the sampling strategy proposed in ISO/WD 19488 has the obvious drawback that it cannot guarantee to have spotted all critical situations: for example a workmanship failure that cannot be detected by visual inspection can be identified only after the measurements. Thus, a sampling criterion based on generic rules cannot find it. However, the con of this strategy is that in general the number of measurements is limited.

### *4.1. Stratified Sampling*

The stratified sampling is the most direct procedure that allows increasing the efficiency of a sampling plan, since it allows reducing the order of magnitude of the sampling error without increasing the sample size.

Stratification is made possible by means of additional information about one or more characters of the population, which is about the structure of the population itself. This allows, based on informed choice, dividing the population into a number of layers as homogeneous as possible, as meaning that within each layer, the considered character has a lower variability. A simple random sample is extracted from each layer; therefore there are as many simple samples as there are layers. These samples are independent of each other and can have different sample sizes. The stratification, due to the way it is implemented, allows obtaining an improvement in the estimates for the same sample size, or to contain the sample size at the same level of efficiency [49].

Considering the above mentioned advantages offered by the stratified sampling, this latter is the solution adopted by UNI 11367 [46] in the case of classification of serial type buildings.

The part of the Italian standard on the classification of buildings and building units that refers to the stratified sampling procedure can be applied in the case of a serial type building. The stratified sampling procedure is based on the concept of homogeneous group. The population of all the building elements that have to be measured for the acoustic classification has to be divided in the homogeneous groups that are defined in the Italian standard on classification. Referring to UNI 11367 [46], generally, a set of test items can be considered homogeneous and therefore subject to a possible sampling (in reference to a specific requirement), if the following conditions are satisfied: item dimensions (with 20% tolerance); dimensions (with 20% tolerance with respect to the volumes) of both transmitting and receiving rooms where the test item is located; the same test methodology; stratigraphy, materials and surface mass; structural constraints (flanking transmissions); presence of equipment passing through the test item; installation techniques. In this section an example is given with reference to the paper presented by the authors at the 38th National Congress of the Acoustical Society of Italy in 2011 [50], concerning the acoustic classification of a building system of a total volume of about 15,000 m$^3$, consisting of two similar buildings, identified as body A1 and body A2, on three floors, with apartments on the ground floor, first and second floor and, in the body A1, a third floor attic. In total, the building system consists of 47 Building Units (BUs), distributed according to their type: six four-room apartments, eight three-room apartments, 25 two-room apartments and eight studios.

The building system was considered a serial type building system, based on the following considerations: it is possible to identify a typical floor (see Figure 4) in which the distribution of BUs is symmetrical with respect to the stairwells; the two-room apartment type is repeated 25 times; the rooms with the same intended use (bedrooms, living rooms, kitchens, *etc.*) have the same shape and size.

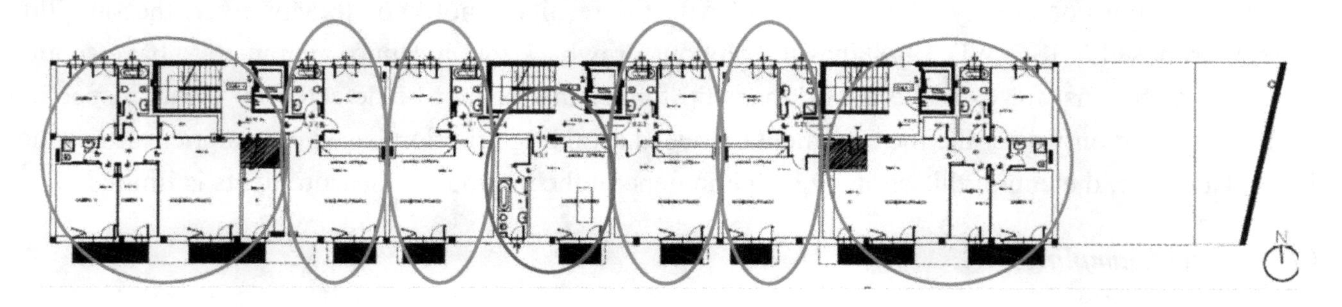

**Figure 4.** Typical floor of the building system considered: the BU typologies are highlighted, in green the four-room apartments, in red the two-room apartments and in blue the studio [50].

For the application of the stratified sampling procedure defined in UNI 11367 [46], it would have been sufficient to use a minimum number of items to be tested equal to at least 10% of the total number of elements of the homogeneous group and not less than three for each homogeneous group. However, in order to obtain the most useful data for a critical examination of the results, the number of items to be tested was higher than the minimum required. In particular, 84% of the vertical partitions were measured. For some requirements, the number of items to be tested of some homogeneous group was equal to two, which is less than the minimum required of three for reasons related to the impossibility to perform further tests (inaccessibility of the rooms). When more measurements than the minimum necessary number were made, in order to simulate the case in which only the minimum sampling number (3) of measurements was performed, the results were reconsidered on the basis of all the combinations without repetition of three elements actually measured and calculating the average. This was done in order to make the choice of the three elements under test as random as possible, and to evaluate the probability of obtaining a specific standard deviation, and a specific class according to the variability and to the randomness of selection of the three elements. The results, obtained on the basis of both the performed measurements and the statistical analysis of the sampling procedure, are as follows. First of all, a methodological indication: a review and a possible redefinition of the homogeneous groups retrospectively (*i.e.*, when the measurement are concluded), this may be useful to formulate an acoustic classification closer to the real situation. This indication comes from the fact that, in the case under study, the values of the impact sound insulation differ greatly and in a systematic manner between the two bodies; the results of body A2 show a worse performance than those of body A1. This difference is due to the installation that, in one case (body A1), was evidently very well done. The influence of workmanship was studied by Craik and Steel [51] who found that workmanship can account for a variation of approximately 2 dB in airborne sound insulation. Within this distinction, the variability of the impact sound insulation values is on the average when compared with the airborne sound insulation of vertical and horizontal partitions.

From the analysis of all the combinations without repetition of the measurements, two possible classifications are found: in one of them, the percentage of BUs of class III (43%) becomes smaller compared to that of class IV, while in the other, the percentage of BUs of class III is rather prevalent (64%); in particular, this is due to the requirement of airborne sound insulation of internal partitions $R'_w$. Actually, the values relating to the airborne sound insulation $R'_w$, for vertical partitions, are in the vast majority of cases very close to the lower limit value (50 dB) of class III; therefore, in the random choice, there is a higher probability that the choice falls on these values straddling the two classes, with the result of moving the larger percentage of BUs from class to class. This analysis makes it clear that it is necessary to adopt, at design level, more conservative design solutions.

Table 7 shows the average, minimum, maximum and standard deviation of the measured performances for each type of technical element; in particular, the variability of the data is described by the standard deviation and it increases with the increase of the latter. The values shown in Table 7 are the net values, as defined in UNI 11367 [46], *i.e.*, the results of a measurement corrected with the measurement uncertainty.

The greatest variability is found for façades; for the building system under classification this is caused mainly by the typical variability of façades, dependent on many types of window frames and the presence of balconies, irrespective of a proper installation and, to a lower extent, also to workmanship.

The lower variability is observed in internal partitions, and in particular floors (horizontal partition), with respect to the sound insulation requirement. The variability of the impact sound insulation is comparable with that of the other requirements. Moreover, the variability of the impact sound insulation for the two bodies separately is comparable, confirming the systematic difference found in body A2.

**Table 7.** Performances variability of technical elements [50].

| Technical Element | Façade | Vertical Partitions | Horizontal Partitions | Horizontal Partitions A1 | Horizontal Partitions A2 |
|---|---|---|---|---|---|
| Quantity used in law requirements | $D_{2mnT,w}$ | R'w | R'w | L'n | L'n |
| number of test elements | 35 | 36 | 21 | 9 | 10 |
| average | 39.6 | 50.1 | 52.6 | 57.6 | 65.5 |
| standard deviation | 3.7 | 1.87 | 1.4 | 2.2 | 2.6 |
| minimum | 30 | 47 | 50 | 54 | 62 |
| maximum | 45 | 54 | 55 | 61 | 70 |

### *4.2. Stratified Sampling Uncertainty*

When a sample survey is used to define the classification of building or BUs, it is necessary to consider the uncertainty associated with the sampling procedure. Moreover, considering that each single measurement result that contributes to the value attributed to a certain requirement has its own measurement uncertainty, it becomes necessary to combine these two uncertainties in a certain way.

In the case of UNI 11367 [46] the representative value of a homogeneous group (Equations (9) and (10)), *i.e.*, the arithmetic mean value of the group with the sampling uncertainty with a one-sided coverage factor, already includes the measurement uncertainty. In fact, the arithmetic mean values $X_{he}$ and $Y_{he}$ are calculated from the net values (*i.e.*, the results of the measurement corrected with the measurement uncertainty) of the homogeneous group itself [46,52] as indicated in UNI 11367 [46] as follows:

$$X_{he} = \frac{\sum_{c=1}^{C_h} X_{hc}}{C_h} \tag{7}$$

$$Y_{he} = \frac{\sum_{c=1}^{C_h} Y_{hc}}{C_h} \tag{8}$$

where $X_{hc}$ is the net value of a sample of a specific requirement (façade sound insulation or airborne sound insulation of the internal partition), $Y_{hc}$ is the net value of a sample of a specific requirement (impact sound insulation or sound pressure level for service equipment), and $C_h$ is the number of samples within a homogeneous group.

The "representative value" $X_h$ and $Y_h$ of each homogeneous group is then obtained as follows [46]:

$$X_h = X_{he} - U_{sh} \tag{9}$$

$$Y_h = Y_{he} + U_{sh} \tag{10}$$

where $U_{sh}$ is the sampling uncertainty equal to the sampling standard deviation $s_{sh}$ times the coverage factor $k$:

$$U_{sh} = s_{sh} \cdot k \tag{11}$$

where $s_{sh}$ is the sampling standard deviation, determined with Equations (12) and (13):

$$s_{shX} = \sqrt{\frac{\sum_{c=1}^{C_h} (X_{he} - X_{hc})^2}{C_h - 1} \frac{(M_h - C_h)}{(M_h - 1)}} \quad (12)$$

$$s_{shY} = \sqrt{\frac{\sum_{c=1}^{C_h} (Y_{he} - Y_{hc})^2}{C_h - 1} \frac{(M_h - C_h)}{(M_h - 1)}} \quad (13)$$

where $s_{shX}$ is the standard deviation referred to the façade sound insulation or to the airborne sound insulation of internal partitions, $s_{shY}$ is the standard deviation referred to the impact sound insulation or to the sound pressure level for service equipment and $M_h$ is the number of all the measurable technical elements within a homogeneous group.

## 5. Conclusions

This study showed that the measurement uncertainty in building acoustics is very high, in particular if the measurements are extended at low frequencies. Therefore, it is extremely important to define the way to manage measurement uncertainty in building acoustics, depending on the different situations.

In the case of a single measurement, the uncertainty that should be associated with this measurement is the reproducibility standard deviation given in ISO 12999-1 [5] multiplied by the appropriate coverage factor to obtain the expanded uncertainty. Such an indication should be independent of the final use of the results, and shall be stated as the expanded uncertainty with a 95% confidence level for a two-sided test. When the single measurement is made in order to verify a requirement, the expanded uncertainty that should be given with the result, should be calculated using a coverage factor for one-sided test and the confidence level should be set to 95%.

When measurements are made to verify the acoustic requirements or the acoustic classification of building units, or buildings, one single measurement might not be enough to this end, and therefore more measurements and more results for the same requirement are necessary. For building acoustics, a sample survey is the best solution in terms of cost and time. There are two main types of sampling strategies used in building acoustics, for the time being: the stratified sampling as stated in UNI 11367 [46] and a sampling procedure taking into account a certain percentage associated with a selection criterion as adopted UNI 11444 [47] and proposed in ISO/WD 19488 [48]. In the former case, the measurement uncertainty is combined with the sampling uncertainty to obtain a reliable classification for BUs, while in the latter case, the sampling uncertainty is not taken into account because the strategy selection includes all the critical acoustic situations. In UNI 11444 [47], the measurement uncertainty is included for each measurement result, as stated in UNI 11367 [46]. In ISO/WD 19488 [48], if classification for different dwellings, rooms or acoustic characteristics varies, the classification assigned is the minimum class obtained, and therefore all the other BUs complied with that class.

In any case, a measurement, whether single or part of a set of measurements for the sampling, should always be associated with its measurement uncertainty. In the case of sampling, either the sampling uncertainty is considered, obtaining a value representative of all the situations considered, or selection criteria of the most critical cases is taken into account, obtaining a value that is not representative of all the situations but is a precautionary value.

## Acknowledgments

We gratefully acknowledge Paolo Cardillo for revising the English text of the manuscript.

## Author Contributions

Chiara Scrosati designed and conceived the study, acquired data, analyzed and interpreted data, supervised the RRTs and drafted the manuscript.

Fabio Scamoni revised the manuscript critically for important intellectual content.

## References and Notes

1. Scrosati, C.; Scamoni, F. Measurement uncertainty in building acoustics Invited and peer reviewed paper. In Proceedings of the 22nd International Congress on Sound and Vibration (ICSV22), Florence, Italy, 12–16 July 2015.
2. *ISO/IEC Guide 98-3:2008 The Guide to the Expression of Uncertainty in Measurement (GUM:1995)*; ISO: Genève, Switzerland, 2008.
3. *ISO 10140 Acoustics—Laboratory Measurement of Sound Insulation of Building Elements*; ISO: Genève, Switzerland, 2010.
4. *ISO 16283 Acoustics—Field Measurement of Sound Insulation in Buildings and of Building Elements*; ISO: Genève, Switzerland, 2014.
5. *ISO 12999-1:2014 Acoustics—Determination and Application of Measurement Uncertainties in Building Acoustics—Part 1: Sound Insulation*; ISO: Genève, Switzerland, 2014.
6. Scamoni. F.; Scrosati, C.; Mussin, M.; Galbusera, E.; Bassanino, M.; Zambon, G.; Radaelli, S. Repeatability and reproducibility of field measurements in buildings. In Proceedings of the Euronoise 2009, Edinburgh, Scotland, 26–28 October 2009.
7. Scrosati, C.; Scamoni, F.; Bassanino, M.; Mussin, M.; Zambon, G. Uncertainty analysis by a round robin test of field measurements of sound insulation in buildings: Single numbers and low frequency bands evaluation—Airborne sound insulation. *Noise Control Eng. J.* **2013**, *61*, 291–306.
8. Scrosati, C.; Scamoni, F.; Zambon, G. Uncertainty of façade sound insulation in buildings by a round robin test. *Appl. Acoust.* **2015**, *96*, 27–38.
9. Decree of the President of the Council of Ministers D.P.C.M. Determinazione Dei Requisiti Acustici Passivi Degli Edifici (Determination of Building Passive Acoustic Requirements). In *G.U. (Official Journal) General Series n.297*; D.P.C.M.: Rome, Italy, 1997.
10. Scrosati, C.; Scamoni, F.; Asdrubali, F.; D'Alessandro, F.; Moretti, E.; Astolfi, A.; Barbaresi, L.; Cellai, G.; Secchi, S.; di Bella, A.; *et al.* Uncertainty of façade sound insulation measurements obtained by a round robin test: The influence of the low frequencies extension, Invited paper. In Proceedings of the 22nd International Congress on Sound and Vibration (ICSV22), Florence, Italy, 12–16 July 2015.

11. *ISO 140-5:1998 Acoustics—Measurement of Sound Insulation in Buildings and of Building Elements—Part 5: Field Measurements of Airborne Sound Insulation of Façade Elements and Façades*; ISO: Genève, Switzerland, 1998.
12. Lang, J. A round robin on sound insulation in building. *Appl. Acoust.* **1997**, *52*, 225–238.
13. Simmons, C. Uncertainty of measured and calculated sound insulation in buildings—Results of a round robin test. *Noise Control Eng. J.* **2007**, *55*, 67–75.
14. Farina, A.; Fausti, P.; Pompoli, R.; Scamoni, F. Intercomparison of laboratory measurements of airborne sound insulation of partitions. In Proceedings of the 1996 International Congress on Noise Control Engineering, (Internoise' 96), Liverpool, UK, 30 July–2 August 1996.
15. Fausti, P.; Pompoli, R.; Smith, R.S. An inter-comparison of laboratory measurements of airborne sound insulation of lightweight plasterboard walls. *J. Build. Acoust.* **1999**, *6*, 127–140.
16. Smith, R.S.; Pompoli, R.; Fausti, P. An Investigation into the reproducibility values of the european inter-laboratory test for lightweight walls. *J. Build. Acoust.* **1999**, *6*, 187–210.
17. Schmitz, A.; Meier, A.; Raabe, G. Inter-laboratory test of sound insulation measurements on heavy walls: Part I—Preliminary test. *J. Build. Acoust.* **1999**, *6*, 159–169.
18. Meier, A.; Schmitz, A.; Raabe, G. Inter-laboratory test of sound insulation measurements on heavy walls: Part II—Results of main test. *J. Build. Acoust.* **1999**, *6*, 171–186.
19. *ISO 140-4:1998 Acoustics—Measurement of Sound Insulation in Buildings and of Building Elements—Part 4: Field Measurements of Airborne Sound Insulation between Rooms*; ISO: Genève, Switzerland, 1998.
20. COST Action TU0901 Integrating and Harmonizing Sound Insulation Aspects in Sustainable Urban Housing Constructions, 2009–2013. Available online: http://www.costtu0901.eu/ (accessed on 15 September 2015).
21. *ISO 717-1:2013 Acoustics—Rating of Sound Insulation in Buildings and of Building Elements—Part 1: Airborne Sound Insulation*; ISO: Genève, Switzerland, 2013.
22. Wittstock, V. On the uncertainty of single-number quantities for rating airborne sound insulation. *Acta Acust. United Acust.* **2007**, *93*, 375–386.
23. Dijckmans, A.; Vermeir, G. Numerical investigation of the repeatability and reproducibility of laboratory sound insulation measurements. *Acta Acust. United Acust.* **2013** *99*, 421–432.
24. Garg, N.; Maji, S. On analyzing the correlations and implications of single-number quantities for rating airborne sound insulation in the frequency range 50 Hz to 5 kHz. *Build. Acoust.* **2015**, *22*, 29–44.
25. *ISO CD 16717-1:2013 Acoustics—Evaluation of Sound Insulation Spectra by Single-Number Values. Part 1: Airborne Sound Insulation*; ISO: Genève, Switzerland, 2013.
26. Mahn, J.; Pearse, J. The uncertainty of the proposed single number ratings for airborne sound insulation. *J Build. Acoust.* **2012**, *19*, 145–172.
27. Hongisto, V.; Keränen, J.; Kylliäinen, M.; Mahn, J. Reproducibility of the present and the proposed single-number quantities of airborne sound insulation. *Acta Acust. United Acust.* **2012**, *98*, 811–819.
28. Machimbarrena, M.; Monteiro, C.R.A.; Pedersoli, S.; Johansson, R.; Smith, S. Uncertainty determination of *in situ* airborne sound insulation measurements. *Appl. Acoust.* **2015**, *89*, 199–210.

29. António; J.; Mateus, D. Influence of low frequency bands on airborne and impact sound insulation single numbers for typical Portuguese buildings. *Appl. Acoust.* **2015**, *89*, 141–151.
30. Grubbs, F.E. Procedures for detecting outlying observation in samples. *Technometrics* **1969**, *11*, 1–21.
31. Grubbs, F.E.; Beck, G. Extension of sample sizes and percentage points for significants tests of outlying observations. *Technometrics* **1972**, *14*, 847–854.
32. Masovic, D.B.; Sumarac Pavlovic, D.S.; Mijic, M.M. On the suitability of ISO 16717-1 reference spectra for rating airborne sound insulation. *J. Acoust. Soc. Am.* **2013**, *134*, EL420–EL425.
33. Rindel, J.H. On the influence of low frequencies on the annoyance of noise from neighbours. In Proceedings of the InterNoise 2003, Seogwipo, Korea, 25–28 August 2003.
34. Park, H.K.; Bradley, J.S. Evaluating standard airborne sound insulation measures in terms of annoyance, loudness, and audibility ratings. *J. Acoust. Soc. Am.* **2009**, *126*, 208–219.
35. Prato, A.; Schiavi, A. Sound insulation of building elements at low frequency: A modal approach, In Proceedings of the Energy Procedia—6th International Building Physics Conference, IBPC 2015, Turin, Italy, 14–17 June 2015.
36. *ISO/DIS 16283-3:2014 Acoustics—Field Measurement of Sound Insulation in Buildings and of Building Elements—Part 3: Façade Sound Insulation*; ISO: Genève, Switzerland, 2014.
37. *ISO 140-2:1991. Acoustics—Measurement of Sound Insulation in Buildings and of Building Elements—Part 2: Determination, Verification and Application of Precision Data*; ISO: Genève, Switzerland, 1991.
38. Berardi, U.; Cirillo, E.; Martellotta, F. Interference effects in field measurements of airborne sound insulation of building façades. *Noise Control Eng. J.* **2011**, *59*, 165–176.
39. Berardi, U. The position of the instruments for the sound insulation measurement of building façades: From ISO 140-5 to ISO 16283-3. *Noise Control Eng. J.* **2013**, *61*, 70–80.
40. Hopkins, C. Revision of international standards on field measurements of airborne, impact and facade sound insulation to form the ISO 16283 series. *Build. Environ.* **2015**, *92*, 703–712.
41. Hopkins, C.; Turner, P. Field measurement of airborne sound insulation between rooms with non-diffuse sound fields at low frequencies. *Appl. Acoust.* **2005**, *66*, 1339–1382.
42. Prato, A.; Ruatta, A.; Schiavi, A. Transmission of impact noise at low frequency: A modal approach for impact sound insulation measurement (50–100 Hz). In Proceedings of the 22nd International Congress on Sound and Vibration (ICSV22), Florence, Italy, 12–16 July 2015.
43. Masovic, D.; Miskinis, K.; Oguc, M.; Scamoni, F.; Scrosati, C. Analysis of façade sound insulation field measurements—Influence of acoustic and non-acoustic parameters. In Proceedings of the INTER-NOISE 2013, Innsbruck, Austria, 15–18 September 2013.
44. Masovic, D.; Miskinis, K.; Oguc, M.; Scamoni, F.; Scrosati, C. Analysis of façade sound insulation field measurements—Comparison of different performance descriptors and influence of low frequencies extension. In Proceedings of the INTER-NOISE 2013, Innsbruck, Austria, 15–18 September 2013.
45. Wittstock, V. Determination of measurement uncertainties in building acoustics by interlaboratory tests. Part 1: Airborne sound insulation. *Acta Acust. United Acust.* **2015**, *101*, 88–98.
46. *UNI 11367:2010 Building Acoustics—Acoustic Classification of Building Units—Evaluation Procedure and In Situ Measurements*; UNI: Milan, Italy, 2010.

47. *UNI 11444:2012 Building Acoustics—Acoustic Classification of Building Units—Guidelines for the Selection of Building Units in not Serial Building Systems*; UNI: Milan, Italy, 2012.
48. *ISO/WD 19488:2015 (Rev. 20 July 2015) Acoustics—Acoustic Classification Scheme for Dwellings*; ISO: Genève, Switzerland, 2012.
49. Scrosati, C.; Pontarollo, C.M.; Scamoni, F.; di Bella, A.; Elia, G. Procedure di verifica delle prestazioni acustiche di edifici: Analisi tramite campionamento, (Acoustic performances of buildings: Sampling procedure analysis). In Proceedings of the 37th AIA National Congress, Siracusa, Italy, 26–28 May 2010.
50. Scamoni, F.; Scrosati, C.; Cera, S. Classificazione acustica in un edificio residenziale a tipologia seriale: Analisi statistica della procedura di campionamento (Acoustic classification of a residential building of serial type: Statistical analysis of sampling procedure). In Proceedings of the 38th AIA National Congress, Rimini, Italy, 8–10 June 2011.
51. Craik, R.J.M.; Steel, J.A. The effect of workmanship on sound transmission through buildings: Part 1. Airborne sound. *Appl. Acoust.* **1989**, *27*, 57–63.
52. Di Bella, A.; Fausti, P.; Scamoni, F.; Secchi, S. Italian experience on acoustic classification of buildings. In Proceedings of the Internoise 2012, New York, NY, USA, 19–22 August 2012.

# 10

# Embedded Ultrasonic Transducers for Active and Passive Concrete Monitoring

**Ernst Niederleithinger *, Julia Wolf †, Frank Mielentz †, Herbert Wiggenhauser † and Stephan Pirskawetz †**

BAM Federal Institute for Materials Research and Testing, Berlin 12200, Germany;
E-Mails: julia.wolf@bam.de (J.W.); frank.mielentz@bam.de (F.M.);
herbert.wiggenhauser@bam.de (H.W.); stephan.pirskawetz@bam.de (S.P.)

† These authors contributed equally to this work.

* Author to whom correspondence should be addressed; E-Mail: ernst.niederleithinger@bam.de

Academic Editor: Thomas Schumacher

---

**Abstract:** Recently developed new transducers for ultrasonic transmission, which can be embedded right into concrete, are now used for non-destructive permanent monitoring of concrete. They can be installed during construction or thereafter. Large volumes of concrete can be monitored for changes of material properties by a limited number of transducers. The transducer design, the main properties as well as installation procedures are presented. It is shown that compressional waves with a central frequency of 62 kHz are mainly generated around the transducer's axis. The transducer can be used as a transmitter or receiver. Application examples demonstrate that the transducers can be used to monitor concrete conditions parameters (stress, temperature, …) as well as damages in an early state or the detection of acoustic events (e.g., crack opening). Besides application in civil engineering our setups can also be used for model studies in geosciences.

**Keywords:** ultrasound; transmission; concrete; damages; cracks; stress

---

## 1. Introduction

Concrete is a complex, multi-phase material. It is made of hydraulic cement, water and aggregates in lots of variations. The first two ingredients start to hydrate and crystallize when in contact with each other. This process is fast in the first hours and days, but can continue for months or years. The aggregates, gravel or crushed stone of various types and sizes (µm-cm range), are used as a filler to save cost and energy. Concrete contains pores, either filled with unbound water or air.

Concrete is the material most produced by mankind. It is considered to be strong, resistive and durable. Some early concrete structures as the cupola of the Roman Pantheon are standing tall after almost 2000 years. However, under certain conditions (hostile environments, adverse load conditions) concrete constructions require attention. For example the apparently ever increasing traffic load (number and individual load of trucks) on bridges may lead to deterioration much earlier than expected at the time of design. Currently, inspections are still mainly based on visual methods, but sophisticated non-destructive methods are used more and more often. The use of bridge instrumentation increases rapidly [1]. However, monitoring is so far limited to local sensors (e.g., strain gauges), which are probing just their close vicinity, or global methods as modal analysis, which looks to the structure as a whole. There is a gap in between. A method which would look at a certain critical volume of concrete with a very limited number of sensors would be of great value.

Ultrasonic transducers with frequencies from 25 kHz to 400 kHz have been used for concrete since decades. They are used in the lab on samples (and sometimes on site) in transmission mode to measure elastic properties and to assess degradation. New point contact transducers have revolutionized the use of echo techniques for thickness measurements and structural imaging. Even the detection of voids in tendon ducts seems to be possible [2]. All transducers used in practice today are for surface mounting. For monitoring this approach shows three strong disadvantages. First, the need for constant coupling, which is hard to realize on the surface in practice. Second, the high influence of surface and external effects (temperature and others) leads to unwanted effects on the measurements. Third, in practical field applications the transducers are prone for accidents or vandalism. Therefore we started to develop a novel transducer, which can be permanently embedded in concrete.

This is not the first or only attempt to embed ultrasonic transducers in concrete. Similar ideas have been proposed e.g., by [3,4], but only in experimental setups for lab applications. For practical applications a much more robust approach would be required. An experiment conducted more than 35 years ago to monitor the hardening of concrete at a massive water dam in Saxony, Germany, by embedded ultrasonic transmitters and receivers recently gave us the opportunity to prove that this kind of sensors might survive in concrete for decades [5].

## 2. A novel Ultrasonic Transducer to Be Embedded in Concrete

### *2.1. Transducer Design and Description*

New ultrasonic transducers ("SO807") have been designed by Acoustic Control Systems, Ltd. (ACS, Moscow, Russia) in cooperation with and exclusively for BAM (Figure 1). The main part is a hollow piezoceramic cylinder of 20 mm diameter and 35 mm length. The electric connections are on the inside. On both ends metallic pieces are clamped to the piezoceramic part. The outer diameter of

15 mm allows stacking of several transducers along a line using standard PVC tubes. Having all cables inside ensures good coupling of the piezo to the concrete and protects the electrical connections during installation. The total length of the transducer is 75 mm.

**Figure 1.** Photograph of the novel ultrasonic transducer for embedment in concrete (Manufacturer: Acsys Ltd., Moscow, Russia. Photo: BAM).

*2.2. Installation*

The transducers can easily be installed at the time of construction. To ensure that they keep their position during casting and vibration, they have to be mounted either direct to the reinforcement or using some kind of stabilizing construction. In our latest bridge installation we have used L-shaped pieces of rebar welded to the reinforcement to hold the transducers at the specified position (Figure 2). Another possibility would be to mount stacked series of transducers using PVC tube segments (similar to the method used for installation in existing structures described below).

**Figure 2.** Photograph of an ultrasonic transducer mounted at a bridge construction site before casting (Photo: Neostrain S.A.).

For existing structures we have developed a method to install one or several transducers at the required depth(s) in a drill hole and ensuring sufficient coupling to the concrete. The hole is drilled with a slightly larger diameter than the transducers and slightly deeper than the installation depth. The transducer(s) and the tube segments are connected with a sealing cap to the structures' surface

(Figure 3). The cap contains an inlet, which is connected to a (liquid) grout reservoir and to the space between the transducers/tubes and the concrete. By connecting a suction pump via an outlet in the cap to the inner hollow space of the transducers and the tubes the grout is sucked into the drill hole and via the crown of the tubes into the inner space (Figure 3c). When the grout appears at the outlet we can be sure the entire space is (at least almost) filled and the transducers are coupled to the concrete. We are using a fast hardening, slightly expanding type of grout.

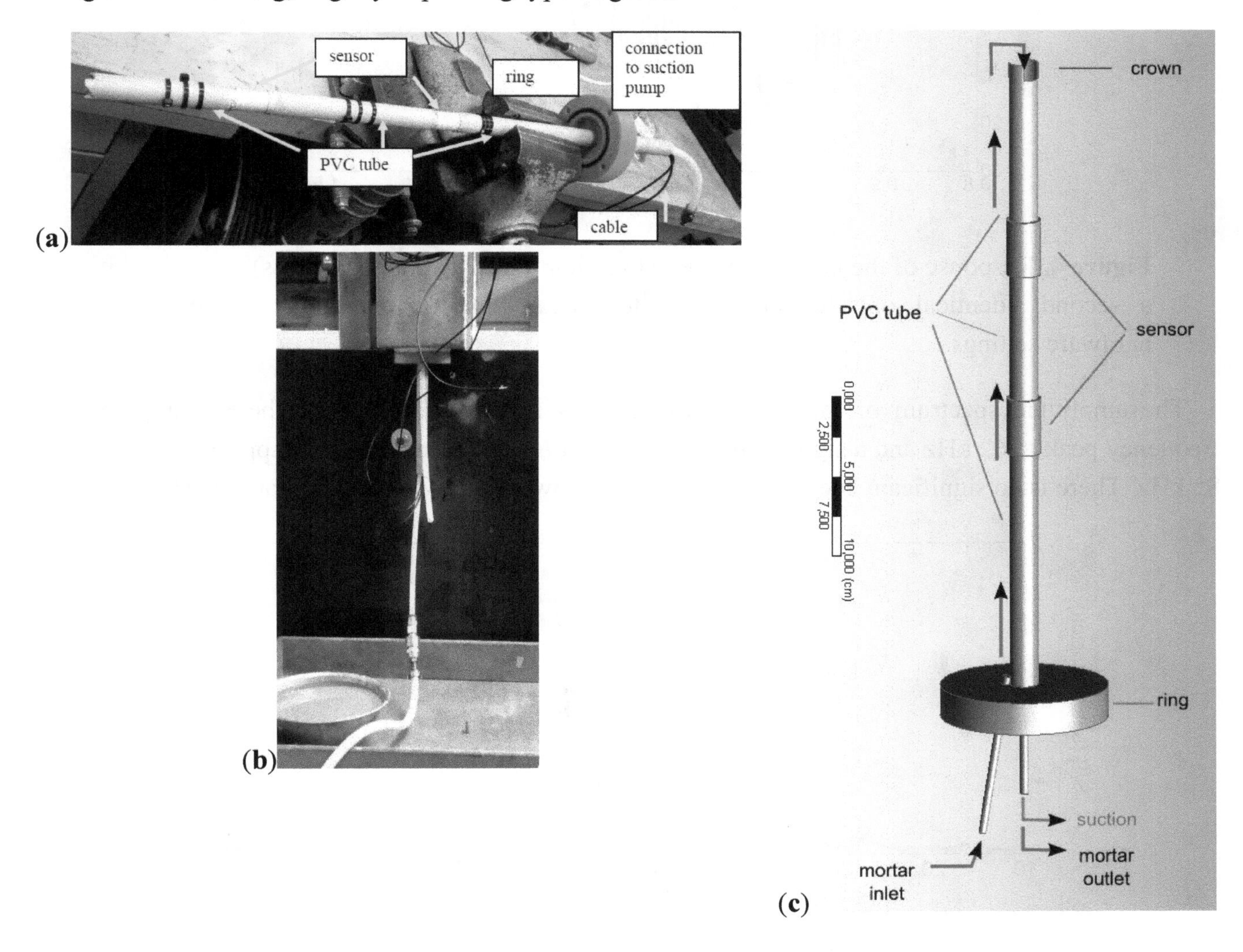


**Figure 3.** (**a**): Two transducers connected by PVC tube segments and equipped with sealing cap for post-concreting installation; (**b**): Installation into a concrete specimen; (**c**): Sketch of grout flow during installation.

## *2.3. Characterization*

The transducers have been characterized by some basic experiments. First, two transducers of the same type have been installed back to back directly without any medium in between to evaluate the frequency spectrum. A short impulse (t = 2 μs) was used as input. The signal recorded by the transducer is displayed in Figure 4. The signal recorded shows a wavelet with about 20 periods and a total duration of ca. 0.3 ms. The reverberations show a lack of damping of the piezo material. As we don't intend to use the transducers for imaging applications, where a sharp response (broadband in frequency domain) would be more beneficial, this behavior is fully satisfactory.

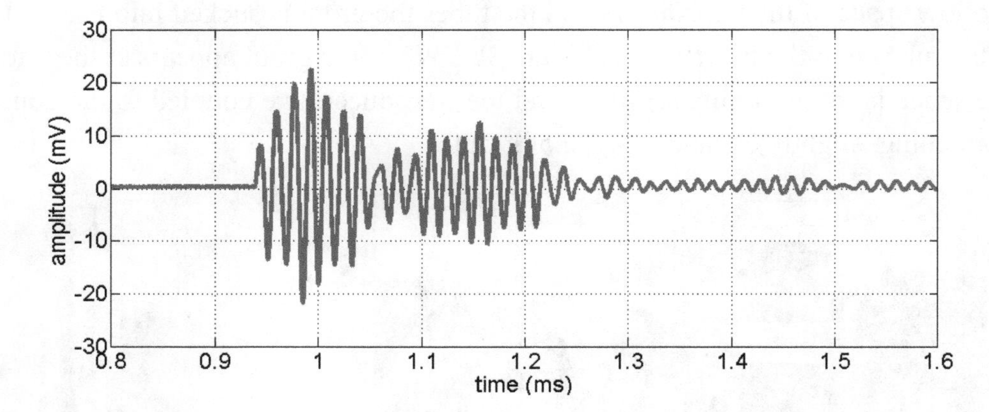


**Figure 4.** Response of the novel transducer to a short electrical impulse (2 µs), recorded by a second, identical one mounted back to back. Time of onset due to recording hardware settings.

The amplitude spectrum of the data of Figure 4 is shown in Figure 5. There is a prominent frequency peak at 62 kHz and a significant second one at 65 kHz. Smaller peaks appear around 50 and 85 kHz. There is no significant energy with frequencies lower than 40 or higher than 90 kHz.

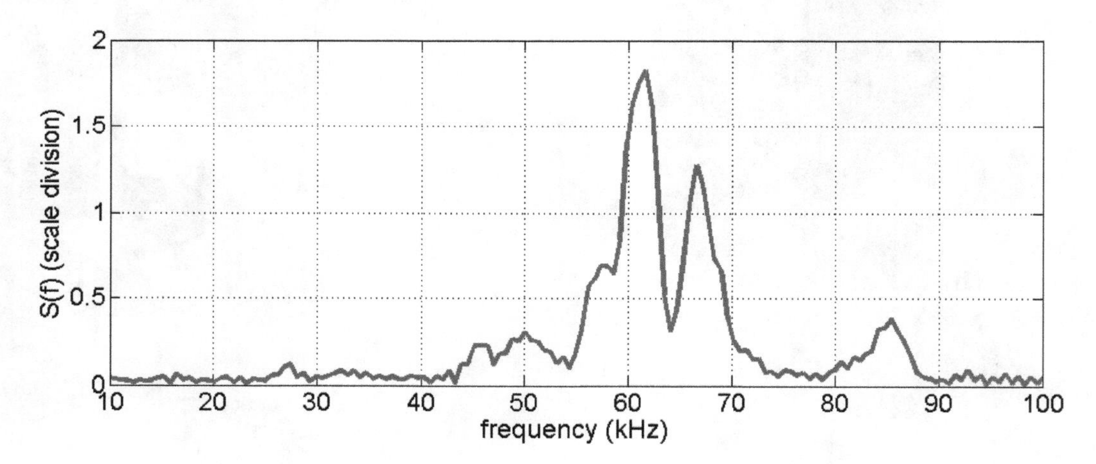


**Figure 5.** Amplitude spectrum of data shown in Figure 4.

A frequency of 62 kHz relates to a wavelength in concrete (compressional wave speed of about 4000 m/s) of ca. 65 mm. This is at least double the size of most aggregates (max aggregate size 32 mm in many types of concrete).

To evaluate the directivity pattern of the new transducer, we have conducted two experiments, the first in a water pool, the second using transducers embedded horizontally or vertically, respectively, in two separate cylindrical concrete blocks (Figure 6). In the first case (water pool) an identical transducer (vertically oriented) was used as receiver. In the second experiment a laser vibrometer was used to record the surface vibrations of the concrete. The laser vibrometer's position was fixed while the block was rotated in 10° intervals before each excitation. In both cases the measurements were taken using (a) vertical and (b) horizontal orientation of the transmitting transducers.

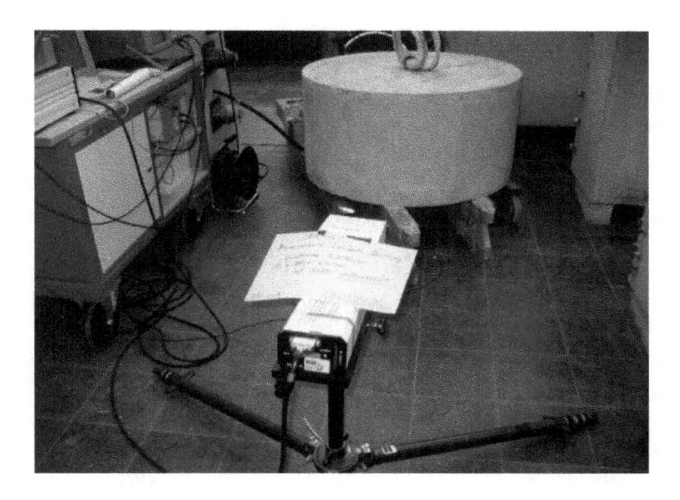

**Figure 6.** Experimental setup for measurement of directivity patterns. Transmitting transducers are embedded in the center of the concrete block. Surface movements are recorded by a laser vibrometer (in front).

The results of the directivity measurements are shown in Figure 7. To produce these plots, the amplitude of the direct arrivals (maximum of the first cycle) have been taken in 10° steps and normalized to the maximum value of all of them. For a vertical transmitter (right in Figure 7) we have recorded an almost perfect circle in water. This had to be expected as the transmitter is rotational symmetric round its vertical axis. In concrete the circular radiation pattern is still recognizable, but far from being perfect, probably due to the inherent inhomogeneities in concrete.

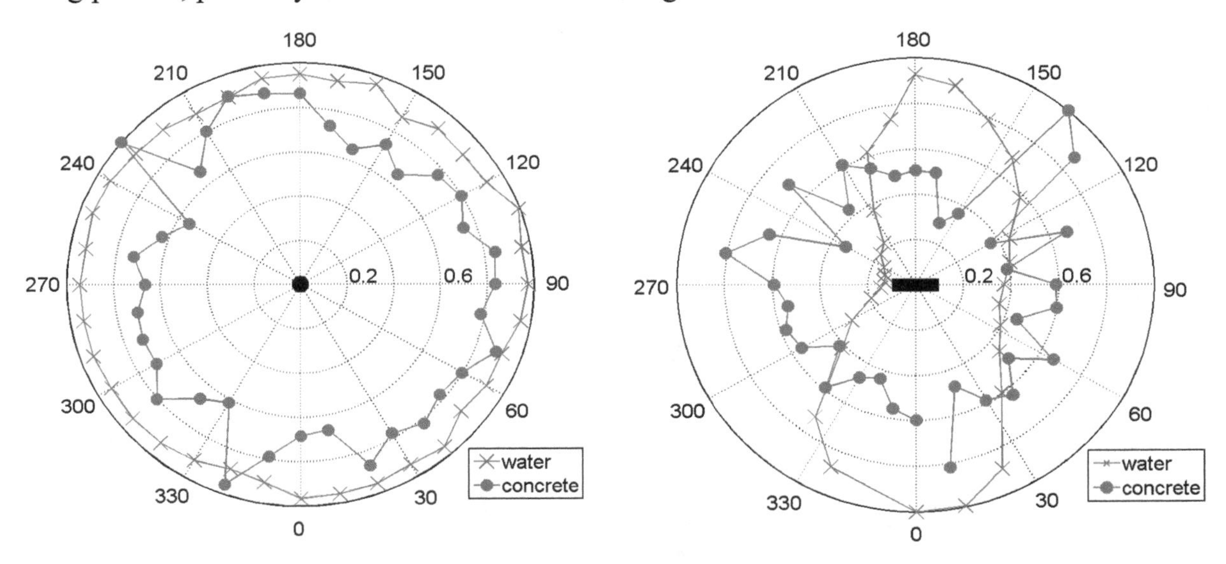


**Figure 7.** Directivity patterns for the new transducer, measured in water (blue crosses) and concrete bodies (green dots). **(Left)** vertical orientation; **(Right)** horizontal orientation. Axis from center to perimeter: relative amplitude. Seen from above.

For horizontal transmitter orientation (left in Figure 7) the directivity pattern in water resembles something close to a number eight. Apparently (and not unexpectedly as applying a voltage to the cylindrical piezoceramic leads mainly to changes in diameter, not in length) the amplitudes emitted in the direction of the transducer's rotational axis are much smaller than in the perpendicular direction.

The experiment with the same configuration in concrete showed that the amplitude is similar in all directions (but has significant variation). Our interpretation is that due to scattering at the concrete's aggregates, pores and cracks, the directivity pattern is somewhat equalized.

Additional experiments have been performed to get an idea on the range of our new transducers. This was done to assess the maximum possible distance between transmitters and receivers in future sensor networks in massive concrete structures. For this purpose a set of transducers was installed in two concrete blocks partly with, partly without reinforcement. Some transducers were installed before casting, some afterwards (see Section 2.2). One block had a maximum aggregate size of 16 mm, the other 32 mm. For all scenarios (aggregate size, reinforcement, and installation type) we have used the same measurement setup. One of the transducers, excited with a 100 V square pulse, was used as transmitter. At two others (distance $r$ = 0.25 and 0.75 m, respectively, from the transmitter) the amplitudes A (maximum of first cycle) of the first arrivals were recorded. The amplitude ratios are shown in Figure 8. It shows, that the attenuation is mainly dependent on the installation type, but not on aggregate size or reinforcement (the latter not shown here). The applicable range of our transducers was estimated from these data to be around three meters [6,7]. For this we calculated the material and frequency specific damping constant α from the amplitudes ($A_1$, $A_2$) of the first arrival after $r_1$ = 0.25 m and $r_2$ = 0.75 m using:

$$-\alpha = \left[ ln\left( \frac{A_2}{A_1} \cdot \frac{r_2}{r_1} \right) \right] \frac{1}{(r_2 - r_1)} \tag{1}$$

As an estimate for the maximum applicable range $r_E$ we have replaced $r_2$, $A_2$ in this equation by $r_E$, $A_E$ and resolved for $r_E$ .Under the assumption for the amplitude $A_E$ at $r_E$ to be at least twice the noise level, which might be different for every measurement environment, we have calculated values between 3 m and 5.5 m for our experimental setups [6,7]. The amplitudes for transducers installed after concreting are larger, probably because the special expanding grout used guarantees perfect coupling.

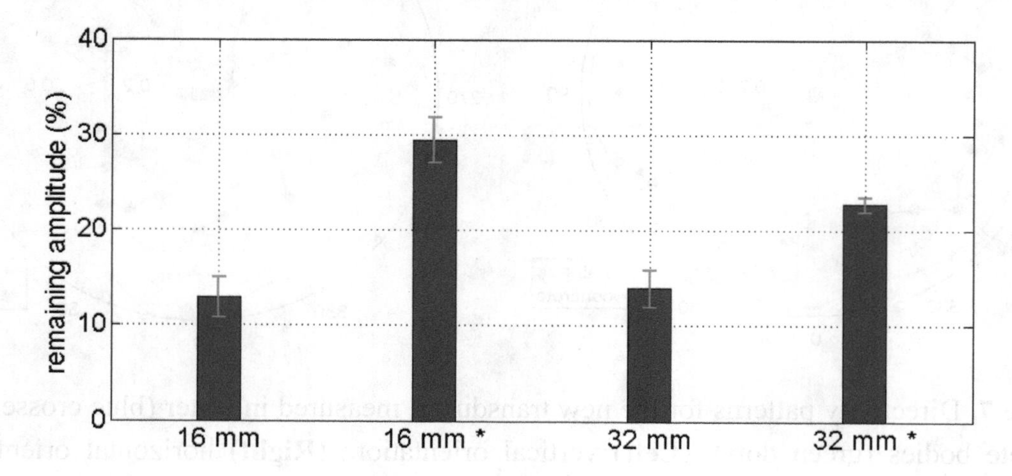


**Figure 8.** Amplitude ratios between near and far receivers in the attenuation experiment for the two specimens (16 and 32 mm max. aggregate size). The * marks values for transducers installed after concreting.

## 3. Short Notes on Ultrasonic Transmission Experiments

### *3.1. Wave Propagation in Concrete*

Piezotransducers induce elastic waves into a concrete body. The propagation of these wave fields depends on the transducer, signal type and frequency as well as the body's material, and geometry. In general, three types of waves are generated: Surface (Rayleigh) waves, which are slightly less important in our case, as we are using embedded transducers, and two types of body waves: compressional and shear. In a homogeneous linear elastic material the velocities of these waves ($c_p$, $c_s$) depend just on a few parameters: Young's modulus E, shear modulus G, Poisson's ratio $\nu$, and density $\rho$:

$$c_p = \sqrt{\frac{E(1-\upsilon)}{\rho(1+\nu)(1-2\nu)}} \tag{2}$$

and:

$$c_s = \sqrt{\frac{G}{\rho}} \tag{3}$$

respectively.

In reality concrete is neither homogeneous nor fully linear elastic. In addition, ultrasonic waves are subject to reflections at internal boundaries or scattering at small (similar or smaller than the wavelength used) objects as aggregates, small cracks or reinforcement rebars. At the same time the amplitudes of the ultrasonic waves are affected by geometrical and intrinsical (material related) attenuation. From literature it is known that the following factors (and potentially also other) have an influence on ultrasonic velocities and the amplitudes:

- Concrete type and compressive strength ([8–10]);
- Stress ([11–15]);
- Temperature ([7,9,10,16,17]);
- Moisture ([18–20]);
- Degradation (microcracking) ([13,21–23]).

Traditional time of flight measurements are considering direct waves only (first arrival at the receiver in most cases). The area of influence is limited to a narrow band ("first Fresnel zone") between transmitter and receiver (Figure 9 left).

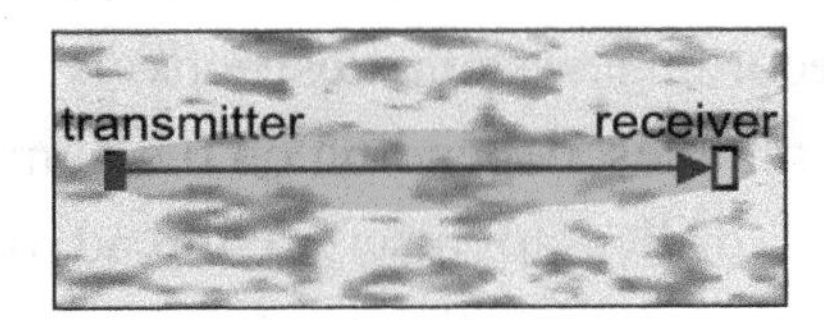


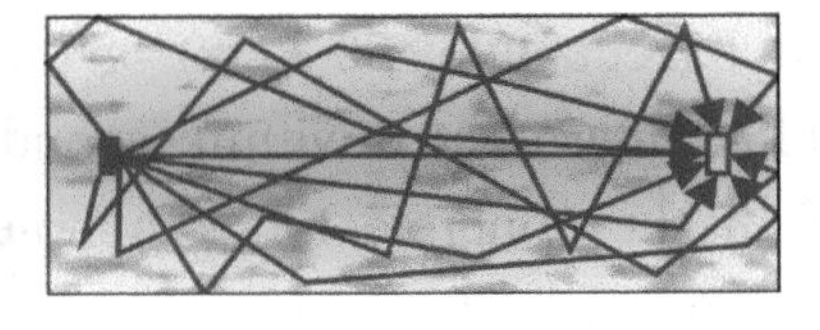

**Figure 9.** Sketch of ray paths (black lines)and area of influence (red) for the direct wave (**left**) and the full signal including coda (**right**).

Lots of other waves, which arrive much later at the receiver, have undergone reflections and scattering, potentially even wave type conversion. It is quite often difficult to evaluate these arrivals separately. However, they contain useful information as they are covering larger areas of the concrete body (Figure 9 right) and are more sensible to velocity changes due to the longer travel paths.

### 3.2. Data Evaluation

The traditional way to interpret transmission data ("time of flight") is described in the corresponding standards on ultrasonic pulse velocity measurements (see e.g., [10,24]). Picking of the first arrival travel times is often done manually. A lot of different automatic picking algorithms have been proposed from simple threshold pickers to more sophisticated ones based on statistical criteria. Based on our experience we prefer the Aikake Information Criterion (AIC) picker as described in [25]. This picker has been successfully used in many of our applications (e.g., [5]). Main disadvantage of the time of flight method is that changes in the medium under investigation may result in very small changes of the arrival times which are hard to detect. In addition, we would need a dense network of transducers in a construction to cover the entire volume by the relatively narrow Fresnel zones.

Recently methods have been introduced to ultrasound investigations in concrete, which are using the entire signal and not just the first arrival. Online monitoring systems can benefit from the use of correlation calculations (of a time series measured during/after load/damage against a reference one). Some more details and preliminary results have been published in [26]. For quantitative evaluation a novel method called coda wave interferometry may be used, which is able to calculate velocity changes from ultrasonic data with a sensitivity of about $2 \times 10^{-5}$ [27]. The method can be expanded to tomographic imaging applications. So far, only a few applications to concrete have been reported [26–29].

## 4. Application Examples

### 4.1. Monitoring of Load Changes

A $1.5 \times 1.5 \times 1.5$ m$^3$ concrete block ("GK32") has been cast in the BAM labs for various tests of the embedded ultrasonic transducers (Figure 10). The lower half contains a certain amount of reinforcement, the upper one is unreinforced. A total number of 18 ultrasonic transducers have been embedded, partly during partly after casting the block. Just ten of those have been used for a load monitoring experiment due to limitation of the available data acquisition equipment. The experiment has already been described in more detail in [26]. A two channel multiplexer had connected the transducers to an ultrasonic transmitter (rectangle, 50 kHz) or data recording system, respectively. All 90 transducer combinations could be interrogated within seconds or minutes, depending on the number of repetitions. In the upper half of the concrete block a hole was drilled to insert a thread bolt. Some of the transducers have been just a few cm away from the center of the load, some almost 1 m. Nuts, $10 \times 10$ cm$^2$ load distributing plates and a piezo load cell provided a way to introduce localized compressional stress in a controlled, repeatable way. Direction of the main compressional load is perpendicular to the front surface shown in Figure 10. However, stresses parallel to the front face are generated as well. Load steps of 5 or 10 kN were applied in various cycles up to a maximum load between 20 and 100 kN, more than one order of magnitude below the compressive strength of the concrete.

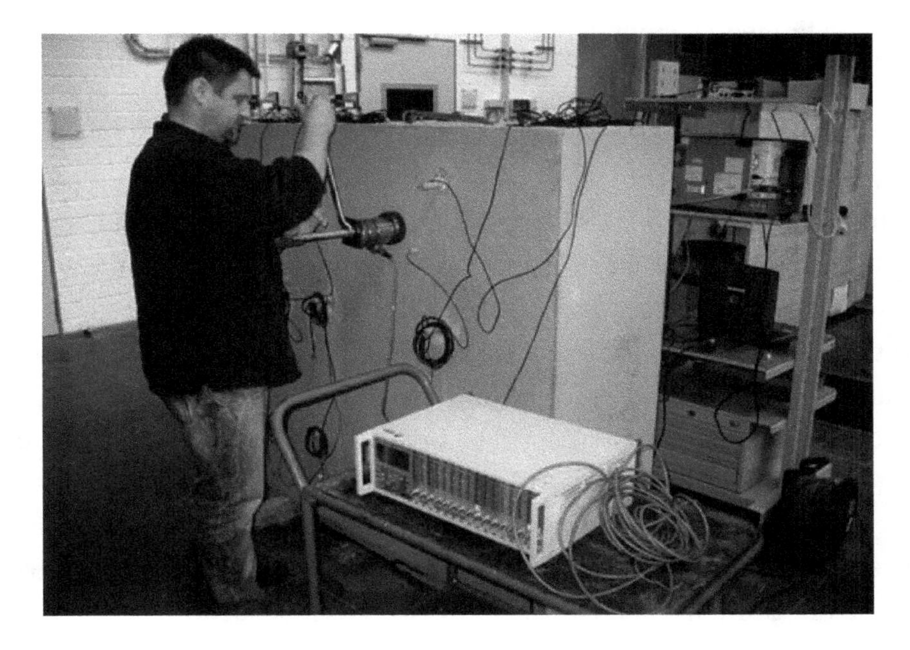

**Figure 10.** Concrete specimen GK32 with embedded transducers and load application system. From [26].

The applied loads, even very small ones, had a clear influence on the ultrasonic signals. A simple but valuable tool to provide a measure for the change is calculating the correlation coefficient between a reference measurement (here: zero load) and all consecutive measurement under various load conditions. Figure 11 shows the development of the correlations coefficient for transmission data between two embedded transducers close to the loading point. Both transducers have the same embedment depth (seen from the front face). Thus, direct waves are traveling perpendicular to the main load direction, but later parts of the signals (reflections, scattering) contain also information from different propagation directions.

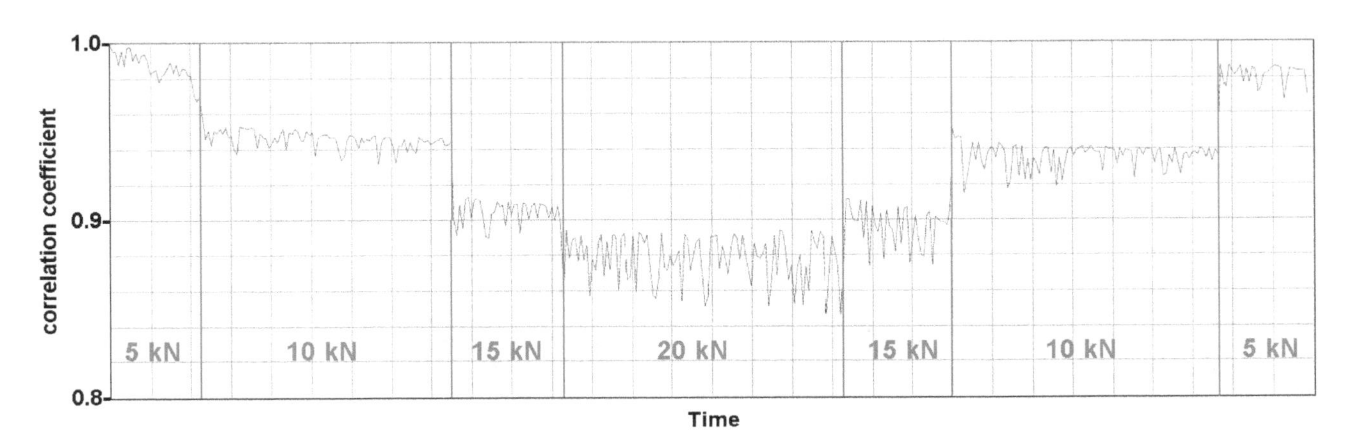


**Figure 11.** Correlation coefficient of 5 ms time series (reference: measurement at zero load) of ultrasonic signals measured by embedded transducers in a concrete block under small local compressional load. Line represents very dense consecutive measurements. From [26].

Even small load changes of 5 kN can clearly be seen, even in the presence of noise. Tomographic coda wave evaluation of the entire embedded transducer data set showed that the area of significant influence of the load is limited to about 0.3 m away from the loading point [26].

### *4.2. Acoustic Emission*

The acoustic emission (AE) technique monitors acoustic waves produced by newly developing micro cracks, opening and closure of existing cracks, friction, *etc.*, all of which are caused by internal stress variations [30]. Based on techniques used in seismology the source of the acoustic emission can be localized. Traditionally the sensors used to detect AE events are arranged on the surface of the monitored element. The distance between them is restricted by the attenuation of the waves within the material. The frequency spectrum of the acoustic events in concrete is below 200 kHz [31]. If a set of embedded transducers would be able to perform passive (AE, localization of active cracks) and active measurements (determination of changes of velocity/attenuation/material parameters) at the same time, this would be a great step forward for structural health monitoring of concrete structures.

Laboratory tests on the use of the embedded transducers for AE were conducted using a similar specimen as discussed in Section 4.1 (“GK16”). All twelve embedded transducers were used with an AMSY6 recording system (Vallen Systeme GmbH, Icking, Germany) to detect acoustic emission events within the specimen and on its surface. All following measurements have been repeated three times. In a first step, the embedded transducers were used one at a time as artificial acoustic emission sources by sending a voltage pulse to each transducer successively. The average velocity of the compressional waves, measured during this automatic sensor test, was around 4500 m/s. Figure 12 shows the corresponding event localizations (green dots) calculated by the location processor option “Planar, plane” of the software Vallen VisualAE. They are coincident with the transducer locations.

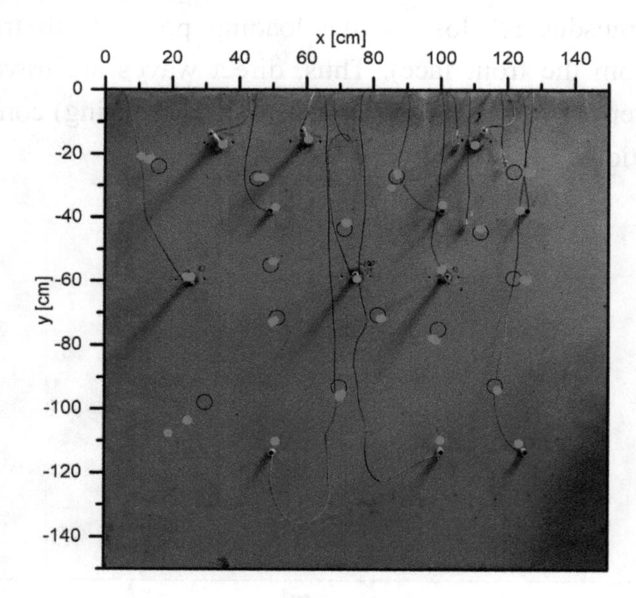


**Figure 12.** Location of acoustic emissions monitored with embedded ultrasonic transducers. Green marks: events localized by AE within the specimen (sources: one of the embedded transducers). Red marks: event localized by AE on the specimen’s surface (source: pencil breaks in black circles).

In a second set of experiments pencil breaks on the specimen's surface (a commonly used test for AE systems according to ASTM E976) have been used. Again, the experiments have been repeated three times at points marked by circles in Figure 12 and the events have been recorded and evaluated using the AMSY6 system. The localization (red dots in Figure 12) in areas with good transducer coverage lies on the mark ±3 cm. Those events outside the transducer array were not located accurately.

The experiment shows that the joint use of the embedded transducers in active and passive measurements is possible. The source needs to be within the sensor network to allow for an accurate localization. Future research will focus on the appropriate distance between the receivers for detecting and localizing acoustic emissions of different frequency and energy.

### *4.3. Time Reversal Experiment*

This application example illustrates the use of the embedded transducers for model experiments in geophysical research. In geophysics, the validation of new approaches for measurement techniques or data evaluation is quite often limited to simulations as the real subsurface (especially in exploration when talking about several kilometers depth) can never fully be explored. So full scale experiments often lack of ground truth. Scaled model experiments might be of help.

Time reversal is a technique to backproject elastic waves, which have been generated by events (e.g., cracks) inside a medium and registered by some sensors inside the medium as well or at the surface, numerically or physically back to their source point. Very few sensors are required for scattering media. In fact, the signals at the sensor, which have started as a spike at the source but then have been reflected, scattered, attenuated and dispersed on the way, can be focused back to the source point (as a perfect spike under ideal conditions). Recently a method to improve the results using deconvolution was proposed [32]. The transducers described in this paper have been used to verify the benefits of this method in case of multiple events overlapping in time [33]. For this purpose three transducers were embedded in a small concrete body (Figure 13) and served as ultrasonic sources (mimicking real acoustic emissions) in the first step. The data recorded at a single external receiver were then processed by the technology proposed in [32] and re-emitted by the external transducer into the model. Our embedded transducers now served as receivers. According to theory energy should focus at the position of a specific transducer at the time corresponding to the original event. This results in distinct sharp peaks in the recorded data (Figure 14). They correspond to the original events in time and shape. It could be shown, that the processed data focused much better in time and space at the transducers positions compared to unprocessed data. This experiment using our new transducers has definitively helped to prove the concept of using deconvolution in time reversal experiments, which would have been difficult under field conditions (events in some km depth) or doubted by practitioners if just done numerically. Details of the methodology, the experiment and the implications of the results are discussed in [33].

### *4.4. Other Applications*

The embedded transducers have been applied in two other fields of applications so far. In several experiments we have used them to instrument concrete lab samples (typical size $15 \times 15 \times 40$ cm$^3$). These samples have been put into climate chambers or CDF test devices to evaluate the influence of temperature [16] and moisture on ultrasonic data. This evaluation is required in real world monitoring application as the influence of loads or deterioration is often covered by environmental influences. Instrumented lab samples may also be used in CDF test for freeze thaw resistance evaluations. The measurement of ultrasonic pulse velocity is already part of these evaluations, but the samples have to be removed from the CDF test chambers so far. Embedded transducers would make these tests much easier.

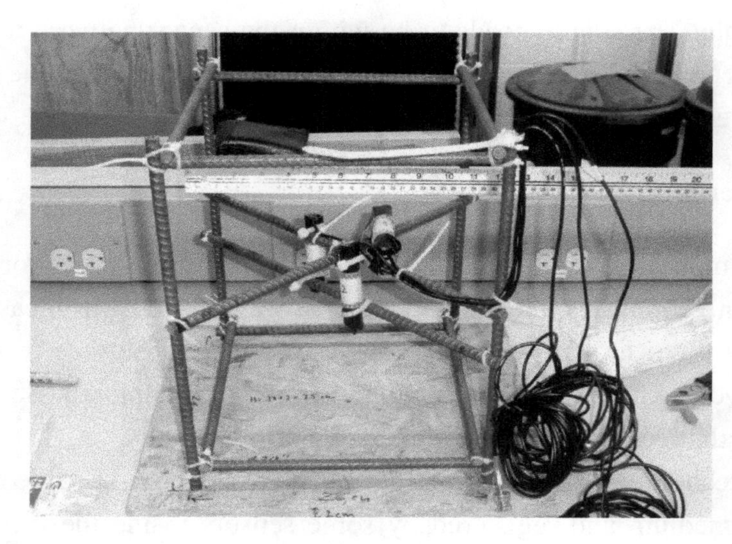

**Figure 13.** Setup of the three embedded transducers on the time reversal model experiment before concreting of the model, with permission from [33].

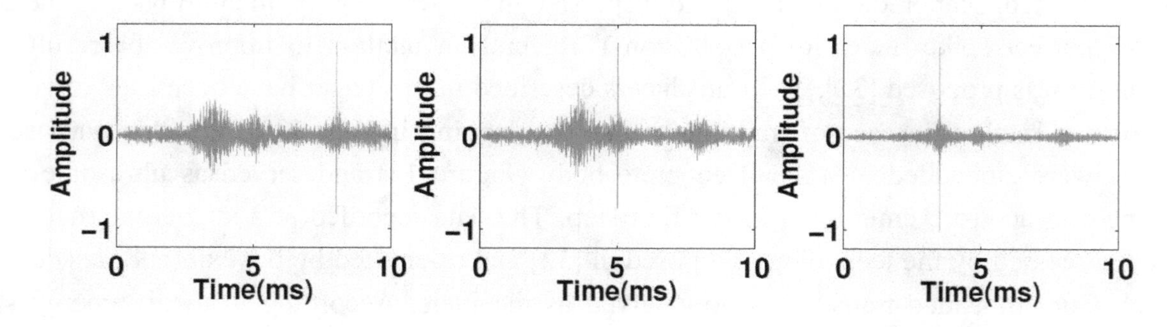


**Figure 14.** Backpropagated signals recorded at the three embedded transducers. Note the sharp peaks at specific times for each transducers reflecting perfectly the original time and place of the acoustic events, with permission from [33].

We have started to use the embedded transducers in full scale monitoring applications as well. An existing, degraded bridge in Turkey is monitored since more than two years by a set of eight transducers. A new bridge in Poland has been equipped with eight transducers before concreting. The hardening of the concrete could be monitored during the first 28 days. Load experiments (a requirement in Poland) will follow.

## 5. Conclusions

The ultrasonic transducers developed for embedment in concrete have shown to be valuable tools for various tasks in structural monitoring. They met our expectations in frequency (around 60 kHz), directivity (almost circular around the main axis) and range (at least three meters). The transducers proved to be very robust. We have developed deployment systems for existing and newly built structures. Early versions are now embedded and used in lab samples and real structures for a few years and are still fully operational.

We have shown that the transducers are useful in a lot of applications. They can be used for active transmission experiments as well as for collecting passive acoustic emission data. They proved to be useful in lab samples, scale experiments and real world monitoring systems. Load changes can be detected and localized as well as environmental influences (temperature, moisture) and various degradation mechanisms. In connection with novel interpretation tools as coda wave interferometry we have very sensible methods for the detection of changes in concrete at hand.

The embedment in concrete has various advantages: The coupling to the concrete is constant, transducers at lab samples have not to be removed before putting them into climate chambers, chemical baths or similar and installations are more vandal proof and less accident prone on real constructions.

## Acknowledgments

The authors wish to acknowledge the help of various colleagues at BAM, the excellent cooperation with the manufacturer of the transducers (Acsys Ltd., Moscow, Russia) as well as the fruitful discussions on coda waves with Sens-Schönfelder (Helmholtz Geoforschungszentrum Potsdam) and Roel Snieder (Colorado School of Mines).

## Author Contributions

Ernst Niederleithinger wrote the main part of paper and worked on the load experiment as well as the time reversal application example.

Julia Wolf performed the experiments on transducer characterization as well as the AE application examples. She designed most of the figures and contributed in writing.

Frank Mielentz is the supervisor of Julia Wolf and contributed greatly to all ultrasonic experiments by designing and evaluating the electronic setups and some of the experiments for transducer characterization.

Herbert Wiggenhauser gave the original idea for the design of the transducers and worked with the manufacturer on the implementation. He also contributed to the experimental design and to data evaluation.

Stephan Pirskawetz performed the experiments and evaluated the data for the AE application example.

## References

1. Karbhari, V.M.; Ansari, M. *Structural Health Monitoring of Civil Infrastructure Systems*; Woodhead Publishing: Cambridge, UK, 2009, ISBN 978-1-84569-392-3.

2. Krause, M.; Mayer, K.; Friese, M.; Milmann, B.; Mielentz, F.; Ballier, G. Progress in ultrasonic tendon duct imaging. *Eur. J. Environ. Civil Eng.* **2011**, *15*, 461–485.
3. Song, G.; Gu, H.; Mo, Y.-L. Smart aggregates: Multi-functional sensors for concrete structures—A tutorial and a review. *Smart Mater. Struct.* **2008**, *17*, doi:10.1088/0964-1726/17/3/033001.
4. Kee, S.H.; Zhu, J. Using piezoelectric sensors for ultrasonic pulse velocity measurements in concrete. *Smart Mater. Struct.* **2013**, *22*, doi:10.1088/0964-1726/22/11/115016.
5. Niederleithinger, E.; Krompholz, R.; Müller, S.; Lautenschläger, R.; Kittler, J. 36 Jahre Talsperre Eibenstock—36 Jahre Überwachung des Eibenstock—36 Years Monitoring the concrete condition by Ultrasound, in German). In Proceedings of 38. Dresdner Wasserbaukolloquium, Germany, 2015. Available online: http://www.researchgate.net/publication/274708685_36_Jahre_Talsperre_Eibenstock__36_Jahre_berwachung_des_Betonzustands_durch_Ultraschall (accessed on 24 April 2015).
6. Wolf, J.; Mielentz, F.; Milmann, B.; Helmerich, R.; Köpp, C.; Wiggenhauser, H. Ultrasound based monitoring system for concrete monolithic objects. In Procedings of the 6th International Conference on Structural Health Monitoring of Intelligent Infrastructure (SHMII-6), Hong Kong, China, 9 December 2013.
7. Wolf, J.; Niederleithinger, E.; Mielentz, F.; Grothe, S; Wiggenhauser, H. Überwachung von Betonkonstruktionen mit eingebetteten Ultraschallsensoren. *Bautechnik* **2014**, *91*, doi:10.1002/bate.201400073.
8. Jones, R.; Facaoaru, I. Recommendations for testing concrete by the ultrasonic pulse method. *Mater. Constr.* **1969**, *2*, 275–284.
9. Crawford, G.I. Guide to Nondestructive Testing of Concrete. *Technical Report FHWA-SA-97-105*; Federal Highway Administration: Washington, DC, USA, 1997.
10. British Standards Institute (BSI). Testing Concrete in Structures. Determination of Ultrasonic Pulse Velocity. BS EN 12504-4:2004 (equivalent German version: DIN EN 12504–4:2004). Available online: http://shop.bsigroup.com/ProductDetail/?pid=000000000030102823 (accessed on 24 April 2015).
11. Sayers, C.M. Stress-induced ultrasonic wave velocity anisotropy in fractured rock. *Ultrasonics* **1988**, *26*, 311–317.
12. Shokouhi, P.; Zoega, A. Surface Wave Velocity-Stress Relationship in Uniaxially Loaded Concrete. *ACI Mater. J.* **2012**, *109*, 141–148.
13. Suaris, W.; Fernando, V. Ultrasonic Pulse Attenuation as a Measure of Damage Growth during Cyclic Loading of Concrete. *ACI Mater. J.* **1987**, *84*, 185–193.
14. Wu, T.T. The Stress Effect on the Ultrasonic Velocity Variations of Concrete under Repeated Loading. *ACI Mater. J.* **1998**, *95*, 519–524.
15. Zhang, Y.; Abraham, O.; Grondin, F.; Loukili, A.; Tournat, V.; Duff, A.-L.; Lascoup, B.; Durand, O. Study of stress-induced velocity variation in concrete under direct tensile force and monitoring of the damage level by using thermally-compensated Coda Wave Interferometry. *Ultrasonics* **2012**, *52*, 1038–1045.
16. Niederleithinger, E.; Wunderlich, C. Influence of small temperature variations on the ultrasonic velocity in concrete. *Rev. Prog. Quant. Nondestruct. Eval.* **2013**, *1511*, 390–397.

17. Zhang, Y.; Abraham, O.; Tournat, V.; Duff, A.L.; Lascoup, B.; Loukili, A.; Grondin, F.; Durand, O. Validation of a thermal bias control technique for Coda Wave Interferometry (CWI). *Ultrasonics* **2013**, *53*, 658–664.
18. Ohdaira, E.; Masuzawa, N. Water content and its effect on ultrasound propagation in concrete—The possibility of NDE. *Ultrasonics* **2000**, *38*, 546–552.
19. Hedenblad, G. *Moisture Permeability of Some Porous Materials*; Report TVBM (Intern 7000-Rapport); Byggnadstysik, DTH, Lyngby, September 1993. Available online: http://lup.lub.lu.se/luur/download?func=downloadFile&recordOId=1653173&fileOId=1653174 (accessed on 24 April 2015).
20. Lencis, U. Moisture Effect on the Ultrasonic Pulse Velocity in Concrete Cured under Normal Conditions and at Elevated Temperature. *Constr. Sci.* **2013**, *14*, 71–78.
21. Payan, C.; Quiviger, A.; Garnier, J.F. Applying diffuse ultrasound under dynamic loading to improve closed crack characterization in concrete. *J. Acoust. Soc. Am.* **2013**, *134*, doi:10.1121/1.4813847.
22. Ramamoorthy, S.K.; Kane, Y.; Turner, J.A. Ultrasound diffusion for crack depth determination in concrete. *J. Acoust. Soc. Am.* **2004**, *115*, 523–529.
23. Payan, C.; Garnier, V. Determination of third order elastic constants in a complex solid applying coda wave interferometry. *Appl. Phys. Lett.* **2009**, *94*, doi:10.1063/1.3064129.
24. ASTM. Standard Test Method for Pulse Velocity through Concrete. ASTM C597-09. Available online: http://www.astm.org/Standards/C597.htm (accessed on 24 April 2015).
25. Tronicke, J. The Influence of High Frequency Uncorrelated Noise on First-Break Arrival Times and Crosshole Traveltime Tomography. *J. Environ. Eng. Geophys.* **2007**, *12*, 173–184.
26. Niederleithinger, E.; Sens-Schönfelder, C.; Grothe, S.; Wiggenhauser, H. Coda Wave Interferometry Used to Localize Compressional Load Effects on a Concrete Specimen. In Proceedings of the 7 th European Workshop on Structiral Health Monitoring (EWSHM 2014), Nantes, France, 8–11 July 2014.
27. Larose, E.; Hall, S. Monitoring stress related velocity variation in concrete with a $2 \times 10^{-5}$ relative resolution using diffuse ultrasound. *J. Acoust. Soc. Am.* **2009**, *125*, 1853–1856.
28. Planes, T.; Larose, E.; Margerin, L.; Rossetto, L.; Sens-Schönfelder, C. Decorrelation and phaseshift of coda waves induced by local changes: Multiple scattering approach and numerical validation. *Waves Random Complex Media* **2014**, *24*, 99–125.
29. Kanu, C. Time-Lapse Monitoring of Localized Changes within Heterogeneous Media with Scattered Waves. Ph.D. Thesis, Colorado School of Mines, Golden, CO, USA, 5 December 2014.
30. Huang, M.; Jiang, L.; Liaw, P.K.; Brooks, C.R.; Seelev, R.; Klarstrom, D.L. Using Acoustic Emission in Fatigue and Fracture Materials Research. *Memb. J. Miner. Metals Mater. Soc.* **1998**, *50*. Available online: http://www.tms.org/pubs/journals/JOM/9811/Huang/Huang-9811.html (accessed on 24 April 2015).
31. Große, C.U.; Schumacher, T. Anwendungen der Schallemissionsanalyse an Betonbauwerken. *Bautechnik* **2013**, *90*, 721–731.

32. Anderson, B.E.; Douma, J.; Ulrich, T.J.; Snieder, R. Improving Spatio-Temporal Focusing and Source Reconstruction through Deconvolution. *Wave Motion* **2015**, *52*, 151–159.
33. Douma, J.; Niederleithinger, E.; Snieder, R. Locating Events Using Time Reversal and Deconvolution: Experimental Application and Analysis. *J. Nondestruct. Eval.* **2015**, in press.

# 11

# Dispersion of Functionalized Silica Micro- and Nanoparticles into Poly(nonamethylene Azelate) by Ultrasonic Micro-Molding

**Angélica Díaz, María T. Casas and Jordi Puiggalí ***

Chemical Engineering Department, Polytechnic University of Catalonia, Av. Diagonal 647, Barcelona E-08028, Spain; E-Mails: angelicadiaz07@hotmail.com (A.D.); m.teresa.casas@upc.edu (M.T.C)

* Author to whom correspondence should be addressed; E-Mail: Jordi.Puiggali@upc.edu

Academic Editors: Dimitrios G. Aggelis and Nathalie Godin

---

**Abstract:** Ultrasound micro-molding technology has proved useful in processing biodegradable polymers with minimum material loss. This makes this technology particularly suitable for the production of biomedical microdevices. The use of silica ($SiO_2$) nanoparticles is also interesting because of advantages like low cost and enhancement of final properties. Evaluation of the capacity to create a homogeneous dispersion of particles is crucial. Specifically, this feature was explored taking into account micro- and nano-sized silica particles and a biodegradable polyester derived from 1,9-nonanodiol and azelaic acid as a matrix. Results demonstrated that composites could be obtained with up to 6 wt. % of silica and that no degradation occurred even if particles were functionalized with a compatibilizer like (3-aminopropyl) triethoxysilane. Incorporation of nanoparticles should have a great influence on properties. Specifically, the effect on crystallization was evaluated by calorimetric and optical microscopy analyses. The overall crystallization rate was enhanced upon addition of functionalized silica nanospheres, even at the low percentage of 3 wt. %. This increase was mainly due to the ability of nanoparticles to act as heterogeneous nuclei during crystallization. However, the enhancement of the secondary nucleation process also played a significant role, as demonstrated by Lauritzen and Hoffmann analysis.

**Keywords:** ultrasound micro-molding technology; functionalized silica nanoparticles; nanocomposites; poly(alkylene dicarboxylate); crystallization kinetics

---

## 1. Introduction

Ultrasonic waves are an energy source that has been employed as a plastic welding procedure for over 40 years because it is clean, efficient, and fast. In fact, absorption of vibration energy leads to polymer friction, which may result in local melting of the sample. The first descriptions of the potential use of ultrasonic waves in plastic powder molding were given by Fairbanks [1] and Crawford *et al.* [2]. Since then different efforts have been made to develop the great potential of this technology. Thus, ultrasonic hot embossing appears as a new, interesting process for fast and low-cost production of microsystems from polymeric materials [3–5]. A wide variety of microdevices (e.g., micro mixers, flow sensors or micro whistles) have been prepared by this technology demonstrating its feasibility.

The use of an ultrasonic source has also been proposed to achieve both the plasticization of the material and the direct injection of the molten material [6–8]. This provides a new micro-molding technology that can be a serious alternative to conventional micro-injection techniques. Micro-molding equipment (Figure 1) is mainly composed of a plasticizing chamber, a controller, a mold, an ultrasonic generator that produces high frequency (30 kHz) from line voltage, and a resonance stack or acoustic unit connected to the generator. This unit consists of: (a) a converter (piezoelectric transducer), where high frequency signals from the generator are transmitted through piezoelectric crystals that expand and contract at the same rate as electrical oscillation; (b) a booster, which amplifies or reduces mechanical oscillation (from 0 to 137.5 μm); and (c) a sonotrode, which transfers this oscillation to the polymer sample in the plasticizing chamber by applying a force (from 100 to 500 N).

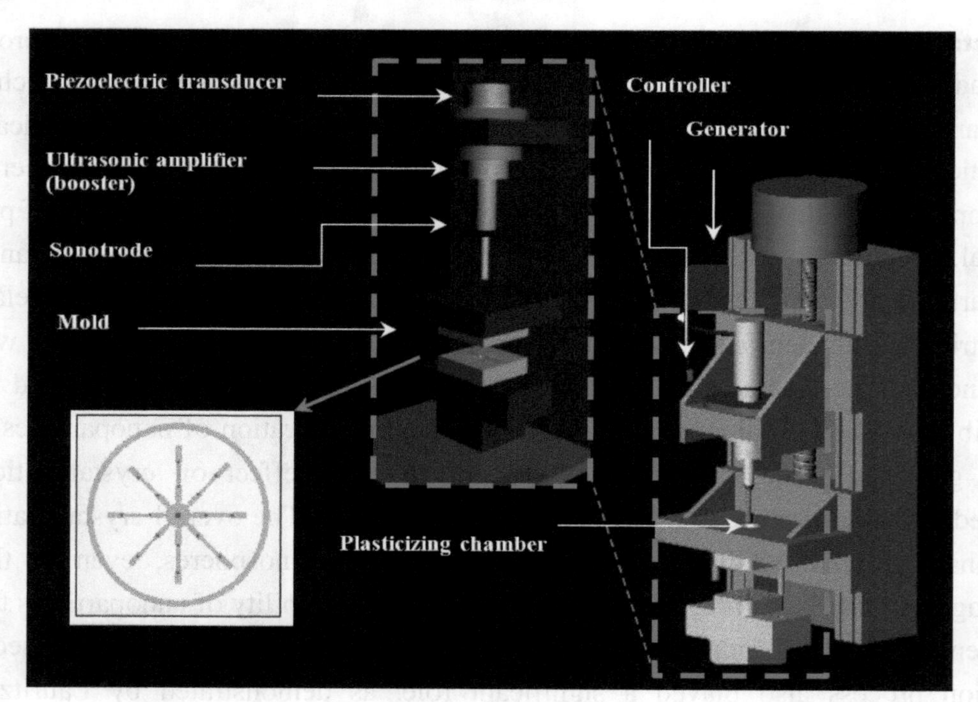


**Figure 1.** Scheme of the main parts of the ultrasound micro-molding machine.

The heat build-up caused by the resulting friction heat melts the polymer, which flows into the mold cavities through their feeding channels under the sonotrode pressure. A controller regulates the main processing parameters: oscillation time (e.g., from 0.5 to 10 s), molding force, and amplitude of the ultrasonic wave.

Despite the potential advantages of using high-intensity ultrasonic waves as an energy source, it should be pointed out that these waves have not only physical effects on the melt rheology of the polymer but also chemical effects on the polymer chain as a result of cavitation and the high temperature that can be achieved inside the plasticizing chamber [9–12]. Therefore, optimization of processing parameters to obtain minimally degraded samples is an essential step in applying the new micro-molding technology. In this sense, we have recently found that it is possible to establish appropriate conditions for biodegradable polymers that could be interesting for the production of biomedical devices such as polylactide (PLA) [13], poly(butylene succinate) (PBS) [14], and poly(nonamethylene azelate) (PE99) [15]. Degradation logically increases with irradiation time and amplitude, but a minimum period is necessary to ensure complete injection into the mold, as well as a minimum amount of vibration energy. Severe degradation can also occur at low ultrasonic amplitude upon application of a high molding force because of chain scissions caused by mechanical shear stress. Nevertheless, a minimum force must be applied to make melt polymer flow and mold cavity filling feasible.

The ability to provide good dispersion is another important reason to evaluate the potential of micro-molding technology. One possible application could involve the incorporation of compounds with added value (e.g., drugs with pharmacological activity) or even of nanoparticles that lead to a large surface area to volume ratio. It is well known that this feature may considerably improve the properties of the material (e.g., physical, chemical, and mechanical) with respect to the properties attained with conventionally filled composites [16–18].

Ultrasound micro-molding was recently found to be effective in obtaining nanocomposites with the final form required for a selected application, a homogeneous clay distribution up to a load of 6 wt. % and, more interestingly, an exfoliated structure without the need for a compatibilizer between the organic polymer and the inorganic silicate clay [17,18]. Furthermore, polymer degradation was minimized by adding pristine clay. These promising results were achieved using PLA, PBS, and PE99 as biodegradable matrices and montmorillonites like neat N757 and C20A, C25A, and N848 organo-modified clays. Incorporation of nanoparticles also had a remarkable influence on the crystallization kinetics of the polymer matrix. Interestingly, significant differences were found depending on the polymer type (e.g., PLA or PE99) despite obtaining similar exfoliated structures [18].

The use of silica ($SiO_2$) nanoparticles to enhance material properties is also of high interest due to their intrinsic advantages (low cost, nontoxicity, high modulus, and ability to modify chemical surface characteristics) [19,20]. Thus, properties can be considerably improved when particles with appropriate surface functionalization to enhance interfacial interactions are well dispersed in the polymer matrix [21–23].

Silica particles with sizes suitable for biomedical applications can be easily prepared from base-catalyzed sol-gel processes [24]. They use organosilane precursors (e.g., tetraethoxysilane), which lead to the formation of a new phase (sol) by hydrolysis and condensation reactions. Subsequently, the condensation of colloidal particles leads to the gel phase. Silica particles can be easily functionalized, giving rise to a high versatility. In fact, functional groups can be attached through covalent bonds during condensation or even after a later grafting process [25].

The present work evaluates the applicability of ultrasound micro-molding technology to prepare dispersion of silica micro- and nanoparticles in a biodegradable polymer matrix that was recently considered for preparation of clay nanocomposites (*i.e.*, PE99). The effect of silica nanoparticle addition on crystallization kinetics is also studied since they should affect nucleation and crystal growth, as recently determined for nanocomposites based on poly(ethylene oxide) [26].

## 2. Experimental Section

### *2.1. Materials*

Poly(nonamethylene azelate) (PE99) was synthesized by thermal polycondensation of azelaic acid with an excess of 1,9-nonanediol (2.2:1 molar ratio), as shown in Figure 2a. Titanium tetrabutoxyde was used as a catalyst and the reaction was first performed in a nitrogen atmosphere at 150 °C for 6 h and then in a vacuum at 180 °C for 18 h. The polymer was purified by precipitation with ethanol of a chloroform solution (10 wt. %). The average molecular weight and polydispersity index determined by GPC and using poly(methyl methacrylate) standards were 13,200 g/mol and 3.1, respectively.

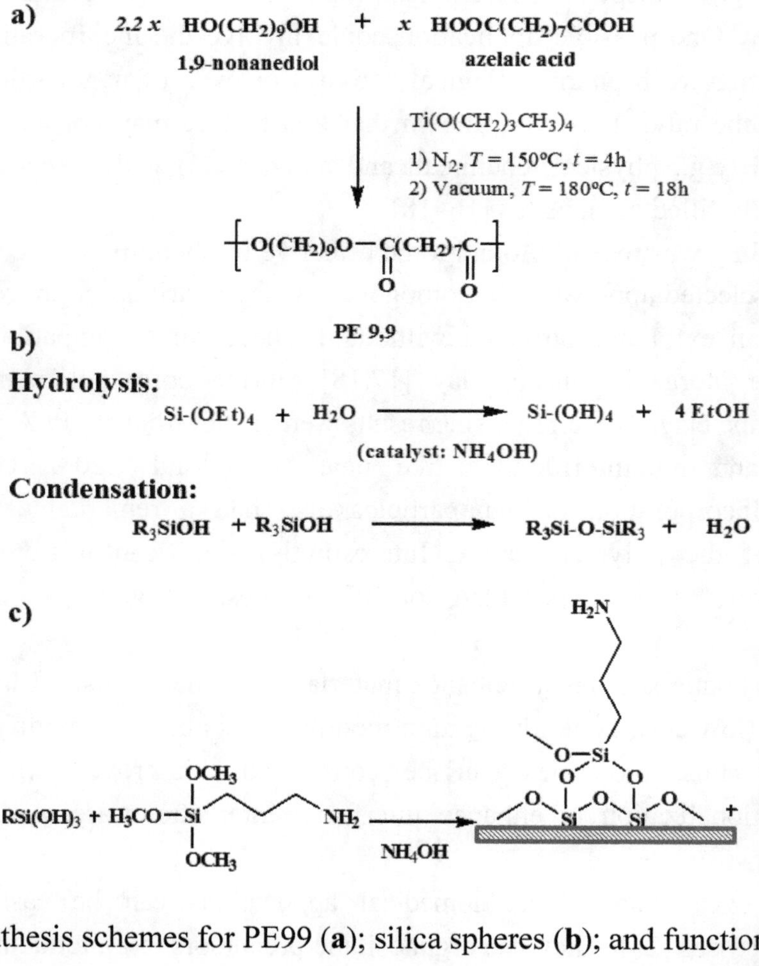

**a)**

$$2.2x\ \mathrm{HO(CH_2)_9OH} + x\ \mathrm{HOOC(CH_2)_7COOH}$$

1,9-nonanediol azelaic acid

$\mathrm{Ti(O(CH_2)_3CH_3)_4}$
1) $N_2$, $T = 150$ °C, $t = 4$h
2) Vacuum, $T = 180$ °C, $t = 18$h

$$\left[\mathrm{O(CH_2)_9O{-}C(=O)(CH_2)_7C(=O)}\right]$$

PE 9,9

**b)**

**Hydrolysis:**

$$\mathrm{Si(OEt)_4 + H_2O \longrightarrow Si(OH)_4 + 4\,EtOH}$$

(catalyst: $NH_4OH$)

**Condensation:**

$$\mathrm{R_3SiOH + R_3SiOH \longrightarrow R_3Si{-}O{-}SiR_3 + H_2O}$$

**c)**

$$\mathrm{RSi(OH)_3 + H_3CO{-}Si(OCH_3)_2{-}(CH_2)_3NH_2 \xrightarrow{NH_4OH} (\text{surface-bound aminopropyl silane}) + 3\,CH_3OH}$$

**Figure 2.** Synthesis schemes for PE99 (**a**); silica spheres (**b**); and functionalization process (**c**).

Preparation of silica micro/nanoparticles (Figure 2b): Deionized water (0.6 mL) and 1.2 mL of a 28% aqueous ammonium hydroxide (1.2 mL) (Aldrich, Madrid, Spain) were added to a flask of 100 mL containing absolute ethanol (50 mL). The mixture was stirred vigorously for 20 min at room temperature using a magnetic agitator. Next 1.2 mL of tetraethoxysilane (Aldrich, Madrid, Spain) were quickly added and the resultant solution was purged with dry nitrogen and stirred for another 16 h. Spheres with a homogeneous diameter close to 600 nm were obtained after filtration. The same protocol except for increasing the stirring speed to 250–1000 rpm was employed to prepare nanoparticles with diameters close to 25 nm.

Functionalization of silica micro/nanoparticles (Figure 2c): 1 g of micro/nanospheres previously dried under a vacuum was introduced into a vessel together with 66 μL of pure ethanol. The mixture was sonicated for 15 min to enhance dispersion using a VWR Ultrasonic cleaner bath (UWR International, New York, NY, USA) at 100 watts, and subsequently 66 μL of ammonium hydroxide was added under stirring for 15 min. Finally, 297 μL of (3-aminopropyl) triethoxysilane (AMPS) was quickly added and the reaction was allowed to progress under stirring for 24 h at room temperature. The resulting solid was centrifuged, repeatedly washed with ethanol, and vacuum dried.

Nanocomposites will be denoted by polymer abbreviation, size (micro or nano), and wt. % of added silica (e.g., PE99-M 6 and PE99-N 6 indicate composites having 6 wt. % of micro- and nanospheres, respectively).

### *2.2. Micro-Molding Equipment*

A prototype Ultrasound Molding Machine (Sonorus®, Ultrasion S.L., Barcelona, Spain) was employed. The apparatus was equipped with a digital ultrasound generator from Mecasonic (1000 W–30 kHz, Barcelona, Spain) a controller (3010 DG digital system, Mecasonic, Barcelona, Spain), a converter, an acoustic unit, and an electric servomotor control (Berneker and Rainer, Barcelona, Spain) dotted with software from Ultrasion S.L. Mold was thermally controlled and designed to prepare eight test specimens. Dimensions of these specimens were $1.5 \times 0.1 \times 0.1$ $cm^3$ and followed IRAM-IAS-U500-102/3 standards.

### *2.3. Measurements*

Molecular weight was estimated by gel permeation chromatography (GPC) using a liquid chromatograph (Shimadzu, model LC-8A, Tokyo, Japan) equipped with an Empower computer program (Waters, Massachusetts, MA, USA). A PL HFIP gel column (Polymer Lab, Agilent Technologies Deutschland GmbH, Böblingen, Germany) and a refractive index detector (Shimadzu RID-10A) were employed. The polymer was dissolved and eluted in 1,1,1,3,3,3-hexafluoroisopropanol containing $CF_3COONa$ (0.05 M) was employed as solvent and elution medium. Flow rate was 0.5 mL/min, the injected volume 100 μL, and the sample concentration 2 mg/mL. Polymethyl methacrylate standards were employed to determine the number and weight average molecular weights and molar-mass dispersities.

A Focused Ion Beam Zeiss Neon40 microscope (Oberkochen, Germany) operating at 5 kV was employed to get SEM micrographs of micro-molded specimens. Carbon coating was accomplished with a Mitec K950 Sputter Coater (Oberkochem, Germany) was used to coat all samples, which were then viewed at an accelerating voltage of 5 kV.

A FTIR 4100 Jasco spectrophotometer dotted with an attenuated total reflectance accessory (Specac MKII Golden Gate Heated Single Reflection Diamond ATR, Jasco International Co. Ltd., Tokyo,

Japan) and a thermal controller was employed to get the FTIR spectra. Samples were placed in an attenuated total reflectance accessory with thermal control and a diamond crystal.

X-ray photoelectron spectroscopy (XPS) was performed with a SPECS system (Berlin, Germany) dotted with an XR50 source of Mg/Al (1253 eV/1487 eV) operating at 150 W. Analyses were performed in a SPECS system equipped with a high intensity twin-anode X-ray source XR50 of Mg/Al (1253 eV/1487 eV) operating at 150 W. A Phoibos 150 MCD-9 XP detector (Berlin, Germany) was employed. The overview spectra were taken with an X-ray spot size of 650 and pass energy of 25 eV in 0.1 eV steps at a pressure below $6 \times 10^{-9}$ mbar. Surface composition was determined through the N 1s and Si 2p peaks at binding energies of 399.2 and 103.2 eV, respectively.

Distribution of micro/nanoparticles in the composites was evaluated by means of a Philips TECNAI 10 electron microscope (Philips Electron Optics, Eindhoven, Holland) at an accelerating voltage of 80 kV. Samples were prepared by embedding the nanocomposite specimens. A low-viscosity, modified Spurr epoxy resin was employed to embed the specimens before curing and cutting in small sections. A Sorvall Porter-Blum microtome (New York, NY, USA) equipped with a diamond knife was employed in this case. The thin sections were collected in a trough filled with water and lifted onto carbon-coated copper grids.

Thermal degradation was performed in a Q50 thermogravimetric analyzer of TA Instruments (TA Instruments, New Castle, DE, USA). Experiments were carried out at a heating rate of 10 °C/min with 5 mg samples and under a flow of dry nitrogen.

Growth rates of spherulites were measured by optical microscopy using a Zeiss Axioscop 40 Pol light polarizing microscope (Oberkochen, Germany). This was equipped with a Linkam temperature control system configured by a THMS 600 heating and freezing stage connected to an LNP 94 liquid nitrogen system. Spherulites were isothermally crystallized at the selected temperature from homogeneous melt thin films. These were prepared by melting 1 mg of the polymer between microscope slides. Subsequently, small sections of the obtained thin films were pressed between two cover slides for microscopy observations. Samples were kept at the hot stage at 90 °C for 5 min to eliminate sample history effects, and then rapidly cooled to the selected temperatures. A Zeiss AxiosCam MRC5 digital camera (Munich, Germany) was employed to follow the diameter evolution of spherulites by taking micrographs at different times. Nucleation was evaluated by counting the number of active nuclei that appeared in the micrographs. The sign of spherulite birefringence was determined by means of a first-order red tint plate placed under crossed polarizers.

Differential scanning calorimetry was performed with a TA instrument Q100 series (TA Instruments, New Castle, DE, USA) with $T_{zero}$ technology and equipped with a refrigerated cooling system (RCS). Crystallization kinetics experiments were conducted under a flow of dry nitrogen with a sample weight around 5 mg. Calibration was performed with indium. Studies were carried out according the following protocol: Samples were firstly heated (20 °C/min) up to 25 °C above their melting temperature, subsequently held at this temperature for 5 min to eliminate the thermal history, and finally cooled to the selected temperature at a rate of 50 °C/min. Samples were kept at the isothermal temperature until baseline was reached. Finally, a new heating run (20 °C) was performed in order to determine the equilibrium melting temperature of samples.

## 3. Results and Discussion

### *3.1. Characterization of Functionalized Silica Particles*

Silica particles were effectively prepared using tetraethoxysilane as a precursor. TEM micrographs clearly revealed a highly regular spherical form (Figure 3a) for silica particles obtained at a low stirring speed. These particles showed a highly homogeneous diameter that varied in a narrow range (*i.e.*, between 570 and 650 nm). Particles were slightly more irregular in form (Figure 3b) when prepared at a higher stirring speed. Diameter size was considerably reduced but a homogeneous distribution could still be observed (*i.e.*, values were always within the 20–30 nm interval). The two kinds of silica preparations will be designated as micro (M) and nano (N) particles.

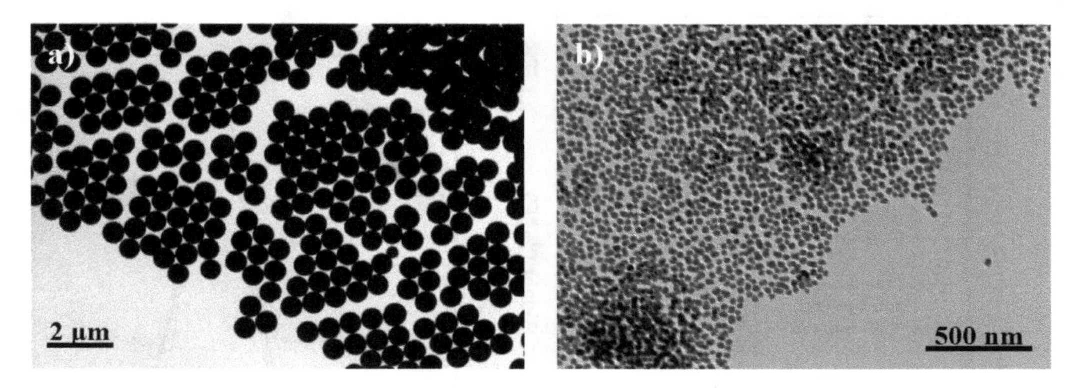


**Figure 3.** TEM micrographs of functionalized microspheres (**a**); and nanospheres (**b**).

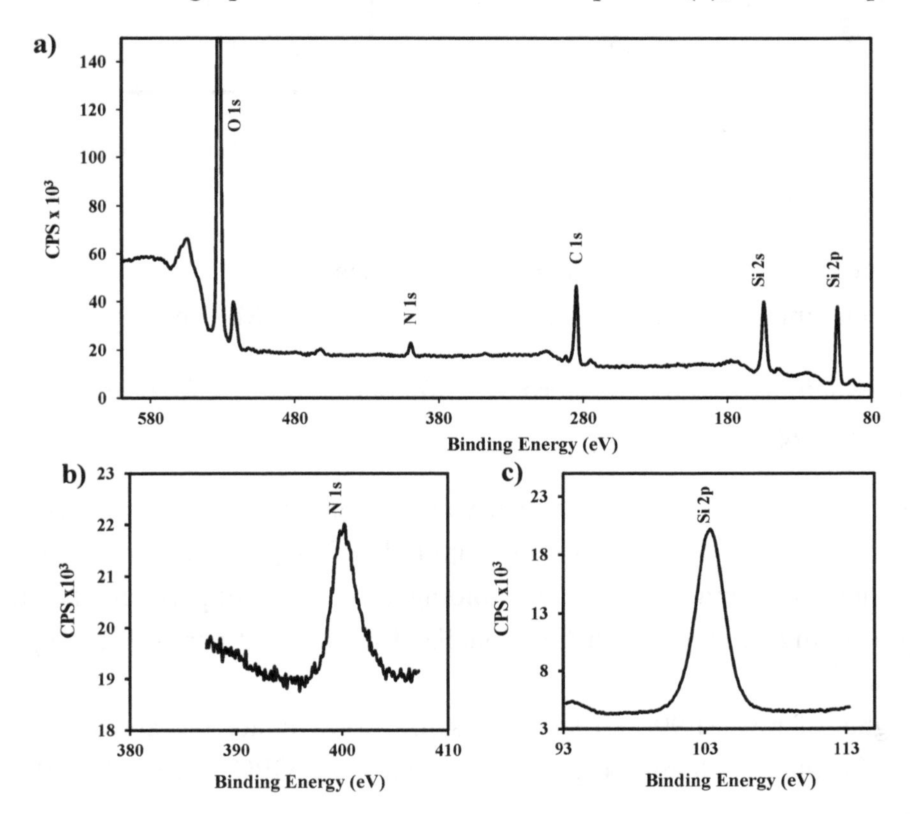


**Figure 4.** Full scale XPS spectra (**a**) and details corresponding to silicon (**b**) and nitrogen (**c**) XPS signals detected in the functionalized silica nanoparticles.

Surface functionalization by reaction with (3-aminopropyl) triethoxysilane was verified by analysis of XPS spectra (Figure 4), which allowed for determining a ratio between N and Si atoms of close to 8% (*i.e.*, 8.8% and 7.3% for micro- and nanoparticles, respectively).

Thermogravimetric analyses (Figure 5) also showed a significant weight loss of functionalized silica particles because of decomposition of grafted AMPS groups, which took place at around 500 °C. This decomposition step corresponded to an approximate weight loss of 9% and, logically, was not detected in non-functionalized particles. Gradual weight loss leading to a value of only 8%–9% at 590 °C was observed for all particles. Thus, at this temperature the total loss was 17% and 8% for functionalized and non-functionalized particles, respectively. Practically no differences were found in the TGA (Figure 5a) and DTGA (Figure 5b) curves of micro- and nanoparticles.

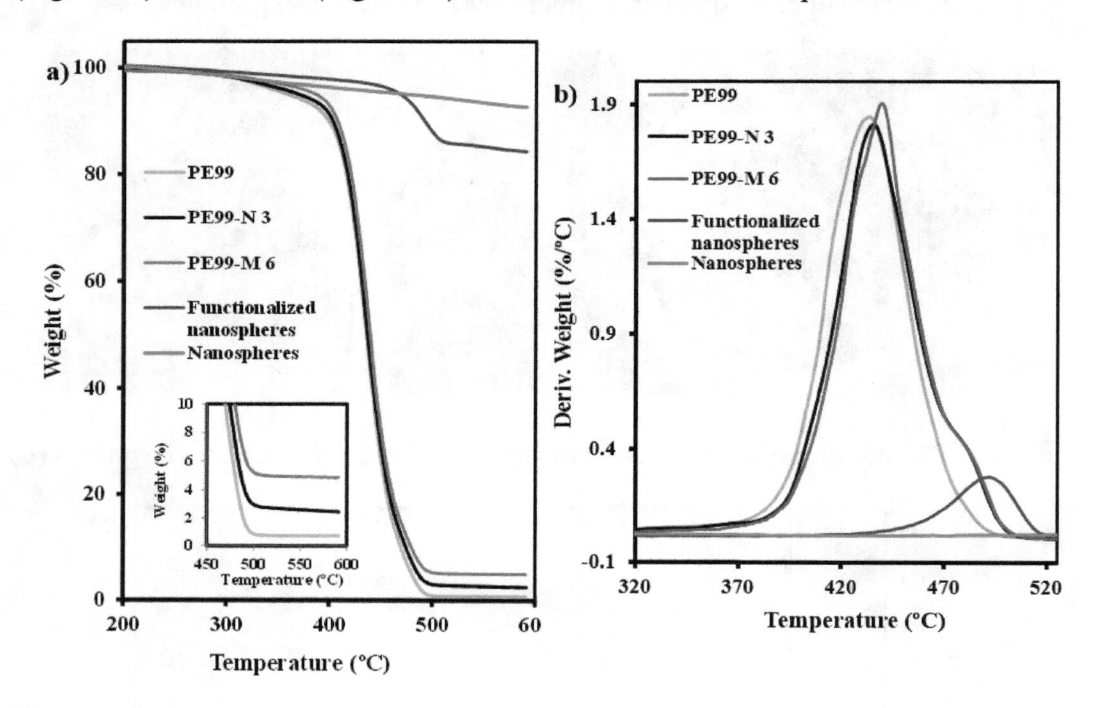


**Figure 5.** TGA (**a**) and DTGA (**b**) curves of silica nanospheres, functionalized silica nanospheres, and micro-molded PE99, PE99-N 3, and PE99-M 6 specimens.

### *3.2. Dispersion of Functionalized Silica Micro- and Nanoparticles by Ultrasound Micro-Molding Technology*

PE99 samples could be processed under relatively mild conditions, as previously established [15]. Thus, a minimum irradiation time of 1.2 s, low amplitude of 24 μm, and a moderate molding force of 300 N were sufficient to guarantee a 100% molding efficiency. Experimental conditions could be maintained for processing mixtures with functionalized micro- and nanospheres up to the maximum test load of 6 wt. %.

Molecular weights of raw PE99 and specimens processed under the above optimized conditions are summarized in Table 1, whereas Figure 6 compares the GPC traces of PE99, PE99-N 3, and PE99-M 6 specimens.

**Table 1.** Molecular weights of processed PE 99 and their mixtures with functionalized micro- and nanoparticles [a].

| Sample | *Mn* (g/mol) | *Mw* (g/mol) | *Mw*/*Mn* |
|---|---|---|---|
| PE99 (raw) | 13,300 | 35,900 | 2.7 |
| PE99 | 14,700 | 37,200 | 2.5 |
| PE99-M 3 | 14,800 | 36,800 | 2.5 |
| PE99-M 6 | 12,600 | 33,800 | 2.7 |
| PE99-N 3 | 15,800 | 39,200 | 2.5 |
| PE99-N 6 | 14,200 | 37,900 | 2.7 |

[a] Micro-molding conditions: Time (s), amplitude (μm) and force (N): 1.2, 10 and 300.

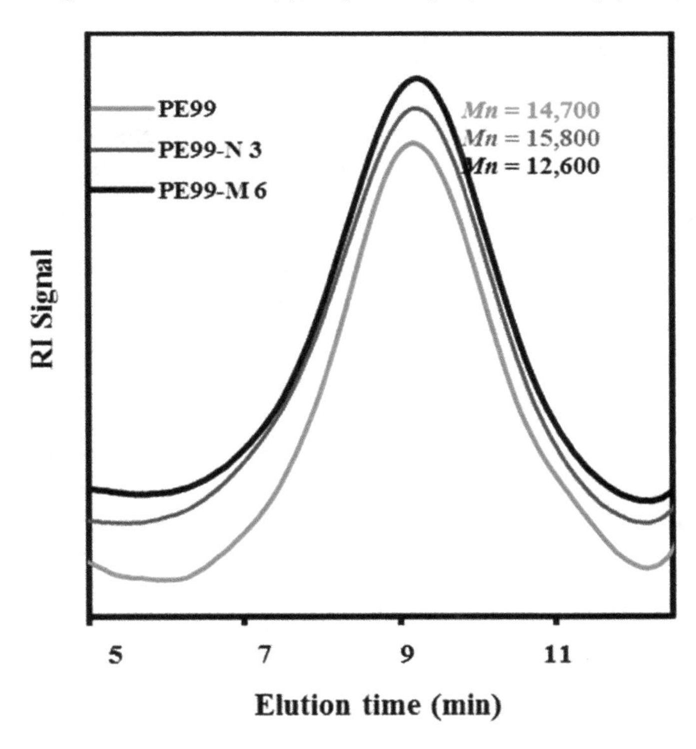


**Figure 6.** GPC molecular weight distribution curves determined for micro-molded PE99, PE-N 3, and PE99-M 6 specimens.

No statistically significant differences were found between samples before and after processing or upon addition of functionalized nanospheres. Only a not highly significant decrease was detected for samples loaded with the maximum percentage (6 wt. %) of microspheres. The new technology appears fully adequate to obtain micropieces with negligible degradation of composites constituted by PE99 and functionalized silica micro/nanoparticles.

Figure 7a,b show optical and SEM images of a representative silica loaded specimen for which a regular texture is detected in the longitudinal section micrographs. All specimens were highly homogeneous, without the presence of cavities that could affect the final properties of the material.

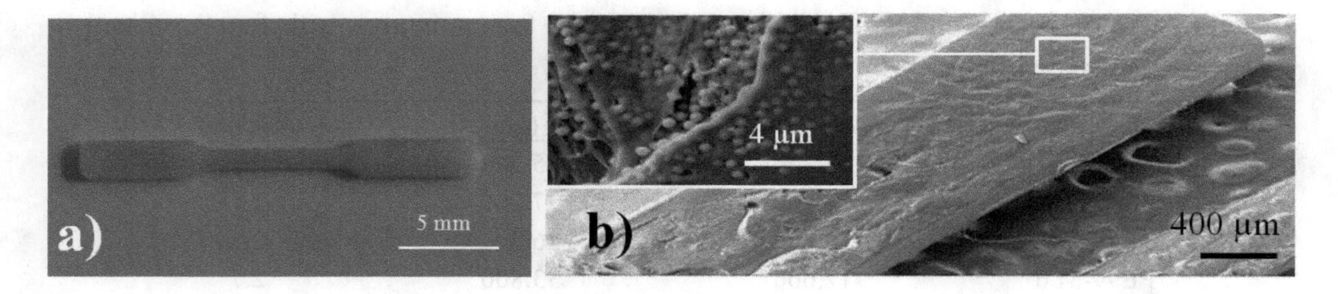


**Figure 7.** Image of a processed PE99 specimen (**a**); SEM micrograph of details of a micro-molded PE99-M 3 specimen (**b**). Microparticles can be observed in the inset.

FTIR spectra were also useful to verify the uniform incorporation of silica particles into micro-molded specimens and to discard concentration in the sprue. Thus, the typical Si–O stretching band at 1075 $cm^{-1}$ was observed in every part of all loaded specimens. Hence, the spectra of specimen zones close to (proximal part) and distant from (distal part) the feeding channel showed a similar ratio between the intensity of the band associated with the C=O stretching band of the polyester and the band associated with the silica particles (see Figure 8 for representative samples containing micro- and nanoparticles).

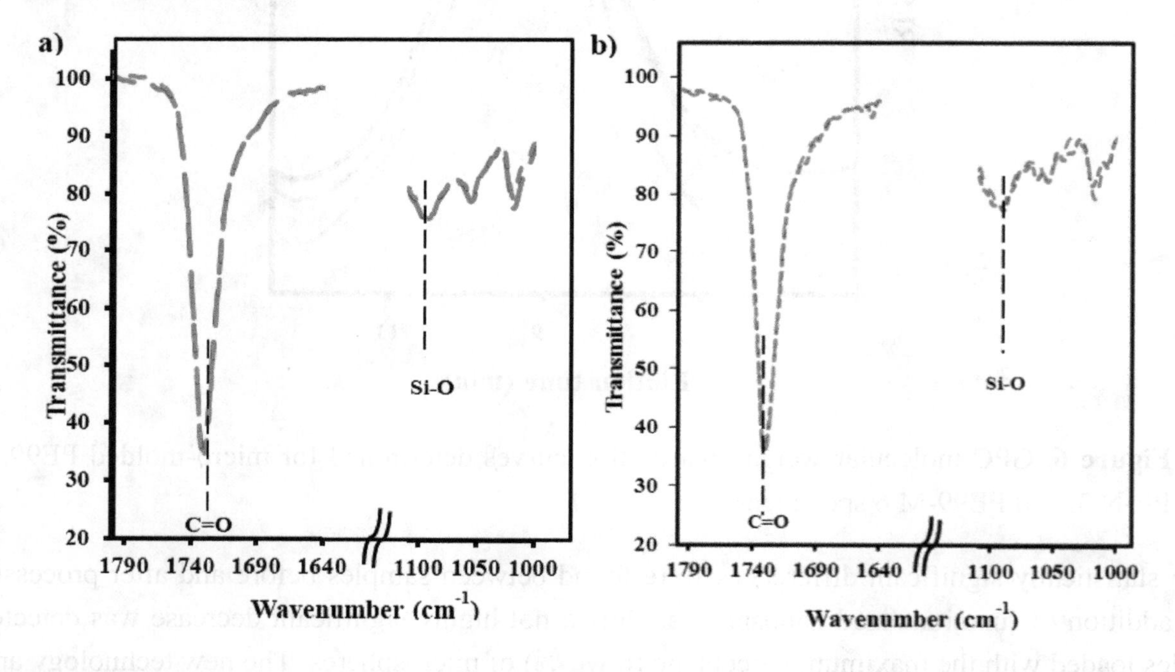


**Figure 8.** FTIR spectra of the characteristic C=O and Si–O stretching bands of micro-molded PE-N 3 (**a**) and PE-M 3 (**b**) specimens. Blue and red lines correspond to the proximal and distal parts of the specimens, respectively.

Analysis of particle dispersion in the processed composites was also carried out by transmission electron microscopy observation of ultrathin sections of micro-molded specimens. Flawless ultrasections could be obtained, as can be seen in the images for representative samples in Figure 9. These micrographs show the distribution of micro- and nanoparticles in the specimen, which is not completely homogeneous since density appears to be locally variable when observed at this scale.

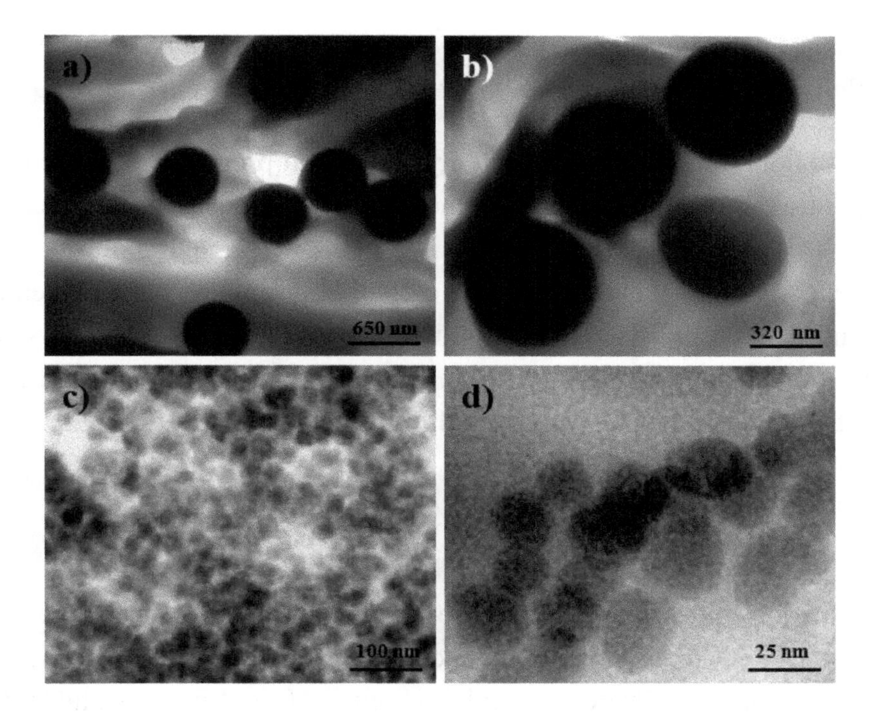


**Figure 9.** TEM micrographs of the dispersion of functionalized silica microspheres (**a**,**b**) and nanospheres (**c**,**d**) in the micro-molded PE99-M 3 and PE99-N 3 specimens.

Thermogravimetric analysis also gave information about the percentage of silica particles incorporated into the polymer matrix and their effect on thermal stability. In all cases, a constant char yield was attained at high temperatures, as shown by representative PE99-N 3 and PE99-M 6 specimens in Figure 5a. Furthermore, the remaining weight percentage was always in full agreement with the expected silica content (*i.e.*, close to 3 and 6 wt. %). The result is meaningful since it has demonstrated again that particles were not generally retained in the sprue. Silica was effectively led through the feeding channels by the molten polymer, giving rise to well-dispersed specimens. A close resemblance between the TGA and DTGA traces of the micro-molded PE99 sample and the silica-loaded composites was also found during the first stages of degradation (probably also as a consequence of the low content of added particles). The added particles even seem to slightly stabilize the sample (Table 2 and Figure 5). In any case, functionalization of silica particles did not have a negative impact on the ultrasound micro-molding process, as was also observed by GPC measurement. Differences were noticeable only at the end of the degradation process; the presence of shoulders around 483 °C in the DTGA plots (Figure 5b) seems to be due to decomposition of grafted AMPS groups.

**Table 2.** Characteristic TGA temperatures and remaining weight percentages for the decomposition of the studied micro-molded specimens.

| Polymer | $T_{onset}$ (°C) | $T_{10\%}$ (°C) | $T_{20\%}$ (°C) | $T_{40\%}$ (°C) | $T_{max}$ (°C) | Remaining Weight (%) |
|---|---|---|---|---|---|---|
| PE 99 | 367 | 397 | 411 | 426 | 433 | 0 |
| PE 99-N 3 | 377 | 405 | 415 | 431 | 436 | 2.7 |
| PE 99-N 6 | 377 | 405 | 415 | 432 | 438 | 5.7 |
| PE 99-M 3 | 377 | 405 | 417 | 431 | 437 | 2.9 |
| PE 99-M 6 | 377 | 404 | 419 | 432 | 439 | 5.8 |

### *3.3. Calorimetric Studies on the Influence of Functionalized Silica Nanoparticles on the Isothermal Crystallization of Poly(nonamethylene Azelate)*

Kinetic analysis was only performed for melt crystallization processes because of the impossibility of obtaining amorphous samples by cooling the melted nanocomposite at the maximum rate allowed by the equipment.

Crystallization experiments were therefore carried out in a narrow temperature interval (*i.e.*, between 56 and 59 °C) due to experimental limitations. Figure 10a shows the crystallization exotherms of the neat polyester and the PE99-N 3 nanocomposite, which allowed for determining the time evolution of the relative degree of crystallinity, $\chi(t)$. The last was calculated according to Equation (1):

$$\chi(t) = \int_{t0}^{t} (\mathrm{d}H/\mathrm{d}t)\mathrm{d}t \Big/ \int_{t0}^{\infty} (\mathrm{d}H/\mathrm{d}t)\mathrm{d}t \quad (1)$$

where $t_0$ is the induction time and d$H$/d$t$ corresponds to the heat flow rate. The evolution of crystallinity always showed a sigmoidal dependence on time for the five melt crystallization experiments performed for the different samples (Figure 10b). Experimental data were analyzed considering the typical Avrami equation [27,28]:

$$1 - \chi(t - t_0) = \exp(-Z\cdot(t - t_0)^n) \quad (2)$$

where $Z$ is a temperature-dependent rate constant and $n$ the Avrami exponent whose value depends on the mechanism and geometry of the crystallization process. A normalized rate constant, $k = Z^{1/n}$, can also be calculated for comparison purposes since corresponding units ($\text{time}^{-1}$) are independent of the specific value of the Avrami exponent.

Plots of $\log(-\ln(1 - \chi(t - t_0)))$ against $\log(t - t_0)$ (Figure 11) allowed for determining the indicated crystallization parameters, which are summarized in Table 3. Avrami exponents for the neat polyester remain in a narrow range (*i.e.*, 2.60–2.94) and have an average value of 2.78. The determined Avrami exponent indicates a predetermined (heterogeneous) nucleation and a spherical growth under geometric constraints. Note that a slight deviation is observed with respect to the theoretical value of 3 and that a value close to 4 should be expected for a sporadic (heterogeneous) and homogeneous nucleations. It should also be pointed out than homogeneous nucleation usually requires high undercooling, which does not correspond with the performed experiments. The exponent slightly decreased for the nanocomposite (*i.e.*, 2.54–2.14, with 2.33 being the average value), suggesting an increase in geometric constraints upon incorporation of the well-dispersed nanospheres. Exponents determined for both samples were found to vary without a well-defined trend within the selected narrow temperature interval.

Reciprocal crystallization half-times ($1/\tau_{1/2}$) are also summarized in Table 3. This parameter is directly determined from DSC isotherms (*i.e.*, it corresponds to the inverse of the difference between thr crystallization start time and half-crystallization time) and can be useful to test the accuracy of the Avrami parameters since an estimated value can be obtained from them (*i.e.*, $1/\tau_{1/2} = (Z/\ln 2)^{1/n}$)).

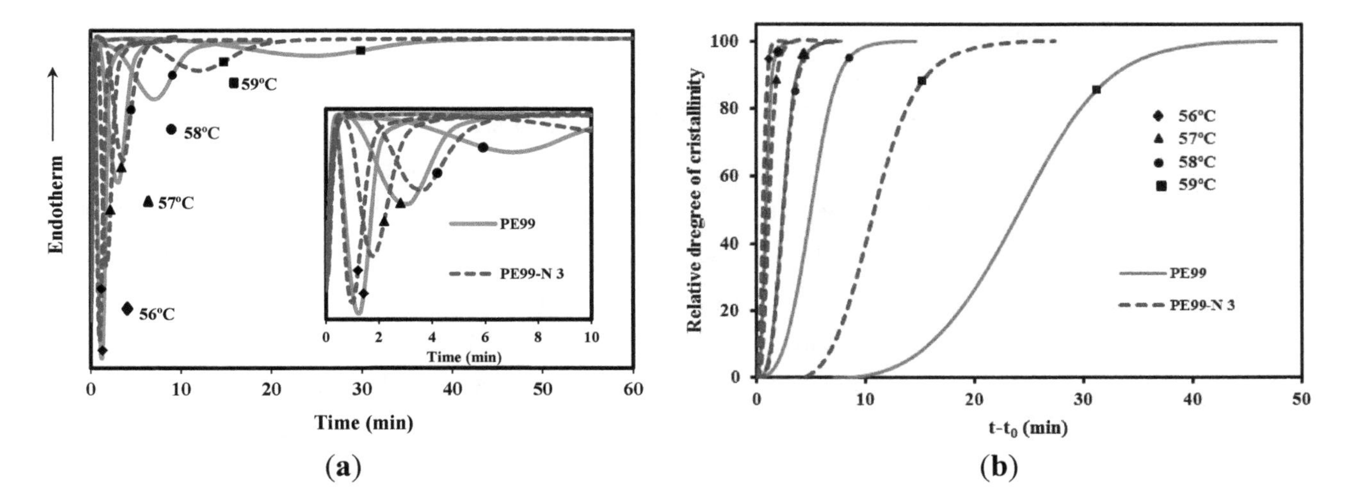


**Figure 10.** (**a**) Exothermic DSC peaks of isothermal crystallizations of PE99 (**garnet**) and PE99-N 3 (**blue**) samples at temperatures between 56 and 59 °C; (**b**) development of the relative degree of crystallinity of PE99 (**garnet**) and PE99-N 3 (**blue**) samples at different crystallization temperatures.

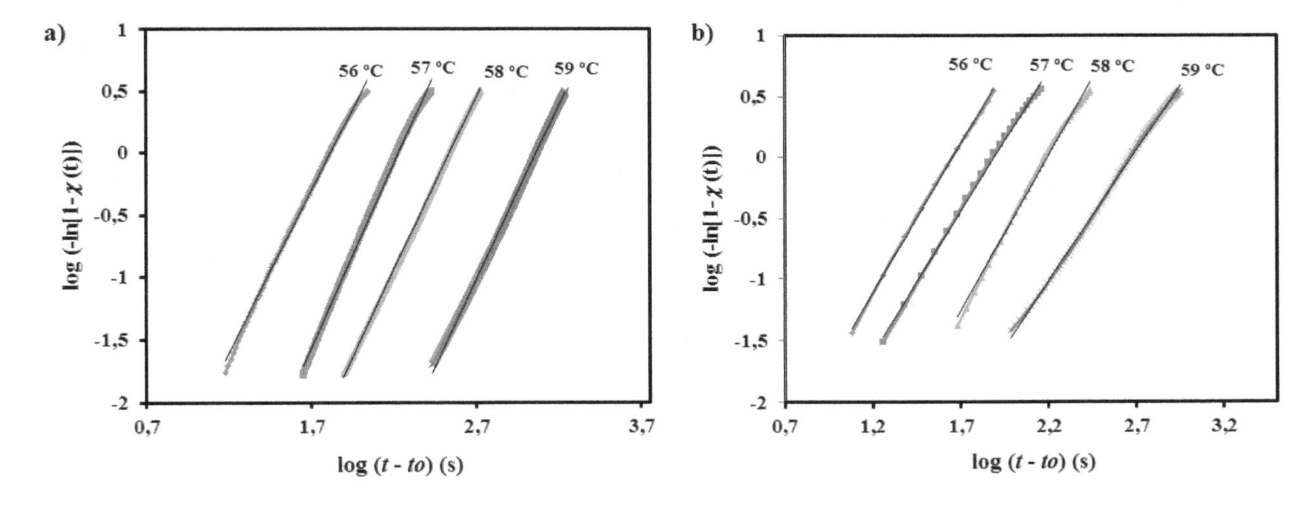


**Figure 11.** Avrami plots of isothermal crystallization of PE99 (**a**) and PE99-N 3 (**b**) at the indicated temperatures.

**Table 3.** Main crystallization kinetic parameters determined by DSC for the neat polyester and its nanocomposite with 3 wt. % of functionalized silica nanospheres.

| Sample | $T$ (°C) | $Z \times 10^6$ (s$^{-n}$) | $n$ | $k \times 10^3$ (s$^{-1}$) | $1/\tau_{1/2} \times 10^3$ (s$^{-1}$) | $(Z/\ln 2)^{1/n} \times 10^3$ (s$^{-1}$) |
|---|---|---|---|---|---|---|
| PE99 | 56 | 18.48 | 2.60 | 15.23 | 17.93 | 17.5 |
| | 57 | 0.28 | 2.94 | 5.89 | 6.82 | 6.67 |
| | 58 | 0.077 | 2.81 | 2.92 | 3.31 | 3.32 |
| | 59 | 0.0028 | 2.79 | 0.86 | 0.97 | 0.99 |
| PE99-N 3 | 56 | 98.56 | 2.42 | 21.98 | 25.97 | 25.58 |
| | 57 | 42.03 | 2.31 | 12.84 | 15.12 | 15.05 |
| | 58 | 2.69 | 2.54 | 6.37 | 7.62 | 7.36 |
| | 59 | 1.89 | 2.14 | 2.12 | 2.54 | 2.52 |

Variation of the overall rate constant with temperature for the neat polymer and its nanocomposites was also evaluated (Table 3). The rate for the nanocomposite increased (*i.e.*, from $2.12 \times 10^{-3}$ $s^{-1}$ to $21.98 \times 10^{-3}$ $s^{-1}$) when crystallization temperature decreased (*i.e.*, from 59 °C to 56 °C), a trend that was also logically observed for PE99. More interestingly, the nanocomposite showed a remarkably higher crystallization rate than the neat polyester at all test crystallization temperatures (e.g., $21.98 \times 10^{-3}$ $s^{-1}$ and $15.23 \times 10^{-3}$ $s^{-1}$ were determined at 56 °C for PE99-N 3 and PE99, respectively). Therefore, incorporation of functionalized silica nanoparticles had a significant influence on the crystallization process and, logically, on the final material properties. It should be pointed out that the ratio between the two kinetic constants decreased (*i.e.*, between 2.5 and 1.4) as crystallization temperature decreased, which justifies further studies on nucleation and crystal growth processes.

### *3.4. Optical Microscopy Studies on the Influence of Functionalized Silica Nanoparticles on the Isothermal Crystallization of Poly(nonamethylene Azelate)*

Crystallization kinetics from the melt state was also studied for micro-molded samples with and without functionalized silica nanospheres by optical microscopy. Spherulite radii grew linearly with time until impingement in both cases, as shown in Figure 12. Crystal growth rates were clearly higher for the nanocomposite at high crystallization temperatures, whereas differences were minimal when this temperature decreased. The relatively high growth rate allowed for collecting experimental data only over a narrow temperature range where crystallization was mainly governed by secondary nucleation (*i.e.*, the typical bell curve of crystal growth rate *versus* temperature could not be obtained). The increase in the crystal growth rate for the nanocomposite is peculiar and suggests favored deposition of molecules onto existing crystal surfaces, as will be discussed. At lower temperatures, this effect seems to be counterbalanced by reduced chain mobility in the presence of silica nanoparticles and spatial constraints imposed by confinement [29,30].

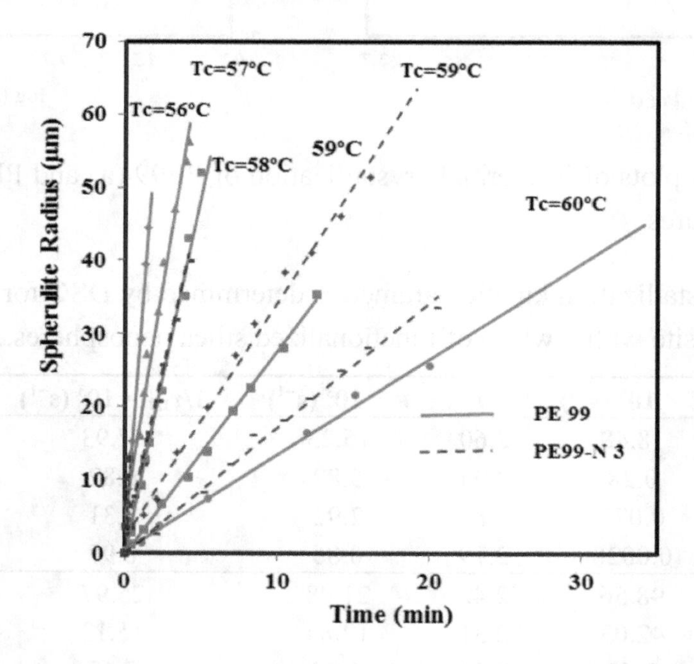


**Figure 12.** Variation of PE99 (**garnet**) and PE99-N 3 (**blue**) spherulite radii with time for isothermal crystallizations performed at the indicated temperatures.

Pristine polyester and nanocomposite spherulites with similar morphological features were formed (Figure 13). Thus, both samples crystallized from the melt into ringed spherulites with negative birefringence. The spacing between rings decreased significantly with decreasing crystallization temperatures (*i.e.*, 13 μm at 60 °C and 5 μm at 56 °C) and decreased slightly upon incorporation of nanospheres (*i.e.*, 5 μm as opposed to 4 μm for crystallizations at 56 °C). A more confusing texture of less defined rings was detected at an intermediate temperature (*i.e.*, 59 °C) for both samples. Note that pristine polyester and nanocomposite spherulites fill the field of view and have a relatively uniform size, suggesting athermal nucleation (*i.e.*, the number of nuclei remains constant during crystallization). More interestingly, this result indicates good dispersion of silica nanoparticles in the polyester matrix. In fact, adsorption of PE99 molecules onto the functionalized surface of silica nanoparticles may hinder particle-particle agglomeration and enhance colloidal stability.

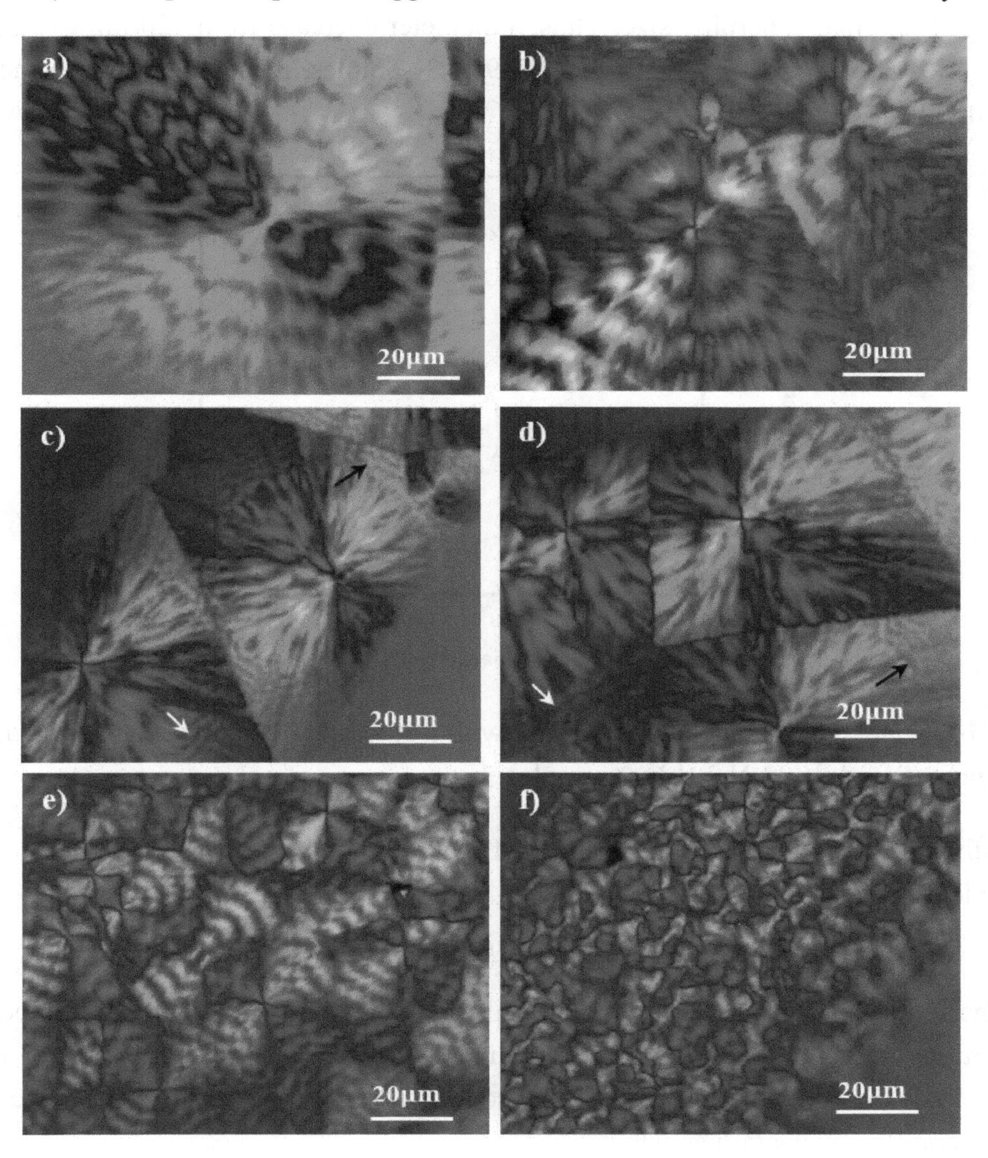


**Figure 13.** Optical micrographs of PE99 (**a,c,e**) and PE99-N 3 (**b,d,f**) spherulites isothermally grown at 60 °C (**a,b**); first step at 59 °C and second step at 54 °C (**c,d**); and 56 °C (**e,f**). Well-defined rings with interspacing between 3 and 2 μm were detected (see arrows) in the outer part of spherulites grown at 54 °C.

Despite the morphological similarities, great differences were detected in the primary nucleation (Figure 14). They were more remarkable at low crystallization temperatures; for example, nucleation densities of 60 and 20 nucleus/mm$^2$ were determined from the optical micrographs taken at 57 °C. Note that crystal growth rates were similar at this temperature; consequently, the differences determined for the overall crystallization rate (e.g., 0.01284 and 0.00589 s$^{-1}$ for PE99 and PE99-N 3 at 57 °C, respectively) were mainly attributed to a nucleation effect. On the contrary, nucleation densities were similar at high temperatures, while crystal growth rates were clearly different. Therefore, the incorporation of nanoparticles had a strong impact on the overall crystallization rate due to differences in crystal growth rate and primary nucleation densities, which became more significant at high and low crystallization temperatures, respectively. Figure 14 also shows that the nucleation density increased exponentially for the two samples at lower crystallization temperatures. These changes in nucleation logically affected the final spherulite size. Thus, a diameter decrease from 120 μm to 35 μm and from 80 μm to 15 μm was observed for PE99 and PE99-N 3 samples, respectively, when the temperature decreased from 60 °C to 56 °C.

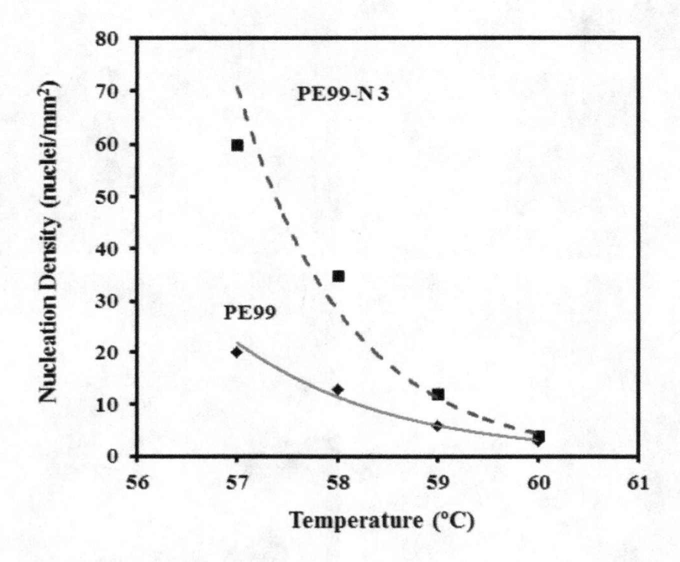


**Figure 14.** Change in the nucleation density with isothermal crystallization temperature for PE99 (**garnet**) and PE99-N 3 (**blue**) samples.

The logarithmic form of the Lauritzen and Hoffman equation [31] was employed to estimate the secondary nucleation constant ($K_g$):

$$\ln G + U^*/R(T_c - T_\infty) = \ln G_0 - K_g/(T_c(\Delta T)f) \tag{3}$$

where $G$ is the radial growth rate, $T_c$ is the crystallization temperature, $T_\infty$ is the temperature below molecular motion ceases, $\Delta T$ is the degree of supercooling, $f$ is a correction factor calculated as $2T_c/(T_m + T_c)$, $U^*$ is the activation energy, $G_0$ is a constant preexponential factor, $R$ is the gas constant, and $K_g$ is the secondary nucleation constant.

The Lauritzen-Hoffman plot was fitted with straight lines ($r^2 > 0.97$) for micro-molded PE99 and PE99-N 3 samples when the "universal" values reported by Suzuki and Kovacs [32] (*i.e.*, $U^*$ = 1500 cal/mol and $T_\infty = T_g - 30$ K) and the experimental $T_g$ of PE99 (*i.e.*, −45 °C) were used in the calculation (Figure 15).

The nucleation term (deduced from the slope of the plot) mainly influenced the kinetic features at low supercoolings in such a way that crystallization rates could become relatively insensitive to the $U^*$ and $T_\infty$ parameters. Therefore, the equilibrium melting temperature was determined since it influenced the degree of supercooling, and consequently the nucleation term. Typical Hoffman-Weeks plots [33] were made with samples crystallized at different temperatures (not shown), leading to equilibrium temperatures of 79.2 °C and 76.4 °C for PE99 and PE99-N 3 samples, respectively. The slight change suggests less perfect lamellae upon addition of functionalized silica nanoparticles. The Lauritzen-Hoffman plot allowed for estimating secondary nucleation constants of $1.30 \times 10^5$ and $1.18 \times 10^5$ $K^2$ for PE99 and PE99-N 3 samples, respectively. These values indicate that the presence of functionalized nanoparticles favored the crystallization process. Thus, enhanced PE99/$SiO_2$ interfacial interaction may decrease the energy involved in the folding of polyester chains and promote their deposition on existing crystal surfaces during the secondary nucleation process [26,34].

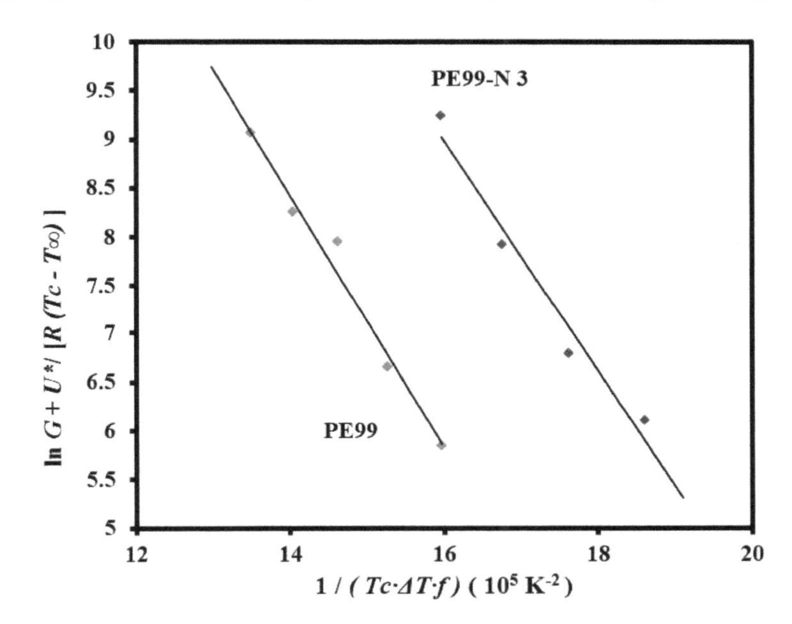


**Figure 15.** Plot of ln$G$ + $U^*/R(T_c - T_\infty)$ *versus* $1/T_c(\Delta T)f$ to determine the $K_g$ secondary nucleation parameter of PE99 (**garnet**) and PE99-N 3 (**blue**) samples.

## 4. Conclusions

Poly(nonamethylene azelate) and its mixtures with functionalized silica micro/nanoparticles could be micro-molded by means of ultrasonic energy and using similar time, amplitude, and force processing parameters. Minimal polymer degradation was detected in the molded specimens, as well as a dispersion of added particles. Furthermore, thermal stability was slightly improved by the addition of silica particles. Decomposition of grafted functional groups was only detected at the end of the thermal degradation process.

Silica nanoparticles had a significant influence on the crystallization process even at a low content of 3 wt. %. This point is important because crystallinity of samples needs to be controlled and specifically gives information about the mold temperature of the ultrasonic equipment that should be established to achieve a determined degree of crystallinity. Avrami exponents slightly decreased compared to the neat polymer as evidence of geometric constraints caused by incorporation of nanospheres. The overall crystallization rate of nanocomposites was always greater than that determined for the neat polymer, with the ratio of the respective constants ranging between 1.4 and 2.5.

Negative and ringed spherulites were obtained at all test temperatures, according to an athermal nucleation process. Primary nucleation was significantly enhanced by the addition of silica nanoparticles, whose functionalized surface was expected to favor the adsorption of polyester molecules. In addition, enhanced PE99/$SiO_2$ interfacial interaction favored molecular deposition on existing crystal surfaces, causing a decrease in the secondary nucleation constant and an increase in the crystal growth rate.

## Acknowledgments

This research was performed in the framework of an INNPACTO project "IPT-2011-0876-420000" and was also supported by grants from MINECO/FEDER and AGAUR (MAT2012-36205, 2014SGR188). We are grateful to Xavier Planta and Jordi Romero from Ultrasion S.L. and the Fundació ASCAMM, respectively, for their technical support. Gonzalo López has also contributed to experimental work concerning the preparation and functionalization of silica particles.

## Author Contributions

Angélica Diáz performed the experiments; María Teresa Casas and Jordi Puiggalí directedthe research; Jordi Puiggalí wrote the manuscript. The manuscript was finalized through contributions from all authors, and all authors also approved the final manuscript.

## References

1. Fairbanks, H.V. Applying ultrasonics to the moulding of plastic powders. *Ultrasonics* **1974**, *12*, 22–24.
2. Paul, D.W.; Crawford, R.J. Ultrasonic moulding of plastic powders. *Ultrasonics* **1981**, *19*, 23–27.
3. Werner, K.S.; Burlage, K.; Gerhardy, C. Ultrasonic Hot Embossing. *Micromachines* **2011**, *2*, 157–166.
4. Khuntontong, P.; Blaser, T.; Maas, D.; Schomburg, W.K. Fabrication of a polymer micro mixer by ultrasonic hot embossing. In Proceedings of the 19th Micro-Mechanics Europe Workshop, MME 2008, Aachen, Germany, 28–30 September 2008.
5. Khuntontong, P.; Blaser, T.; Schomburg, W.K. Fabrication of molded interconnection devices by ultrasonic hot embossing of thin films. *IEEE Trans. Electron. Packag. Manuf.* **2009**, *32*, 152–156.
6. Michaeli, W.; Spennemann, A.; Gartner, R. New plastification concepts for micro injection moulding. *Microsyst. Technol.* **2002**, *8*, 55–57.

7. Michaeli, W.; Starke, C. Ultrasonic investigations of the thermoplastics injection moulding process. *Polym. Test.* **2005**, *24*, 205–209.
8. Michaeli, W.; Kamps, T.; Hopmann, C. Manufacturing of polymer micro parts by ultrasonic plasticization and direct injection. *Microsyst. Technol.* **2011**, *17*, 243–249.
9. Chen, J.; Chen, Y.; Li, H.; Lai, S.Y.; Jow, J. Physical and chemical effects of ultrasound vibration on polymer melt in extrusion. *Ultrason. Sonochem.* **2010**, *17*, 66–71.
10. Chen, G.; Guo, S.; Li, H. Ultrasonic improvement of rheological behavior of polystyrene. *J. Appl. Polym. Sci.* **2002**, *84*, 2451–2460.
11. Kang, J.; Chen, J.; Cao, Y.; Li, H. Effects of ultrasound on the conformation and crystallization behavior of isotactic polypropylene and β-isotactic polypropylene. *Polymer* **2010**, *51*, 249–256.
12. Cao, Y.; Li, H. Influence of ultrasound on the processing and structure of polypropylene during extrusion. *Polym. Eng. Sci.* **2002**, *42*, 1534–1540.
13. Sacristán, M.; Plantá, X.; Morell, M.; Puiggalí, J. Effects of ultrasonic vibration on the micro-molding processing of polylactide. *Ultrason. Sonochem.* **2014**, *21*, 376–386.
14. Planellas, M.; Sacristán, M.; Rey, L.; Olmo, C.; Aymamí, J.; Casas, M.T.; del Valle, L.J.; Franco, L.; Puiggalí, J. Micro-molding with ultrasonic vibration energy: New method to disperse nanoclays in polymer matrices. *Ultrason. Sonochem.* **2014**, *21*, 1557–1569.
15. Díaz, A.; Casas, M.T.; del Valle, L.J.; Aymamí, J.; Olmo, C.; Puiggalí, J. Preparation of micro-molded exfoliated clay nanocomposites by means of ultrasonic technology. *J. Polym. Res.* **2014**, *21*, 584–596.
16. Usuki, A.; Kojima, Y.; Kawasumi, M.; Okada, A.; Fukushima, Y.; Kurauchi, T.; Kamigaito, O. Mechanical properties of nylon 6-clay hybrid. *J. Mater. Res.* **1993**, *8*, 1185–1189.
17. Kojima, Y.; Usuki, A.; Kawasumi, M.; Okada, A.; Kurauchi, T.; Kamigaito, O. Synthesis of nylon 6-clay hybrid by montmorillonite intercalated with ε-caprolactam. *J. Polym. Sci. Part A* **1993**, *31*, 983–986.
18. Kojima, Y.; Usuki, A.; Kawasumi, M.; Okada, A.; Kurauchi, T.; Kamigaito, O. One-pot synthesis of nylon 6-clay hybrid. *J. Polym. Sci. Part A* **1993**, *31*, 1755–1758.
19. Burgaz, E. Poly(ethylene-oxide)/clay/silica nanocomposites: Morphology and thermomechanical properties. *Polymer* **2011**, *52*, 5118–5126.
20. Choi, M.; Kim, C.; Jeon, S.O.; Yook, K.S.; Lee, J.Y.; Jang, J. Synthesis of titania embedded silica hollow nanospheres via sonication mediated etching and re-deposition. *Chem. Commun.* **2011**, *47*, 7092–7094.
21. Yang, F.; Ou, Y.; Yu, Z. Polyamide 6/silica nanocomposites prepared by *in situ* polymerization. *J. Appl. Polym. Sci.* **1998**, *69*, 355–361.
22. Xu, X.; Li, B.; Lu, H.; Zhang, Z.; Wang, H. The effect of the interface structure of different surface-modified nano-$SiO_2$ on the mechanical properties of nylon 66 composites. *J. Appl. Polym. Sci.* **2008**, *107*, 2007–2014.
23. Vassiliou, A.A.; Papageorgiou, G.Z.; Achilias, D.S.; Bikiaris, D.N. Non-isothermal crystallisation kinetics of *in situ* prepared poly(ε-caprolactone)/surface-treated $SiO_2$ nanocomposites. *Macromol. Chem. Phys.* **2007**, *208*, 364–376.

24. Busbee, J.D.; Juhl, A.T.; Natarajan, L.V.; Tongdilia, V.P.; Bunning, T.J.; Vaia, R.A.; Braun, P.V. $SiO_2$ Nanoparticle Sequestration via Reactive Functionalization in Holographic Polymer-Dispersed Liquid Crystals. *Adv. Mater.* **2009**, *21*, 3659–3662.
25. Li, Z.; Barnes, J.C.; Bosoy, A.; Stoddart, J.F.; Zink, J.I. Mesoporous silica nanoparticles in biomedical applications. *Chem. Soc. Rev.* **2012**, *41*, 2590–2605.
26. Lee, E.; Hong, J.Y.; Ungar, G.; Jang, J. Crystallization of poly(ethylene oxide) embedded with surface-modified $SiO_2$ nanoparticles. *Polym. Int.* **2013**, *62*, 1112–1122.
27. Avrami, M. Kinetics of phase change. I General Theory. *J. Chem. Phys.* **1939**, *7*, 1103–1112.
28. Avrami, M. Kinetics if phase change. II Transformation time relations for random distribution of nuclei. *J. Chem. Phys.* **1940**, *8*, 212–224.
29. Kennedy, M.; Turturro, G.; Brown, G.R.; St-Pierre, L.E. Silica retards radial growth of spherulites in isotactic polystyrene. *Nature* **1980**, *287*, 316–317.
30. Nitta, K.H.; Asuka, K.; Liu, B.; Terano, M. The effect of the addition of silica particles on linear spherulite growth rate of isotactic polypropylene and its explanation by lamellar cluster model. *Polymer* **2006**, *47*, 6457–6463.
31. Lauritzen, J.I.; Hoffman, J.D. Extension of theory of growth of chain folded polymer crystals to large undercoolings. *J. Appl. Phys.* **1973**, *44*, 4340–4352.
32. Suzuki, T.; Kovacs, A.J. Temperature dependence of spherulitic growth rate of isotactic polystyrene. A critical comparison with the kinetic theory of surface nucleation. *Polym. J.* **1970**, *1*, 82–100.
33. Hoffman, J.D.; Weeks, J.J. Melting process and the equilibrium melting temperature of polychlorotrifluoroethylene. *J. Res. Natl. Bur. Stand.* **1962**, *66*, 13–28.
34. Wang, K.; Wu, J.; Zeng, H. Radial growth rate of spherulites in polypropylene/barium sulfate composites. *Eur. Polym. J.* **2003**, *39*, 1647–1652.

# 12

# Electromagnetic Acoustic Transducers Applied to High Temperature Plates for Potential use in the Solar Thermal Industry

**Maria Kogia [1,*], Liang Cheng [1], Abbas Mohimi [1,2], Vassilios Kappatos [1], Tat-Hean Gan [1,2,*], Wamadeva Balachandran [1] and Cem Selcuk [1]**

[1] Brunel Innovation Centre (BIC), Brunel University, Cambridge CB21 2AL, UK; E-Mails: liang.cheng@brunel.ac.uk (L.C.); abbas.mohimi@brunel.ac.uk (A.M.); vassilis.kappatos@brunel.ac.uk (V.K.); wamadeva.balachandran@brunel.ac.uk (W.B.); cem.selcuk@brunel.ac.uk (C.S.)

[2] TWI Ltd., Granta Park, Great Abington, Cambridge CB21 6AL, UK

† This paper is an extended version of paper published in the 6th International Conference on Emerging Technologies in Non-destructive Testing (ETNDT6), Brussels, Belgium, 27–29 May 2015.

* Author to whom correspondence should be addressed; E-Mail: maria.kogia@brunel.ac.uk (M.G.), tat-hean.gan@brunel.ac.uk (T.-H.G.)

Academic Editor: Dimitrios G. Aggelis

---

**Abstract:** Concentrated Solar Plants (CSPs) are used in solar thermal industry for collecting and converting sunlight into electricity. Parabolic trough CSPs are the most widely used type of CSP and an absorber tube is an essential part of them. The hostile operating environment of the absorber tubes, such as high temperatures (400–550 °C), contraction/expansion, and vibrations, may lead them to suffer from creep, thermo-mechanical fatigue, and hot corrosion. Hence, their condition monitoring is of crucial importance and a very challenging task as well. Electromagnetic Acoustic Transducers (EMATs) are a promising, non-contact technology of transducers that has the potential to be used for the inspection of large structures at high temperatures by exciting Guided Waves. In this paper, a study regarding the potential use of EMATs in this application and their performance at high temperature is presented. A Periodic Permanent Magnet (PPM) EMAT with a racetrack coil, designed to excite Shear Horizontal waves ($SH_0$), has been theoretically and experimentally evaluated at both room and high temperatures.

**Keywords:** Electromagnetic Acoustic Transducer (EMAT); guided waves; solar thermal plants; non-destructive testing; high temperatures

## 1. Introduction

Solar thermal industry is an environmentally friendly means of power generation, converting solar energy into electricity. Concentrated Solar Plants (CSPs) are employed in the solar thermal industry for collecting sunlight and converting it firstly into heat and later into electricity. There are several types of CSPs such as solar towers, dish concentrators, and parabolic trough CSPs; the latter is the most widely used [1] and is shown in Figure 1a.

**Figure 1.** (**a**) Parabolic Trough Concentrated Plants [2]; (**b**) reflector and absorber tube [3]; (**c**) absorber Tube [4].

The two essential components of parabolic trough CSPs are the long, parabolic trough shaped reflectors and the absorber tubes. The reflector collects and focuses the sunlight to the absorber tube, which is located at the focal point of the reflector and runs its whole length. The absorber tubes are composed of long, thin, stainless steel tubes covered by a glass envelope under vacuum, as shown in Figure 1b,c. Their entire surface should be exposed to sunlight and therefore there are few access points for any extra hardware to be attached to them. Inside the tubes, working fluid, either water/molten salt or synthetic oil, is flowing, absorbing the heat of the sunlight and transmitting it into heat/steam engines for the final stages of the power generation procedure.

The high temperatures (400–500 °C), the contraction/expansion endured due to the fast cooling of the tube at high temperatures, and the vibrations can result in the appearance of defects either on the surface or inside the absorber tube. Problems with absorber tubes have been reported [5–7], such as creep, thermo-mechanical fatigue, and hot corrosion. The turbulent mixing of hot and cold flow streams of the working fluid may result in temperature variations along the pipe and consequently in thermomechanical fatigue [8–11]. Corrosion is another common problem absorber tubes may suffer from [12–14]; local pitting corrosion can cause the initiation of stress corrosion cracking or result in small-scale leaks. Most stainless steel pipes are vulnerable to pitting corrosion and stress corrosion cracking as well [15]. The low flow of the working fluid may also lead the absorber tube to overheat; consequently, the absorber tube may be subject to creep damage, thermal oxidation, softening, and/or stress rupture [16].

Non Destructive Testing (NDT) techniques are suggested to be applied for the structural health monitoring or inspection of the absorber tubes. Several NDT techniques can be used for either the inspection or the monitoring of high temperature structures; Acoustic Emission (AE), Eddy Current (EC), Holographic Interferometry, Laser Ultrasonic, Guided Wave Testing (GWT), and Infrared Thermography (IR) are some of those. These techniques have been reported for operating at temperatures up to 300 °C, though with shortcomings. Some of the drawbacks are qualitative results, sensitivity to noise, laboratorial utilization, and the need for coupling media [17–20]. AE is a passive NDT technique that has been widely deployed in Structural Health Monitoring (SHM); it monitors the elastic waves generated after the initiation or propagation of a crack [21]. Nevertheless, AE is sensitive to noise and gives qualitative results. EC is a non-contact technique that can be employed for the inspection of any electrically conductive material; however, it is subject to the skin effect, leading it to be mainly used for the detection of surface and subsurface defects [22]. Holographic Interferometry can give detailed results and be used for the detection of small defects, but it is mainly used for laboratorial tests as its setup is complicating and it is sensitive to vibrations [23]. Laser Ultrasonic has a small and adjustable footprint; therefore, it can be used for the inspection of irregular surfaces and samples of small and complex geometry. It induces high frequency ultrasound and thus very small defects can be detected as well. However, its setup is complicated and it is mainly used for laboratorial tests. GWT is used for the inspection or monitoring of large structures; mainly piezoelectric transducers are employed upon the structure being tested exciting/receiving guided waves [24]. Piezoelectric transducers require direct access to the specimen and a coupling medium (usually a water-based gel), and their response cannot travel through a vacuum. IR can be used in this application mainly for overheating identification; however, the length of the pipes makes this technique practically inefficient. The camera needs to scan the whole length of the pipe, which is time-consuming and may not be possible while the absorber tube is operating [25].

The monitoring of absorber tubes requires a non-contact NDT technique that can be applied at high temperatures without the need for a couplant and can inspect the whole length of the tube from a single point. Therefore, GWT can be applied with the use of non-contact transducers that can withstand high temperatures and excite/receive guided waves and more particularly a T(0,1) wave mode (or $SH_0$) [26,27]. The EMAT can be used for this application, since it is a non-contact technology that has been used in GWT and can be applied for the inspection of structures under hostile conditions such as moving specimen and elevated temperatures.

In the following section the main operating principles of EMATs are described and emphasis is also placed on the potentials and limitations of EMATs on guided waves and high temperatures. A brief description of this study follows. The results obtained from the theoretical study are shown in Section 4, where the dispersion curves of the absorber tubes were calculated and processed and a PPM EMAT with racetrack coil has been evaluated mainly regarding its guided wave purity characteristics at both room and high temperatures. In Sections 5 and 6 the experimental setup and the results of the experimental evaluation of a pair of PPM EMATs are presented. The EMATs were tested as far as their efficiency to excite $SH_0$ is concerned and the parameters that may affect their performance at both room and high temperatures. The EMATs were validated for their wave purity, their sensitivity to noise, their lift-off limitations, their power requirements, and their performance at high temperatures while they were exciting/receiving guided waves. Hence, a comparison between the simulated and the experimental results is finally presented. A high temperature EMAT exciting $SH_0$ wave and operating up to 400–550 °C for either monitoring or long-term inspection is also to be accomplished in the near future.

## 2. EMAT Technology

EMAT is a non-contact technology of transducers that can be used for the inspection of moving structures or a specimen that operates at elevated temperatures. Their response can travel through a vacuum as well, making them even more suitable for this application compared to piezoelectric transducers. EMATs can excite/receive guided waves and thus they can be employed in GWT for either the inspection or the monitoring of large structures [27–30]. Hence, high-temperature EMATs may be used for the inspection or monitoring of absorber tubes. Nevertheless, the high-temperature EMATs that have been reported so far have been designed for thickness measurements [31–33]. Therefore, a high-temperature EMAT that excites/receives guided waves for the long-term inspection of high-temperature structures is still required.

A typical EMAT transducer is composed of either a permanent magnet or an electromagnet for the generation of a static magnetic field, and a coil. The coil is driven by an alternating current that generates a dynamic magnetic field. This dynamic magnetic field induces an eddy current in the specimen placed below the coil. If the specimen being tested is an electrical conductor and non-ferromagnetic, mainly Lorentz force is exerted upon the material particles; Equation (1) shows that Lorentz force is equal to the product of eddy current density and the overall magnetic field:

$$F_L = J_e \times (B_{st} + B_{dyn}) \tag{1}$$

where $F_L$ is Lorentz force, $J_e$ is the eddy current density, $B_{st}$ stands for the static magnetic field, and $B_{dyn}$ refers to the dynamic magnetic field. In this application, the absorber tubes are made of 316 L stainless steel, which is paramagnetic; therefore the main force generated in them would be Lorentz force.

The EMAT configuration and the material properties of its main components affect EMAT performance at both room and high temperatures. The material the coil is made of influences the EMAT performance. The impedance of the coil affects the energy transmitted to the specimen and the SNR/quality of the signal received; Equation (2) shows how the resistance increases with temperature rise and Equation (3) demonstrates the relationship between the temperature and the noise level of the signal received:

$$R(T) = R_0[1 + \alpha(T - T_0)] \tag{2}$$

where $R_0$ is the resistance of a single turn coil, $\alpha$ is the temperature coefficient of resistance, and $T_0$ is the room temperature.

$$V_{Noise} = (4K\beta T R_{EMAT})^2 \tag{3}$$

where $K$ is the Boltzmann constant, $T$ is the temperature measured in Kelvin, $\beta$ is the bandwidth, and $R_{EMAT}$ is the resistance of the EMAT coil per turn. Consequently, the coil should be preferably made of a low-resistance material such as copper. However, copper is known to oxidize at high temperatures and, therefore, other materials could be used for this application such as silver, platinum, nickel, or constantan. Hernandez-Valle (2011) has already made an electromagnet EMAT that can operate at up to 600 °C without any cooling system for thickness measurements and apart from designing a high temperature electromagnet he has also tested several coil designs at high temperatures [32].

The permanent magnets also have limitations regarding their maximum operating temperature. The Maximum Operating Temperature (MOT) of a magnet is equal to half of its Curie Point; beyond Curie Temperature, the strength of the magnet decreases rapidly. Consequently, high Curie Point magnets are preferable for high-temperature applications. Nevertheless, the magnetic strength of the main two types of high-temperature magnets, Alnico (with a maximum operating temperature of 500 °C) and SmCo (with a maximum operating temperature of 300 °C), is smaller than the magnetic strength of Neo (NdFeB) magnets. However, NdFeB magnets cannot be used at temperatures higher than 200 °C.

A cooling system may also be required so that both the magnets and the coil will operate efficiently. Idris *et al.* have designed and tested a water-cooled EMAT that can obtain signals up to 1000 °C for thickness measurements. The EMAT was exposed to the heat source for as much time as it needed for the signal to be recorded and then it was removed [33]. Oil- and air-cooled EMATs have also been designed. Generally, the reported high-temperature EMATs seem to operate at high temperatures for short periods of time, which are suitable for inspection, but they cannot be used in long-term condition monitoring. Consequently, an EMAT operating at high temperatures (500 °C) for long periods of time is required and could be used for the structural health monitoring of high-temperature structures such as the absorber tubes.

## 3. Our Methodology

This study is divided into two main parts, the theoretical validation of the PPM EMAT design and the experimental evaluation of a pair of PPM EMATs. Figure 2 shows the schematic of the methodology followed in this study. In the theoretical part, the dispersion curves of a 3 mm thick, 316 L stainless steel plate were calculated for both room and high temperatures. The resonant frequency of EMAT and the wave velocity of $SH_0$ at the resonant frequency of EMAT for room temperature, 60 °C, 100 °C and 180 °C are also calculated. Electromagnetic simulations in COMSOL were also carried out for the theoretical validation of this EMAT design regarding its wave purity characteristics at both room and high temperatures. Experiments were conducted with a pair of PPM EMATs with racetrack coil on a stainless steel plate. During the experiments the EMATs were tested regarding their wave purity potentials, their lift-off limitations, their power requirements, and their performance at high temperatures (up to 180 °C). The results obtained from the experimental procedure were compared with the results from the theoretical validation of the EMAT and the conclusions of this study are summarized and presented in the last part of this paper.

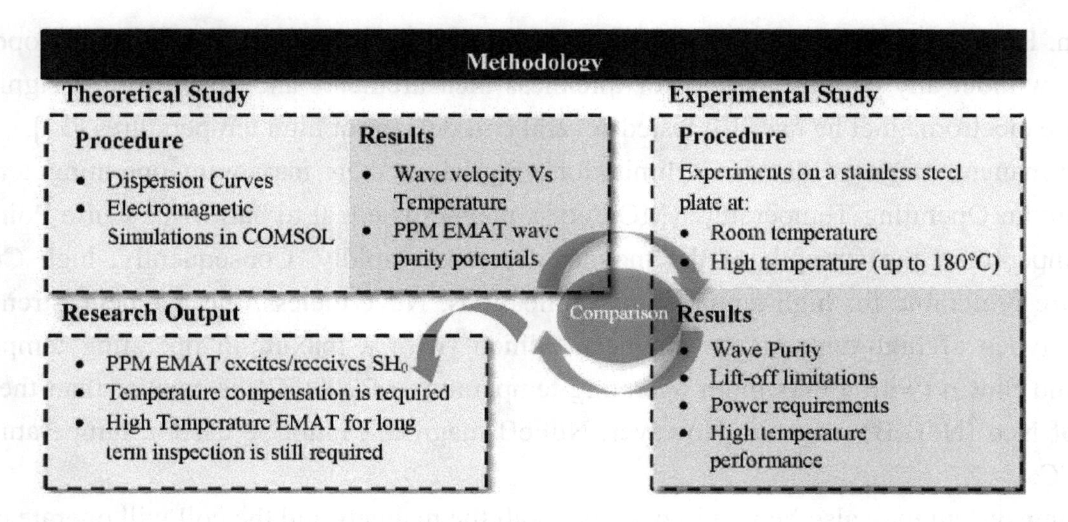


**Figure 2.** Methodology schematic.

## 4. Theoretical Study

### *4.1. Dispersion Curves*

GWT is an NDT technique that can be employed in this applicaton. Low-frequency waves can travel big distances without being significantly attenuated; however, their relatively large wavelength limits the size of defect that can be detected. A 10 mm length of defect was set as the minimum size defect that should be detected; considering that the wavelength should be smaller than the double of the minimum size defect, in this case the wavelength should be smaller than 20 mm. A $SH_0$ wave of 12 mm wavelength was introduced to this structure.

As the wave propagates inside the specimen, it strikes at the boundaries of the medium, resulting in the change of its waveform. Consequently, there is an infinite number of possible wave modes that may appear in the material; their velocity changes in respect not only to the material properties of the specimen but to its geometry as well. Hence, the same mode at a different frequency may propagate with a different mode shape and velocity. This phenomenon is called dispersion; most of the wave modes in guided waves are dispersive Nevertheless, T(0,1) (or, alternatively, $SH_0$ on plates) is not dispersive, making the interpretation of the signal received less complicated. Its displacement is also in-plane, making it more suitable for this application, since the wave will propagate all along the pipe without it being affected by the working fluid that flows inside the pipe; T(0,1) cannot propagate in liquids. The number of wave modes that may appear in a plate is also smaller compared to a pipe and therefore the interpretation of the signal received from a plate may be less complicated as well.

In this preliminary study, the EMAT is simulated and designed for exciting $SH_0$ waves, which will propagate axially on a 316 L stainless steel plate. The dispersion curves of a 3 mm thick, 316 L stainless steel plate of 8000 kg/m$^3$ density, 195 GPa Young's modulus, and 0.285 Poisson ratio have been calculated and demonstrated in Figure 3. This figure shows that the wavelength curve crosses the $SH_0$ curve at 256 kHz frequency and thus this should be the resonant frequency of the EMAT. The wavelength curve also crosses the $S_0$, $A_0$, and $A_1$ curves; Table 1 shows at which frequency the wavelength curve crosses each wave mode curve and their wave velocity at these frequencies. Hence, an EMAT designed to excite $SH_0$ of 12 mm wavelength should be driven with an AC current of

256 kHz frequency. Otherwise if the EMAT is tuned to any other frequency, it will be likely for it to excite Lamp waves instead of $SH_0$. This is more possible if the coil design is meander instead of racetrack, for meander coils are used for Lamp waves as well. At 256 kHz, the $SH_0$ and the $A_0$ have the same group velocity. As a result, both wave modes can be excited/received from the EMAT, simultaneously resulting in a more complicated signal, since $A_0$ is dispersive at this frequency. Nevertheless, the orientation of the displacement of each wave mode is different; the $SH_0$ has an in-plane displacement while the $A_0$ has an out-of-plane displacement. Thus, a further study should be conducted regarding the wave mode, and more particularly the displacement to which the PPM EMAT is sensitive. Consequently, an electromagnetic simulation calculating and showing the amplitude and the direction of the wave modes generated by a PPM EMAT is needed.

Material properties such as the Young's Modulus, Poisson ratio, and density of a specimen change with temperature rise as well, resulting in variations in the velocity of the propagating wave mode. The Young's modulus and density of the specimen decrease with an increase in temperature, while the Poisson ratio increases. These changes result in a decrease in the velocity of the $SH_0$. Table 1 summarizes the wave velocity of $SH_0$, $S_0$, $A_0$, and $A_1$ at room temperature, 60 °C, 100 °C, and 180 °C. Hence, as the temperature rises and the wave velocity decreases, any reflections received from the EMAT should shift in time. Hence, the temperature should be kept stable during the inspection or temperature compensation should take place during the signal interpretation.

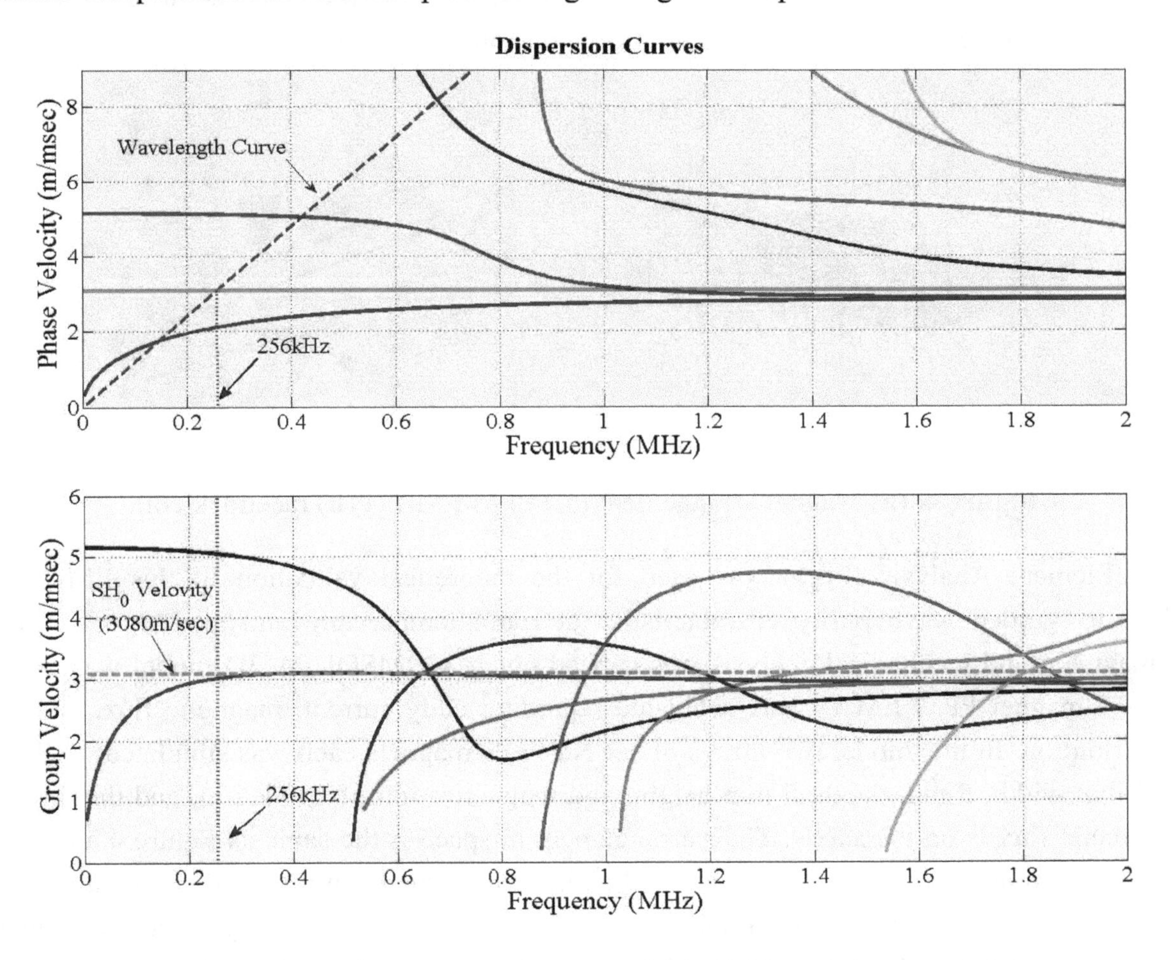


**Figure 3.** Dispersion curves of a 3 mm thick 316 L stainless steel plate.

**Table 1.** Dispersion Curves, wave velocity, and frequency.

| Dispersion Curves—$SH_0/S_0/A_0/A_1$ Frequency & Wave Velocity | | | | | | | | |
|---|---|---|---|---|---|---|---|---|
| **Value** | **Frequency (kHz)** | | | | **Velocity (m/s)** | | | |
| Temperature (°C) | 22 | 60 | 100 | 180 | 22 | 60 | 100 | 180 |
| $SH_0$ | 256 | 255 | 254.2 | 250.6 | 3080 | 3067 | 3051 | 3008 |
| $S_0$ | 420 | 419 | 415.5 | 405 | 5044 | 5022 | 5003 | 4944 |
| $A_0$ | 143 | 140 | 139 | 136 | 1736 | 1720 | 1711 | 1685 |
| $A_1$ | 676 | 673 | 668 | 657 | 8243 | 8217 | 8203 | 8129 |

### *4.2. Electromagnetic Simulations*

Both the coil design and the magnet configuration affect the distribution of Lorentz force in space, its direction, and its amplitude. As a result, the wave mode the EMAT is to generate and/or receive gets affected by the configuration of its two main components. An EMAT configuration that can be used for the $SH_0$ is the PPM EMAT with racetrack coil [26–40].

In a PPM EMAT, the distance between two adjacent magnets whose magnetic field has the same direction (pitch) is equal to the wavelength; the arrangement of the magnets is illustrated in Figure 4. Racetrack coil is also used in this EMAT configuration. In this coil design there are no gaps within its turns, resulting in its broadband response in frequency. Hence, only the magnets' arrangement and the frequency of the AC current can affect the frequency response of EMAT.

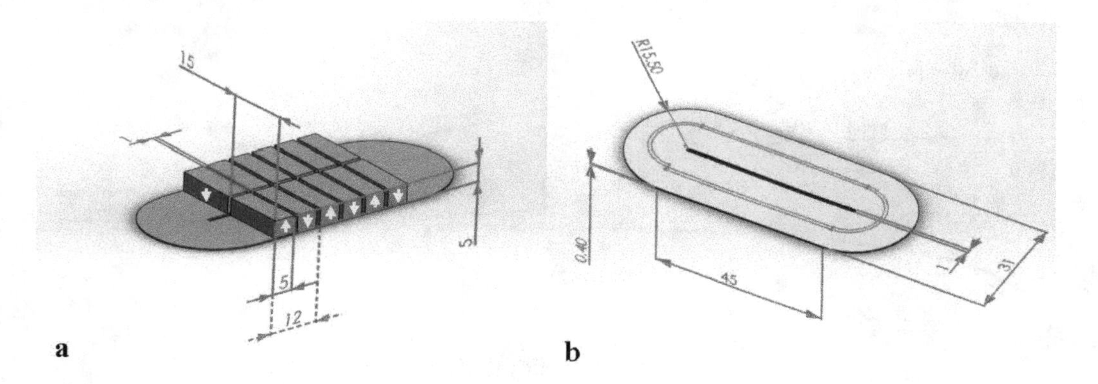


**Figure 4.** (**a**) Magnet arrangement in a PPM EMAT; (**b**) racetrack coil.

Finite Element Analysis (FEA) was used for the theoretical validation of this EMAT design regarding its guided wave purity characteristics at room temperature and at 180 °C. A coupled electromagnetic and mechanical analysis was carried out in COMSOL. A 3D model was created, in which a 12-magnet PPM EMAT was evaluated regarding eddy current, magnetic flux, and Lorentz force distribution. In this model two arrays of six Nd-Fe-B magnets each was simulated; each magnet has a 15 mm width, 5 mm depth, 5 mm height, and magnetization of 750 kA/m, and the direction of their magnetic flux is on the $z$ axis. Their arrangement in space is the same as Figure 4 and, thus, the distance between one another is 1 mm. Due to the high requirements in time and computational power these complicated models have, the copper coil has been simplified and designed as two rectangular blocks of 35 mm width, 15 mm depth, and 0.4 mm height. The AC current driven to the coil is a two-cycle, Hanning windowed sinusoidal wave of 20 A amplitude and 256 kHz central frequency with direction on the $x$ axis. The orientation of the excitation current flowing inside one rectangular coil is opposite to the orientation of the current inside the other coil. The EMAT has a lift-off of 0.6 mm from the specimen, which is a 316 L stainless steel plate of 3 mm thickness, 750 mm width, and 750 mm depth and of the same material properties as the plate simulated in the previous section at room temperature and at 180 °C. The material, magnetic, and electrical properties of the EMAT did not change with temperature.

Figure 5a,b show the excitation/eddy current at 4 μs and the static magnetic flux distribution, respectively. Both the electric and the magnetic field are uniformly distributed. The orientation of the eddy current alters between the two sides of the coil and it is on the $x$–$y$ plane. The orientation of the magnetic field alters as it is depicted in Figure 5b; its orientation is mainly on the $z$ axis and its maximum strength is observed at the center of each magnet separately, as is expected. Thus, the Lorentz force generated should mainly result in an in-plane displacement. A probe was also placed 30 cm away from the EMAT for obtaining the in-plane ($x$–$y$ plane) and out-of-plane ($x$–$z$ plane) displacement. Figure 5c shows that the in-plane displacement maximizes at 98 μs while the out-of-plane displacement maximizes at 100 μs; the Time of Flight (ToF) of the in-plane displacement matches with the wave velocity of the $SH_0$, as was calculated from the dispersion curves. while the out-of-plane displacement can be the $A_0$. Hence, this EMAT configuration may excite the $A_0$ wave mode as well. However, the maximum value of the out-of-plane displacement is significantly smaller than the in-plane displacement and therefore experimental validation of this EMAT design regarding its guided wave purity characteristics is still required.

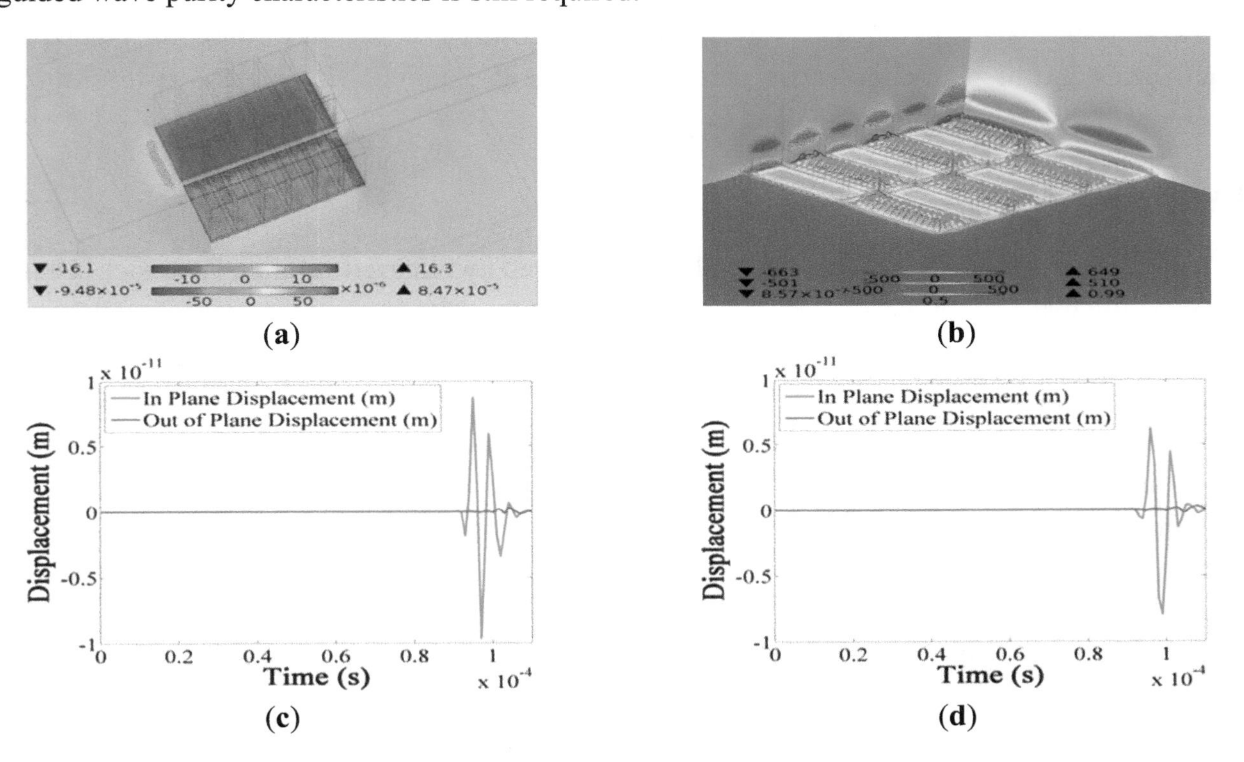


**Figure 5.** (**a**) Excitation/eddy current distribution at 12 μs; (**b**) magnetic flux distribution of the 12 Nd-Fe-B; (**c**) in-plane/out-of-plane displacement 30 cm away from the EMAT at room temperature; (**d**) in-plane/out-of-plane displacement 30 cm away from the EMAT at 180 °C.

Figure 5d shows the in-plane and out-of-plane displacement 30 cm away from the EMAT at 180 °C; both displacements are shifted in time, as was expected since the wave velocity changes with temperature rise and the ToF matches with the wave velocity as it was calculated from the dispersion curves. The amplitude of the in-plane displacement decreased; however, the amplitude of the out-of-plane displacement did not decrease with the temperature rise. As was mentioned, in the FEA model only the material properties of the specimen changed so that the effect of the temperature rise would be simulated. Nevertheless, the temperature rise affects the components of the EMAT as well, as was mentioned in the introduction, and thus a further decrease in the amplitude of both displacements may be observed in the experimental results, which the current model does not take into account.

Consequently, this EMAT configuration can mainly generate in-plane displacement ($SH_0$); however, it may also generate a small out-of-plane displacement ($A_0$) at this specific frequency. This can result in complicating signal analysis, since the out-of-plane displacement can be dispersive and mode conversion can also occur, which can lead to incorrect conclusions regarding the structural integrity of the specimen. In GWT, EMATs should excite a single wave mode, so that the signal received from the specimen can provide valid information regarding the structural integrity of the specimen. Also, the temperature rise affected the ultrasonic response of the EMAT, as was expected; amplitude attenuation and shifting in time were the main two changes in the signal received as the temperature increased. Hence, the wave purity characteristics of the PPM EMAT for GWT are of great importance for this application as well as the effect of temperature on the ultrasonic response of EMATs and thus an experimental validation of this EMAT design regarding its guided wave characteristics and its high-temperature performance is still required. The simulation results will be compared with the experimental results.

## 5. Experimental Setup

A pair of PPM EMATs with racetrack coil, manufactured by Sonemat Limited (Warwick, UK), was experimentally evaluated at both room and high temperatures. The pitch of the magnets is 12 mm and equal to the wavelength of the $SH_0$ [41]. During these experiments, the EMATs were tested regarding defect detection using guided waves at both room and high temperatures. Their sensitivity to noise was also experimentally evaluated; the influence of common mode noise on their performance and the effect of common ground connection between the EMAT and the specimen on the noise reduction were tested. Their lift-off limitations were validated; the thickness of the glass envelope, under which is the stainless steel pipe of the absorber tubes, will attenuate the signal generated by the EMAT and will perform as a fixed lift-off between the EMAT and the stainless steel pipe. Hence, the maximum lift-off EMATs can reach when they are used for GWT needs to be known. Their power requirements were also investigated, since only Lorentz force is generated in stainless steel, making the EMATs less efficient and more power demanding.

The experimental setup used is shown in Figure 6. The specimen used is a 316Ti stainless steel square plate of 1.25 m length and 3 mm thickness. The EMAT coil is made of lacquered copper wires of 0.315 mm diameter and the maximum operating temperature of the EMATs is 250 °C. Ritec RAM 5000 SNAP pulser/receiver (RITEC Inc., Warwick, RI, USA) was used for driving the EMAT transmitter with a six-cycle, Hanning-windowed pulse of 256 kHz frequency. Ritec was also used for amplifying the signal received with a gain of 80 dB and filtering it within the bandwidth of 10 kHz and 20 MHz. The signal is finally collected, averaged, and recorded in a 2-channel Agilent oscilloscope (Keysight Technologies Inc., Santa Rosa, CA, USA).

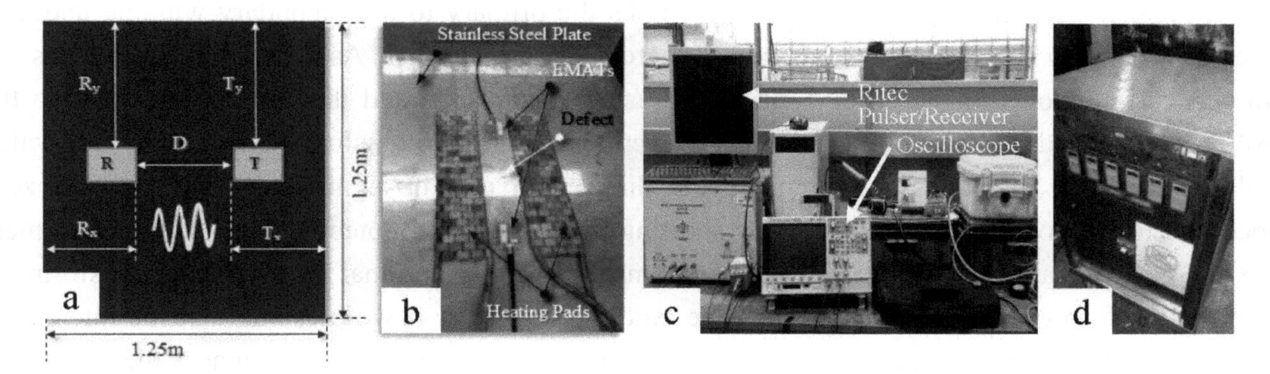


**Figure 6.** (**a**) Schematic of experimental setup; (**b**) experimental setup (EMATs, specimen, heating pads); (**c**) pulser/receiver—oscilloscope; (**d**) heating unit.

During the room-temperature experiments both defect-free and defective areas were tested. The distance between the transmitter and the receiver was equal to 30 cm. Five defects were created with different length and mass loss each and located 10 cm away one from the other. The defect tested was 20 mm long and with 66.6% mass loss; the transmitter was 15 cm away from one edge of the defect and the receiver was 15 cm away from the other edge of the defect. The effect of the voltage difference between the EMAT and the specimen on the quality of the signal received has been also investigated; in fact, an additional thin, stainless steel cover was also placed all around the transducers, touching both the EMATs and the specimen, for establishing a common ground connection. The influence of lift-off on EMAT response was investigated from zero to 1 mm lift-off with a step of 0.1 mm. A study regarding the power supply requirements of these EMATs was also accomplished by gradually decreasing the power output of Ritec with a step of 5% starting from its maximum power level (5000 W) and stopping at 20% of its maximum power, where no useful information could be retrieved anymore from the signal received. For the high-temperature experiments, the distance between the transmitter and the receiver was equal to 30 cm with the defect located 15 cm away from the transmitter and 15 cm away from the receiver as well (similar to the room temperature setup). The temperature was increased from ambient to 180 °C with a step of 10 °C, using a three-phase heating unit. During the high-temperature experiments the EMATs were continuously in contact with the specimen.

## 6. Experimental Results and Discussion

### *6.1. Room Temperature Experiments*

During this set of experiments, the EMATs were tested for their sensitivity to common mode noise. Four case studies were investigated; the EMATs were employed in both a defect-free and a defective area, with and without common ground connection with the specimen (shielding). More particularly, the connection/interaction between the EMAT and the specimen is differential, since the specimen induces to the EMAT coil an alternating, differential mode current. Parasitic capacitance exists between the specimen and the EMAT as a result of their physical spacing and the presence of dielectric between them [35]. Tranformers perform in a similar way and they also suffer from common mode noise; in real transformers, a small capacitance links the primary to the secondary winding and also serves as a path for the common mode current across the transformer. As a result, in both cases the common current flows to the ground via the parasitic capacitance and thus no current flows to the EMAT coil/specimen or secondary winding. Nevertheless, an autotransformer acts as a high-value parallel impedance that does not attenuate the differential current significantly but presents zero impedance to the common mode signals by shorting them to ground potential [42]. Autotransformers have smaller resistance and leakage reactance compared to conventional two winding transformers; therefore, the former is more efficient than the latter [43]. Hence, if we presume that the interaction between the EMAT and the specimen is equivalent to a transformer and we connect them so as to perform as an autotransformer, then the common mode noise should be cancelled and the EMAT receiver should work more efficiently. In this case, the alternating, differential mode current will be induced to the EMAT coil and the signal received will have an enhanced SNR and valid information would be retrieved from it. The autotransformer connection is also more robust, reasulting in an increase of the amplitude of the "wanted" signal. If an extra layer of stainless steel is attached to EMATs, touching both the EMAT housing and the specimen, then a common ground connection (autotransformer) is established. Figure 7 shows the equivalent electrical circuit of the EMAT/specimen connection.

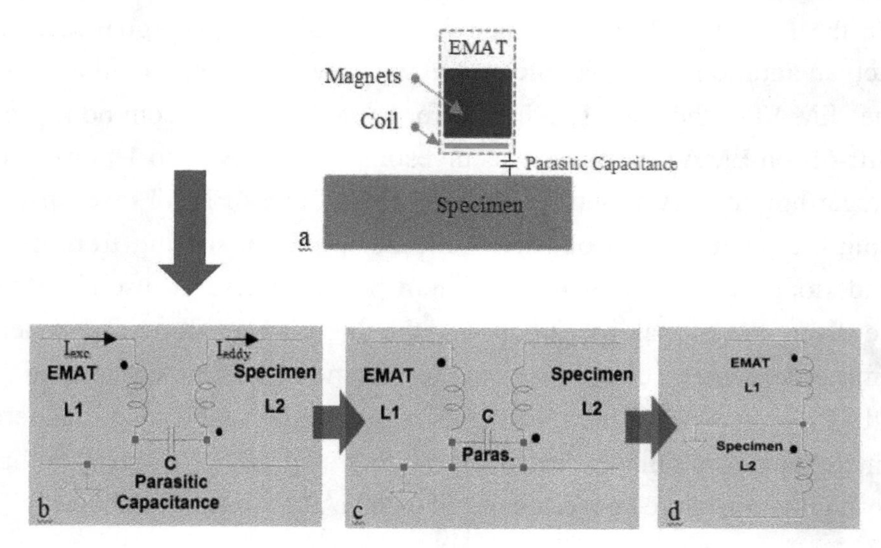


**Figure 7.** (**a**) Schematic of EMAT/specimen connection; (**b**) equivalent electrical circuit of EMAT/specimen without common ground connection; (**c**) common ground connection; (**d**) EMAT/specimen—autotransformer schematic.

Figure 8a shows the signal received from a defect-free area when the EMAT and the specimen do not have any common ground connection. In this figure the first reflection is the signal transmitted from the transmitter to the receiver and the other three are coming from the edges of the plate. In this case the noise level is high and the amplitude of all the reflections is low, leading to a low SNR.

Figure 8b illustrates the signal received when the defective area was tested without the common ground connection. Similar to Figure 8a, the signal transmitted and the three reflections from the edges of the plate are clearly obvious in the signal received. However, no reflections from the defect are obvious. Figure 9a demonstrates the signal received when the EMATs test the defect-free area while a common ground connection between the EMATs and the specimen has been established. In this case, the amplitude of the reflections increased, the noise level decreased, and as a result the SNR increased by five times. Figure 9b shows the signal received from the defective area when the EMATs and the specimen had a common ground connection. Similar to Figure 9a, the SNR of the signal increased four times more compared to the signal shown in Figure 8b. Actually, in Figure 9b both reflections from the crack are clearly obvious.

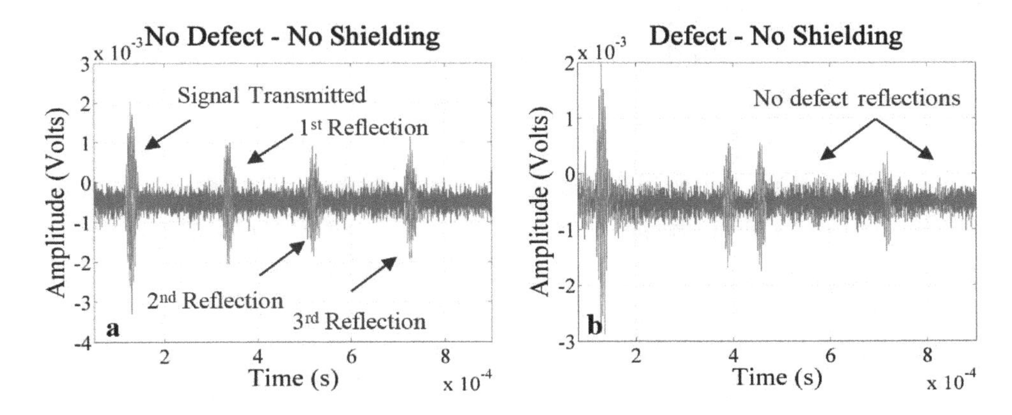


**Figure 8.** (**a**) Signal received from the defect-free area; (**b**) signal received from the defect.

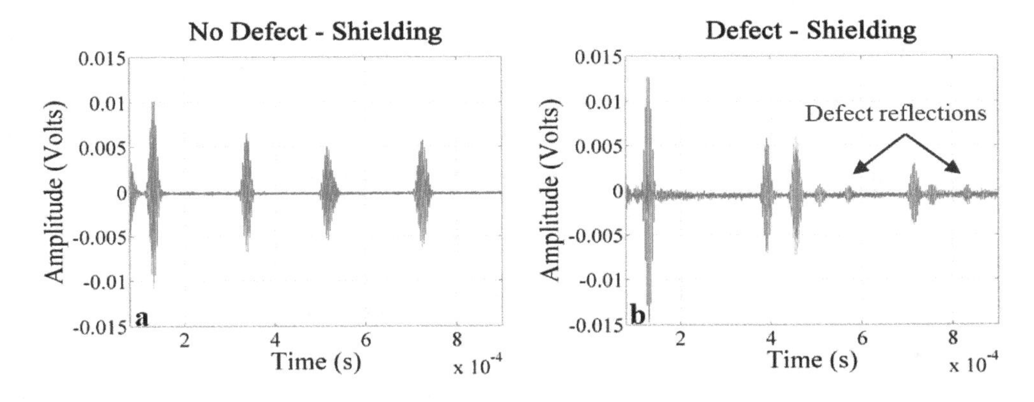


**Figure 9.** (**a**) Signal received from the defect-free area with shielding; (**b**) signal received from the defect with shielding.

Hence, the shielding has resulted in an enhanced SNR due to the noise cancellation. The electromagnetic coupling between the specimen and the EMAT receiver is weak when they are connected as a two-winding transformer (no shielding); in this case, the electromagnetic losses between the EMAT and the specimen are greater than in a conventional two-winding transformer, since no ferrite connects the EMAT and the specimen. The air between them increases the noise level and attenuates the electromagnetic coupling/"wanted" signal (there is no ferrite to drive the electromagnetic wave; on the contrary, the electromagnetic wave scatters when air is between the EMAT and the specimen). When the specimen/EMAT connection performs as an autotransformer, the noise level decreases and the amplitude of the "wanted" signal (ultrasonic response of the specimen) increases; no noise interferes with the EMAT receiver and thus more current is induced to the coil. The autotransformer is more efficient than a two-winding transformer and as a result more current is induced in the coil when the autotransformer connection is established between the EMAT and the specimen. The unshielded EMAT has greater losses than a conventional transformer would have due to the weak connection between the EMAT and the specimen, while an autotransformer has smaller losses compared to both the unshielded EMAT and a conventional transformer. Thus, a significant increase in the amplitude of the "wanted" signal is observed when the EMAT is shielded. Consequently, the voltage difference between the EMAT and the specimen significantly affects the quality of the signal received and when there is no voltage difference and both components are connected to the ground, the probability of defect detection increases as well. Also, the frequency selected based on the size defect and the dispersion curves is proven to be correct for the detection of the 20 mm long defect.

The attenuation of the signal and the ToF of the main four reflections from the edges of the plate remain features that enable us to distinguish the defective from the defect-free areas when there was no common ground connection. Although the velocity of $SH_0$ was not to change during the experiments, for no temperature rise occurred, the ToF of both the first and the second reflection from the edges of the plate change from one case study to the other. It is likely that the defect causes this time delay by trapping a portion of the energy/spectrum of the wave propagating inside the plate.

From all the above figures and more especially from Figure 9a,b, it is obvious that only the $SH_0$ wave mode had been received from the EMAT; the time of arrival of the reflections matches with the $SH_0$ velocity, as it was calculated from the dispersion curves and the electromagnetic model. Hence, the experimental results match with the theoretical. However, further experimental investigation should be conducted regarding the wave purity characteristics of this EMAT design. In this set of experiments the EMAT receiver was mainly evaluated regarding its ultrasonic potentials; laser interferometry tests, during which the actual displacement generated from the EMAT transmitter can be observed, may also be required for further experimental evaluation of the EMAT transmitter regarding its wave purity characteristics.

Figure 10a illustrates how the amplitude of the signal transmitted changes (%) with respect to the lift-off. According to the literature, EMATs are sensitive to lift-off [28] and therefore their efficiency decreases with the increase of lift-off. The amplitude of the signal transmitted decreases almost linearly with the lift-off increase; when the lift-off is equal to 1 mm only the signal transmitted can be clearly observed. This confirms the high sensitivity of EMATs to lift-off; more particularly, the lift-off limitations of EMATs differ depending on the application the EMAT is designed for. EMATs for thickness measurements can still operate efficiently when the lift-off exceeds 1 mm; however, EMATs for guided waves are more sensitive to lift-off. A parameter that influences the performance of EMAT regarding lift-off is the impedance of the coil. The impedance changes with lift-off as well as with the material properties of the specimen. The inductance due to the magnetization of the specimen is smaller when the EMAT is employed on paramagnetic materials compared to ferromagnetic materials. Consequently, an alternative coil design should be used so that EMATs can be efficiently used in the inspection of absorber tubes, since the thickness of the glass envelope is 7 mm; stacked coils may be more efficient.

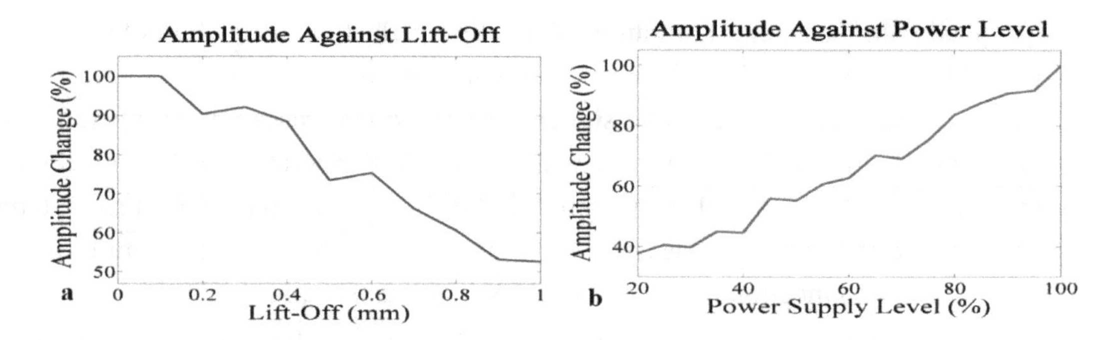


**Figure 10.** Amplitude change of signal transmitted against (**a**) lift-off increase; (**b**) power increase.

An experimental evaluation of these EMATs regarding their power requirements was also conducted. The power level decreased gradually from 100% to 20% with a step of 5%; we stopped there as no useful information could be retrieved from the signal received when the power level was smaller than 20%. Figure 10b shows how the amplitude of the signal transmitted increases with power supply increase; it can be observed that the amplitude increases almost linearly with the power increase. Similar to lift-off, the impedance of the coil affects the power requirements of EMATs. If the pulser/receiver unit drives the EMAT transmitter through an output resistor, the magnitude of the impedance of the coil should be equal to the output resistor of the pulser unit, so that the voltage drop in the coil will be minimized. Therefore, impedance matching is always required, which means that either a unit with zero output impedance should be chosen or an impedance matching network should be added between the pulser and the EMAT transmitter. Impedance matching should be used for the coil to be driven with the maximum power possible so that strong signals will be obtained. Nevertheless, the power supply level of EMATs remains high, leading to the conclusion that EMATs are considered to be more efficient as receivers rather than transmitters.

### *6.2. High-Temperature Experiments*

As was mentioned in the theoretical section, high-temperature EMATs have been designed so far only for thickness measurements and thus an EMAT that can withstand high temperatures and excite guided waves is still required. A first approach for that would be the selection of the suitable high Curie magnets and high-temperature coil or the design of a cooling system so that the EMAT would be as efficient as possible at high temperatures. However, all of the above may result in a more complicated design. Hence, a further study about room temperature, guided wave EMATs, and their performance at high temperatures should be conducted prior to the design of a new EMAT.

Hence, the EMATs were tested from ambient temperature to 180 °C with a step of 10 °C; the maximum operating temperature of these EMATs is 250 °C and therefore they were tested up to 180 °C only, so that any serious and irreversible damage will be avoided. The EMATs were continuously exposed to the heat source with zero lift-off during the rise in temperature, while the overall time they were exposed to the heat was equal to 15 min. This set of experiments was conducted three times. Figure 11a–d show the signal received at room temperature, 60 °C, 100 °C, and 180 °C, respectively. Firstly, in the signal obtained at room temperature, both the reflections from the plate edges and the first two reflections from the defect are clear. However, it is obvious that the amplitude of the second reflection from the plate diminishes greatly after 60 °C, while at 100 °C and 180 °C it can hardly be noticed. Similarly, the amplitute of the first reflection from the defect and the forth reflection from the plate decreases with the increase in temperature.

Figure 12 shows how the amplitude of the signal transmitted decreases with temperature rise; it is clear that the amplitute dwindles almost linearly with the rise in temperature. However, the amplitude of the signal transmitted in 30 °C and 40 °C was slightly larger than the amplitude at room temperature in the areas marked in red in Figure 12. Also, the amplitude error alters with temperature rise. A reason for that may be the ground connection between the EMAT and the specimen. The thermal conductivity of stainless steel is low and therefore the specimen was not heated up uniformly; as a result, the plate bended and the mechanical connection between the EMAT ground and the specimen altered with temperature rise due to the gradient of the bend. This mechanical/electrical connection significantly influences the amplitude of the signal transmitted and thus may be the reason for the amplitude increase at 30 °C and 40 °C.

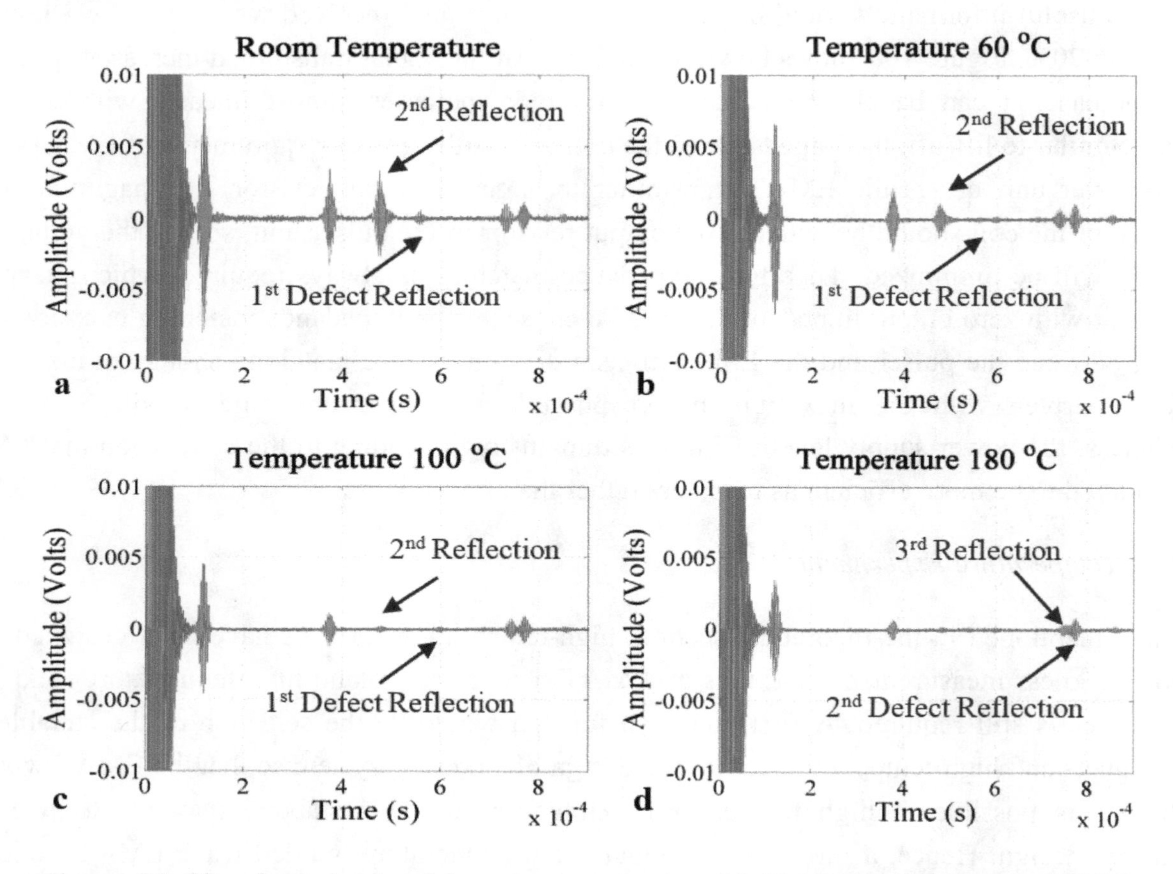


**Figure 11.** Signal received at (**a**) room temperature; (**b**) 60 °C; (**c**) 100 °C; (**d**) 180 °C.

Time shifting is also observed, as was expected due to the change in the wave velocity at high temperatures, presented in the theoretical study. The third reflection from the plate shifts in time and starts coming closer to the second reflection from the crack, leading to an increase of the magnitude of the latter. Consequently, the experimental results match the theoretical. Temperature compensation should take place and both the ToF of the reflections as well as the drop of their amplitude may be two features that can be further processed and used for the identification of the temperature of the structure being tested. Room-temperature EMATs cannot be used for the monitoring of high-temperature structures (>100 °C); however, a high-temperature EMAT specifically designed for the monitoring of high-temperature structures (>200 °C) should be compared with the performance of this EMAT up to 100 °C.

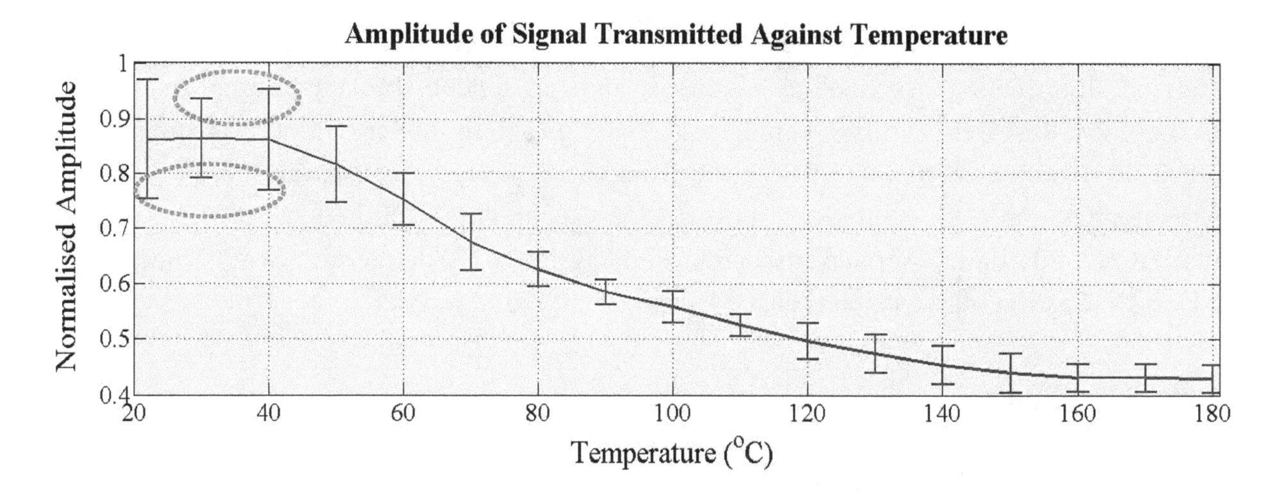


**Figure 12.** Amplitude of the signal transmitted against temperature.

## 7. Conclusions and Future Work

The absorber tube is an essential part of Parabolic Trough CSPs and is very likely to get damaged due to its hostile operating conditions. Hence, NDT techniques are required for their monitoring and/or inspection. A promising technique for this application is the use of high-temperature EMAT transducers for the excitation of guided waves and more particularly of the T(0,1) wave mode. A theoretical study about the GWT of absorber tubes, their dispersion curves, and the wave mode that should be applied for their inspection were calculated. The ultrasonic response of a PPM EMAT and its wave purity characteristics were also presented. A pair of PPM EMATs was experimentally evaluated regarding its wave purity, its sensitivity to noise, its lift-off limitations, its power requirements, and its performance at high temperatures while it was exciting/receiving guided waves. It was found that a PPM EMAT receiver can mainly detect $SH_0$ and thus it can be used for the GWT of the absorber tubes in terms of wave purity characteristics, as the theoretical study also showed. However, laser interferometry tests are also required for the transmitter to be validated. Also, a common ground connection between the EMAT and the specimen can significantly enhance the SNR of the signal received. The current EMAT design is not efficient enough to inspect a stainless steel pipe with a lift-off larger than 1 mm and therefore it cannot be employed for the inspection of absorber tubes. A room-temperature EMAT cannot be applied at high temperatures (<200 °C) and thus a high-temperature EMAT is still required for the guided wave monitoring of high-temperature structures.

The design and manufacturing of a high-temperature PPM EMAT with racetrack coil operating efficiently at temperatures higher than 300 °C for long-term inspection or monitoring is our next step. Several thermal, electromagnetic, and mechanical simulations have already been carried out and have given encouraging results. Moreover, a further experimental investigation regarding the impedance of the EMAT and its relationship with the lift-off and the material properties of the specimen being tested is also another part of our future research.

## Acknowledgments

The authors are indebted to the European Commission for the provision of funding through the INTERSOLAR FP7 project. The INTERSOLAR project is coordinated and managed by Computerized Information Technology Limited and is funded by the European Commission through the FP7 Research for the benefit of SMEs program under Grant Agreement Number: GA-SME-2013-1-605028. The INTERSOLAR project is a collaborative research project between the following organizations: Computerized Information Technology Limited, PSP S.A., Technology Assistance BCNA 2010 S.L., Applied Inspection Limited, INGETEAM Service S.A., Brunel University, Universidad De Castilla—La Mancha (UCLM), and ENGITEC Limited.

## Author Contributions

M.K. performed the literature review and the finite element simulations, conducted all the experiments and wrote the paper; A.M., V.K. and C.S. reviewed the manuscript; T.-H.G., W.B. and L.C. supervised the research.

## References

1. Poullikkas, A. Economic analysis of power generation from parabolic trough solar thermal plants for the Mediterranean region—A case study for the island of Cyprus. *Renew. Sustain. Energy Rev.* **2009**, *13*, 2474–2484.
2. Staff, C.W. Industry Experts Trumpet Untold Solar Potential. 2013. Available online: http://www.constructionweekonline.com/article-23701-industry-experts-trumpet-untold-solar-potential/ (accessed on 12 March 2015).
3. Taggart, S. Parabolic troughs: Concentrating Solar Power (CSP)'s Quiet Achiever. 2008. Available online: http://www.renewableenergyfocus.com/view/3390/parabolic-troughs-concentrating-solar-power-csp-s-quiet-achiever (accessed on 12 March 2015).
4. Steinfeld, A. Topics for Master/Bachelor Thesis and Semester Projects: Spectral Optical Properties of Glass Envelopes of Parabolic trough Concentrators. Available online: http://www.pre.ethz.ch/teaching/topics/?id=87 (accessed on 18 August 2015).
5. Guillot, S.; Faika, A.; Rakhmatullina, A.; Lambert, V.; Verona, E.; Echegut, P.; Bessadaa, C.; Calvet, N.; Py, X. Corrosion effects between molten salts and thermal storage material for concentrated solar power plants. *Appl. Energy* **2012**, *94*, 174–181.
6. Herrmann, U.; Kearney, D.W. Survey of Thermal Energy Storage for Parabolic Trough Power Plants. *J. Sol. Energy Eng.* **2002**, *124*, 145–152.
7. Papaelias, M.; Cheng, L.; Kogia, M.; Mohimi, A.; Kappatos, V.; Selcuk, C.; Constantinou, L.; Muñoz, C.Q.G.; Marquez, F.P.G.; Gan, T.H. Inspection and structural health monitoring techniques for concentrated solar power plants. *Renew. Energy* **2015**, *85*, 1178–1191.

8. Lee, J.L.; Hu, L.W.; Saha, P.; Kazimi, M.S. Numerical analysis of thermal striping induced high cycle thermal fatigue in a mixing tee. *Nucl. Eng. Des.* **2009**, *239*, 833–839.
9. Noguchi, Y.; Okada, H.; Semba, H.; Yoshizawa, M. Isothermal, thermo-mechanical and bithermal fatigue life of Ni base alloy HR6W for piping in 700 °C USC power plants. *Procedia Eng.* **2011**, *11*, 1127–1132.
10. Jinu, G.R.; Sathiya, P.; Ravichandran, G.; Rathinam, A. Comparison of thermal fatigue behaviour of ASTM A 213 grade T-92 base and weld tubes. *J. Mech. Sci. Technol.* **2010**, *24*, 1067–1076.
11. Sunny, S.; Patil, R.; Singh, K. Assessment of thermal fatigue failure for BS 3059 boiler tube experimental procedure using smithy furnace. *Int. J. Emerg. Technol. Adv. Eng.* **2012**, *2*, 391–398.
12. Posteraro, K. Thwart Corrosion under Industrial Insulation. *Chem. Eng. Prog.* **1999**, *95*, 43–47.
13. Halliday, M. Preventing corrosion under insulation-new generation solutions for an age old problem. *J. Prot. Coat. Linings* **2007**, *24*, 24–36.
14. De Vogelaere, F. Corrosion under insulation. *Process. Saf. Prog.* **2009**, *28*, 30–35.
15. Kumar, M.S.; Sujata, M.; Venkataswamy, M.A.; Bhaumik, S.K. Failure analysis of a stainless steel pipeline. *Eng. Fail. Anal.* **2008**, *15*, 497–504.
16. Stine, W.B.; Harrigan, R.W. *Solar Energy Systems Design*; John Wiley and Sons, Inc.: New York, NY, USA, 1986.
17. Hutchins, D.A.; Saleh, C.; Moles, M.; Farahbahkhsh, B. Ultrasonic NDE Using a Concentric Laser/EMAT System. *J. Nondestruct. Eval.* **1990**, *9*, 247–261.
18. Kirk, K.J.; Lee, C.K.; Cochran, S. Ultrasonic thin film transducers for high-temperature NDT. *Insight* **2005**, *47*, 85–87.
19. Momona, S.; Moevus, M.; Godina, N.; R'Mili, M.; Reynauda, P.; Fayolle, G. Acoustic emission and lifetime prediction during static fatigue tests onceramic-matrix-composite at high temperature under air. *Compos. A Appl. Sci. Manuf.* **2010**, *41*, 913–918
20. Shen, G.; Li, T. Infrared thermography for high temperature pipe. *Insight* **2007**, *49*, 151–153.
21. Beattie, G. Acoustic emission principles and instrumentation. *J. Acoust. Emiss.* **1983**, *2*, 95–128.
22. Hagemaier, D.J. *Fundamentals of Eddy Current Testing*; American Society for Nondestructive, Testing: Columbus, OH, USA, 1990.
23. Malmo, J.T.; Jøkberg, O.J.; Slettemoen, G.A. Interferometric testing at very high temperatures by TV holography (ESPI). *Exp. Mech.* **1998**, *28*, 315–321.
24. Silk, M.G.; Bainton, K.F. The propagation in metal tubing of ultrasonic wave modes equivalent to Lamb waves. *Ultrasonics* **1979**, *17*, 11–19.
25. Pfander, M.; Lupfert, E.; Pistor, P. Infrared temperature measurements on solar trough absorber tubes. *Sol. Energy* **2007**, *81*, 629–635.
26. Cawley, P.; Alleyne, D. The use of Lamb waves for the long range inspection of large structures. *Ultrasonics* **1996**, *34*, 287–290.
27. Hirao, M.; Ogi, H. A SH-wave EMAT technique for gas pipeline inspection. *NDT E Int.* **1999**, *32*, 127–132.
28. Bottger, W.; Schneider, H.; Weingarten, W. Prototype EMAT system for tube inspection with guided ultrasonic waves. *Nucl. Eng. Des.* **1986**, *102*, 369–376.

29. Wilcox, P.; Lowe, M.; Cawley, P. Omnidirectional guided wave inspection of large metallic plate structures using an EMAT array. *IEEE Trans. Ultrason. Ferroelectr. Freq. Control* **2005**, *52*, 653–665.
30. Andruschak, N.; Saletes, I.; Filleter, T.; Sinclair, A. An NDT guided wave technique for the identification of corrosion defects at support locations. *NDT E Int.* **2015**, *75*, 72–79.
31. Burrows, S.E.; Fan, Y.; Dixon, S. High temperature thickness measurements of stainless steel and low carbon steel using electromagnetic acoustic transducers. *NDT E Int.* **2014**, *68*, 73–77.
32. Hernandez, V.F. Pulsed-Electromagnet EMAT for High Temperature Applications. Ph.D. Thesis, University of Warwick, Warwick, UK, 2011.
33. Idris, A.; Edwards, C.; Palmer, S.B. Acoustic wave measurement at elevated temperature using a pulsed laser generator and an Electromagnetic Acoustic Transducer detector. *Nondestruct. Test. Eval.* **1994**, *11*, 195–213.
34. Dixon, S.; Edwards, C.; Palmer, S.B. High-accuracy non-contact ultrasonic thickness gauging of aluminium sheet using electromagnetic acoustic transducer. *Ultrasonics* **2001**, *39*, 445–453.
35. Jian, X.; Dixon, S.; Edwards, R.S.; Morrison, J. Coupling mechanism of an EMAT. *Ultrasonics* **2006**, *44*, 653–656.
36. Ribichini, R.; Cegla, F.; Nagy, P.B.; Cawley, P. Study and Comparison of Different EMAT Configurations for SH Wave Inspection. *IEEE Trans. Ultrason. Ferroelectr. Freq. Control* **2011**, *58*, 2571–2581.
37. Gaultier, J.; Mustafa, V.; Chahbaz, A. EMAT Generation of Polarized Shear Waves for Pipe Inspection. In Proceeding of the 4th PACNDT, Toronto, ON, Canada, 14–18 September 1998.
38. Rose, J.L.; Lee, C.M.; Hay, T.R.; Cho, Y.; Park, I.K. Rail Inspection with Guided Waves. In Processing of the 12th A-PCNDT—Asia-Pacific Conference on NDT, Auckland, New Zealand, 5–10 November 2006.
39. Jackel, P.; Niese, F. EMAT Application: Corrosion Detection with Guided Waves in Rod, Pipes and Plates. In Proceedings of the 11th European Conference on Non-Destructive Testing (ECNDT 2014), Prague, Czech, 6–10 October 2014.
40. Hübschen, G. Generation of Horizontally Polarized Shear Waves with EMAT Transducers. *e-J. Nondestruct. Test.* **1998**, *3*, 1–7.
41. Kogia, M.; Mohimi, A.; Liang, C.; Kappatos, V.; Selcuk, C.; Gan, T.H. High temperature Electromagnetic Acoustic Transducer for the inspections of jointed solar thermal tubes. In Proceedings of the First Young Professionals International Conference, Budapest, Hungary, 17–20 September 2014.
42. Pulse Electronics, Understanding Common Mode Noise. Available online: www.pulseelectronics.com/download/3124/g204/pdf (accessed on 10 November 2015).
43. Winders, J. Autotransformers and Three-Winding Transformers. In *Power Transformers: Principles and Applications*; 1st ed.; Marcel Dekker: New York, NY, USA, 2002.

# 13

# Enhancement of Spatial Resolution using a Metamaterial Sensor in Nondestructive Evaluation

**Adriana Savin [1,*], Alina Bruma [2,*], Rozina Steigmann [1,3], Nicoleta Iftimie [1] and Dagmar Faktorova [4]**

[1] Nondestructive Testing Department, National Institute of R&D for Technical Physics, Iasi 700050, Romania; E-Mails: steigmann@phys-iasi.ro (R.S.); niftimie@phys-iasi.ro (N.I.)
[2] CRISMAT Laboratory, Ecole Nationale Superieure d'Ingenieurs de Caen, Universite de Caen Basse Normandie, 6 Blvd Marechal Juin, Caen 14050, France
[3] Faculty of Physics, University Al.I. Cuza, 11 Carol I Blvd, Iasi 700506, Romania
[4] Faculty of Electrical Engineering, University of Žilina, Univerzitná 1, Žilina 010 26, Slovakia; E-Mail: dagmar.faktorova@fel.uniza.sk

† This paper is an extended version of paper published in the 6th International Conference on Emerging Technologies in Non-destructive Testing, ETNDT6 held in Brussels, 27–29 May 2015.

* Authors to whom correspondence should be addressed: E-Mails: asavin@phys-iasi.ro (A.S.); bruma.alina@outlook.com (A.B.)

Academic Editor: Dimitrios G. Aggelis

---

**Abstract:** The current stage of non-destructive evaluation techniques imposes the development of new electromagnetic methods that are based on high spatial resolution and increased sensitivity. Printed circuit boards, integrated circuit boards, composite materials with polymeric matrix containing conductive fibers, as well as some types of biosensors are devices of interest in using such evaluation methods. In order to achieve high performance, the work frequencies must be either radiofrequencies or microwaves. At these frequencies, at the dielectric/conductor interface, plasmon polaritons can appear, propagating between conductive regions as evanescent waves. Detection of these waves, containing required information, can be done using sensors with metamaterial lenses. We propose in this paper the enhancement of the spatial resolution using electromagnetic methods, which can be accomplished in this case using evanescent waves that appear in the current study in slits of materials such as the spaces between carbon fibers in Carbon Fibers Reinforced Plastics or in materials of interest in the nondestructive evaluation field with industrial applications, where microscopic cracks are present. We propose herein a unique design of the metamaterials for use in nondestructive evaluation based on Conical Swiss Rolls configurations, which assure the robust concentration/focusing of the incident electromagnetic waves (practically impossible to be focused using classical materials), as well as the robust manipulation of evanescent waves. Applying this testing method, spatial resolution of approximately λ/2000 can be achieved. This testing method can be successfully applied in a variety of applications of paramount importance such as defect/damage detection in materials used in a variety of industrial applications, such as automotive and aviation technologies.

**Keywords:** nondestructive evaluation; metamaterials lens; metallic strip gratings; fiber reinforced plastic composites; evanescent waves

## 1. Introduction

In recent years, several nondestructive evaluation (NDE) techniques have been developed for detecting the effect of damages/embedded objects in homogeneous media.

The electromagnetic nondestructive evaluation (eNDE) of materials consists in the application of an electromagnetic (EM) field with frequencies ranging from tens of Hz to tens of GHz, to the examined object and evaluating the interaction between the field and the eventually material discontinuities. A generic NDE system is presented in Figure 1.

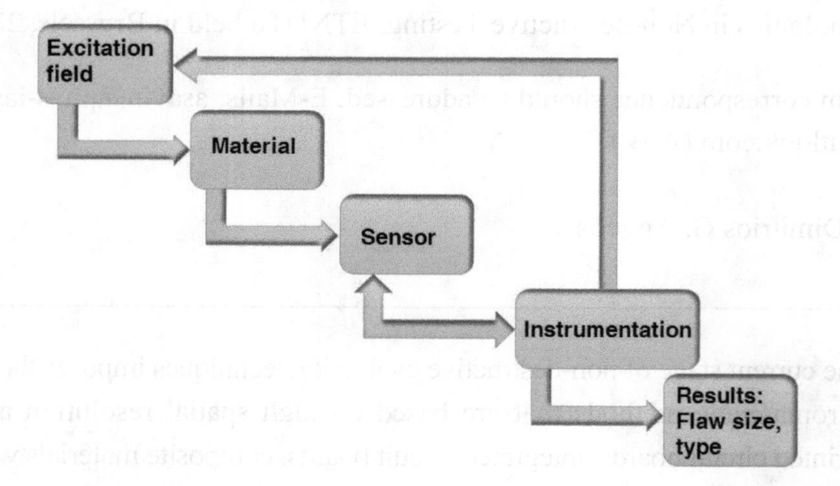


**Figure 1.** Principles of nondestructive evaluation operation.

Effective data acquisition and display capabilities have led to developments in extraction and recording information about discontinuities and material properties. Fundamentally, these methods were involved with the evolution of reflected and/or transmitted waves after interacting with the test part. If the examined object is electrically conductive, under the action of the incident EM field, eddy currents will be induced in the material, according to Faraday's Law [1].

If the incident EM field has low frequency, usually created by coils circulated by alternating current [2–9] or current impulses [10–13], different shapes of coils [14] assure the detection of the secondary EM field. In order to obtain a better signal/noise ratio [15–19], it is necessary to use the smallest possible lift-off (distance between the EM transducer and the controlled piece). This requires working in the near field because the generated and scattered EM waves are evanescent waves (waves that are rapidly attenuated with distance [20] and are difficult to be focalized using classical materials).

Recently, a new approach to the design of EM structures has been introduced, where the paths of EM waves are controlled by modifying constitutive parameters through a variation of space features. The interaction of the EM field with periodical metallic structures constitutes an interesting domain, both from a theoretical point of view as well as experimentally [21–24]. The metallic strip gratings (MSG) can act as filters and polarizers [25], also representing the basic elements in rigid and flexible printed circuits boards, serving to supply and to the transmission of the signals. MSGs present special properties when they are excited with a transversal electric/magnetic along the $z$ axis (TEz/TMz) polarized EM field. These structures are intensively studied from a theoretical point of view [26,27], for obtaining complex information about their behavior in electronic applications [28] and the design of new types of metamaterials (MMs) starting from the existence of surface plasmons polaritons (SPPs) [29]. These applications impose a rigorous and rapid quality control.

Extending the study to composite materials, carbon fiber reinforced plastics (CFRP) with uniaxial symmetry [30], as well in the case of reinforcement with carbon fibers woven [7,31], the possibility of manipulating the evanescent waves that appear in the space between fibers indicates the necessity to improve the spatial resolution of the sensors, to emphasize microscopic flaws.

The composite materials found wide usage in modern technologies due to the development of industries [32], ranging from medicine and sporting goods to aeronautics components. These composite components can have various dimensions, from very small panels used in satellite structures to large ones used in miniature naval vessels as hulls, 30 m long and 25 mm thick.

Fiber-reinforced polymer composites (FRPC) have superior mechanical properties to metallic structures, assuring simple manufacturing of layered products [33]. FRPCs are classified by the type of reinforcement fibers (carbon, glass, or aramids) or the type of matrix (thermoset or thermoplastic). Despite the fact that thermoplastic composites are more expensive, they have been preferred in the construction of complex structures, due to the advantages of thermoplastic matrix such as recyclability, aesthetic finishing, high impact resistance, chemical resistance, hard crystalline or rubbery surface options, and eco-friendly manufacturing, making the entire process cost less [34]. Woven carbon fibers are recommended due to their high strength-to-weight ratio. Polyphenylene sulfide (PPS) has excellent properties [35], tying into the advantages described above. Woven carbon fibers/PPS laminates are characterized by reduced damages but are susceptible to impacts with low energies, leading to delaminations, desbonding of carbon fibers, and/or matrix cracking [36]. The aerospace industry requires the highest quality control and product release specifications [37,38]. The raw material standard, prepreg materials, require mechanical property testing (*i.e.*, interlaminar strength and tensile strength) from specimens. Final parts also require NDE [39] and structural health monitoring (SHM) [40,41]. The presence of different types of defects such as voids, inclusions, desbondings, improper cure, and delaminations are common during the manufacture and use of composite materials.

Alternative techniques based on phenomenological changes in the composite materials were developed to measure damage due to impacts: acoustic emission [42,43], infrared imaging [44,45], electrical resistance [46,47], or non-contact techniques such as digital image correlation [48] and X-ray tomography [49,50], but these methods have their limitations, being complementary. Nowadays, NDE methods are developed for post-damage inspection and assessment such as thermography [51,52], ultrasonic C-scan [53,54], eddy current [55,56], and optical fiber [57,58], offering quantitative results. Other techniques can be used for online health monitoring of CFRP structures, embedding external sensors or additional fiber input in CFRPs.

This paper proposes the possibility to enhance the spatial resolution of eNDE methods applied to MSG and to CFRP, using a sensor with MM lenses. For this, special attention is granted to the sensor based on a lens [59,60] with conical Swiss rolls, which allows for the manipulation of evanescent waves created in slits and in the dielectric insulating the carbon fibers, respectively, and can reach a spatial resolution for visualization of carbon fibers' layout and eventually of flaws such as delamination created by impact.

## 2. Metamaterial Sensor for eNDE and Theory

MMs, EM structures with distinguished properties, have started to be studied especially in the last few years. EM MMs belong to the class of artificially engineered materials, which can provide an engineered response to EM radiation, not available from naturally occurring materials. These are often defined as the structures of metallic and/or dielectric elements, periodically arranged in two or three dimensions [61]. The size of the structure is typically smaller than the free space wavelength of incoming EM waves. Nowadays, multitudes of MM structural elements type are known, conferring special EM properties. Depending on the frequency of the incident EM field, the type and geometrical shape of the MM may have a high relative magnetic permeability, either positive or negative [61]. Also, MM lenses allow the amplification of evanescent waves [62]. These properties strongly depend on the geometry of MM rather than their composition [63], and were experimentally demonstrated [64]. MMs have started to interest engineers and physicists due to their wide application in perfect lens [65], slow light [66], data storage [67], *etc.*

In order to find the effective permittivity and permeability of a slab of MM, the material has to be homogeneous. The permittivity and permeability can be found from the $S$ parameters data. For a MM slab characterized by effective permittivity $\varepsilon_{eff}$ and effective magnetic permeability $\mu_{eff}$, the refractive index is:

$$n = \sqrt{\varepsilon_{\text{eff}} \mu_{\text{eff}}} \tag{1}$$

and the impedance is given by:

$$Z = \sqrt{\frac{\mu_{\text{eff}}}{\varepsilon_{\text{eff}}}} \tag{2}$$

The relationship between $\varepsilon$ and $\mu$ for a medium as well as the wave propagation through it can categorize the media into the following classes [68,69]:

(a) Double positive medium (DSP) when $\varepsilon > 0$ and $\mu > 0$; Only propagating waves;

(b) Single negative medium-electric negative (ENG), when $\varepsilon < 0$ and $\mu > 0$; only evanescent waves;

(c) Double negative medium (DNG) when $\varepsilon < 0$ and $\mu < 0$; propagating waves and evanescent waves;

(d) Single negative medium-magnetic negative (MNG), when $\varepsilon > 0$ and $\mu < 0$; only propagating waves.

When the effective electrical permittivity $\varepsilon_{eff}$, and the effective magnetic permeability $\mu_{eff}$ of a MM slab are simultaneously −1, the refractive index of the slab is $n = -1$ [70]. Therefore, the surface impedance of such an MM is $Z = 1$, there is no mismatch and consequently no reflection at the slab-air interface [64]. This MM slab forms a perfect lens [62] and is not only focusing the EM field, but also focuses the evanescent waves [62].

Due to experimental difficulties in obtaining a perfect lens, the manipulation of the evanescent modes can be made with an EM sensor with MM lenses that have, at the operation frequency, either $\varepsilon_{eff} = -1$ when electric evanescent modes can be manipulated, or $\mu_{eff} = -1$ when the lens can focus magnetic evanescent modes [61]. Moreover, working at a frequency that ensures $\mu_{eff} = -1$ for the same lens, the magnetic evanescent modes can be manipulated [20,71].

According to the above classes, as shown in Ref. [30], the electric evanescent modes can be manipulated with a sensor made from a special MM, named Conical Swiss Roll (CSR) [31], functioning in a frequency range that ensures the maximum $\mu_{eff}$ [60]; it did not accomplish the conditions imposed for a perfect lens but will lead to the substantial enhancement of spatial resolution. The spatial resolution of the system (the distance between two distinctively visible points) was verified according to [59] and the analysis of data obtained shows that the realization of MM lenses in the RF range is possible using the CSR, whose distortions are minimal and whose calculation is made based on Fourier optic principles.

EM sensors with MM lenses have been made using two CSRs, the operation frequencies depending both on the constitutive parameters of MM and the polarization of the incident EM field ($TE_z$ or $TM_z$). Figure 2 shows the developed sensor with MM [7,59,60]. A CSR consists of a number of spiral-wound layers of an insulated conductor on a conical mandrel [31]. The EM sensor is absolute send-receive type; it has the emission part made from one-turn rectangular coil with 35 × 70 mm dimensions, using Cu wire with a 1 mm diameter. When an EM $TM_z$ polarized waves acts, at normal incidence, the magnetic field being parallel with the y axis such that $H_x = H_z = 0$ and $H_y \neq 0$, in very near field, between carbon fibers similar with MSG, evanescent waves can appear. In the focal image point a reception coil with MM lens is placed, having one turn with 1 mm average diameter made of Cu wire with 0.1 mm diameter, to convert localized energy in electromotive force (e.m.f.). An optimized work frequency of 476 MHz assures magnetic effective permeability of 22. At this frequency, the property of CSR to act as an alternative magnetic flux concentrator has been verified.

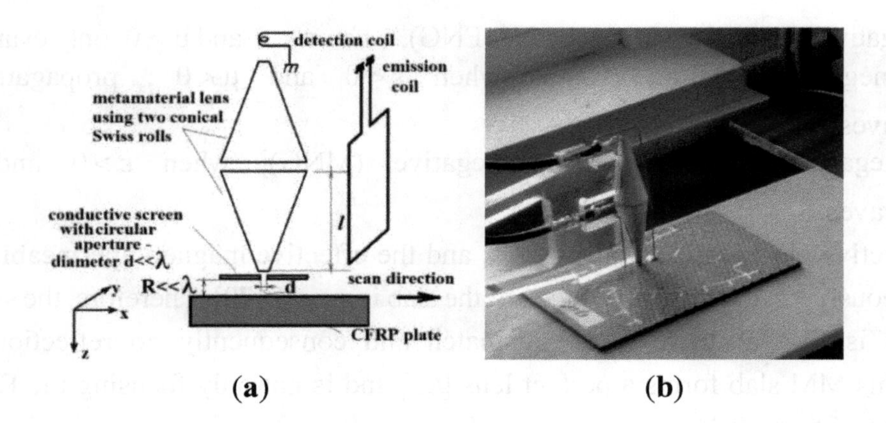


**Figure 2.** Sensor with MM lens: (**a**) schematic representation; (**b**) photograph.

The calculus of the MM lens based on CSR was presented in Ref. [31]; the field in the focal plane of the lens is given by

$$H\left(x,y,z_0+2l\right)=\frac{H_0}{\pi^2}e^{j\left(\frac{k_b+k_a}{2}x+\frac{k_b+k_a}{2}y\right)}\frac{\sin\left(\frac{k_b-k_a}{2}x\right)}{x}\frac{\sin\left(\frac{k_b-k_a}{2}y\right)}{y},\tag{3}$$

where $z_0 = R$ (Figure 2a) with $R << \lambda$, and $H_0$ is the amplitude of incident magnetic field.

The principle scheme of sensor for the evanescent wave's detection is shown in Figure 2a. In order to enhance the spatial resolution of the sensor, a conductive screen having a circular aperture made from perfect electric conductor (PEC) material with a very small diameter is placed in front of the lens. The circular aperture serves for the diffraction of the evanescent waves that can occur on slits. This ensures paraxial incident beam. The diameter of focal spot provided by MM lens is given by [72]:

$$D=\frac{4\pi}{k_b-k_a}\tag{4}$$

and is equal with the diameter of the small basis of the conical Swiss roll, *i.e.*, 3.2 mm. The MM lens with CSR will be displaced along the *x*-axis (Figure 2a). When $k_a = 0$ and this value is inserted into Equation (4), $k_b$ is obtained and the field in the focal plane is calculated with Equation (3). The detection principle is similar with near-field EM scanning microscopy (NFESM). NFSEM imaging is a sampling technique, *i.e.*, the sample (in our case MSG or carbon fibers) is probed point by point by raster scanning with the sensor over the sample surface and recording for energy image pixel a corresponding EM signature.

Selecting a region with dimensions $\left(x_c+x_d\right)\cdot\left(x_c+x_d\right)$, (where $x_c$ and $x_d$ are coordinates for conducting/dielectric material), using the Fourier optics methods [72,73], an object $O(x,y)$ that can represent the eigenmodes in function of the polarization of an incident EM field, has, while passing through the circular aperture and the lens, an image $I(x', y')$, given by [68]:

$$I(x',y')=\frac{1}{\lambda^2 d_1 d_2}\int_{-\infty}^{\infty}\int_{-\infty}^{\infty}\exp\left[i\frac{k\left((x'-x_1)^2+(y'-y_1)^2\right)}{2d_2}\right]P(x,y)\exp\left[i\frac{k\left(x_1^2+y_1^2\right)}{2f}\right]\times \left(\int_{-\infty}^{\infty}\int_{-\infty}^{\infty}O(x,y)\exp\left[i\frac{k\left((x_1-x)^2+(y_1-y)^2\right)}{2d_1}\right]\mathrm{d}x\mathrm{d}y\right)\mathrm{d}x_1\mathrm{d}y_1 \tag{5}$$

where $P(x,y)$ is the pupil function defined as [73**Error! Bookmark not defined.**]:

$$P(x,y)=\begin{cases}1 & x^2+y^2\le d^2\\ 0 & \text{otherwise}\end{cases} \tag{6}$$

$O(x,y)$ is the object defined as:

$$O(x,y)=\begin{cases}e_v(x,y) & \text{for TEz polarized incident waves}\\ h_v(x,y) & \text{for TMz polarized incident waves}\end{cases} \tag{7}$$

$f$ is the focal distance of the lens equal with the height of CSR; $\lambda$ is the wavelength in a vacuum; $k=2\pi/\lambda$ is the wave number; $d_1=R+l$ is the distance from the object to the center of the lens; and $d_2=l$ is the distance from the center of the lens to the detecting coil.

The most convenient method from an experimental point of view is to measure $S$ parameters for the MM that fill a waveguide or in free space, using an emission and reception antennas. The relation between $S_{11}$ and $S_{21}$, applying the effective method [74] using a 4395A Network/Spectrum/Impedance Analyzer Agilent (Agilent Technologies, Santa Clara, CA, USA) coupled with an Agilent 87511A S Parameters Test kit and effective refractive index $n$, is given by [75,76]:

$$S_{11}=\frac{R_{0_1}\left(1-\mathrm{e}^{j2nk_0d}\right)}{1-R_{0_1}^2\mathrm{e}^{j2nk_0d}}\quad S_{21}=\frac{\left(1-R_{0_1}^2\right)\mathrm{e}^{j2nk_0d}}{1-R_{0_1}^2\mathrm{e}^{j2nk_0d}} \tag{8}$$

where $R_{0_1}=\dfrac{Z-1}{Z+1}$ and the impedance $Z$ is obtained by inverting Equation (8) yielding:

$$Z=\pm\sqrt{\frac{(1+S_{11})^2-S_{21}^2}{(1-S_{11})^2-S_{21}^2}}\quad e^{jnk_0d}=X\pm j\sqrt{1-X^2} \tag{9}$$

where $X=\dfrac{1}{2S_{21}(1-S_{11}^2+S_{21}^2)}$.

The focal distance of the lens using an MM is given in Ref. [31] with $f\cong l$, where $l$ is the height of a CSR. Assuming that the passive MM slab has an effective refractive index $n$ and impedance $Z$, according to Ref. [74] the effective permittivity $\varepsilon_{\text{eff}}$ and permeability $\mu_{\text{eff}}$ are directly $\mu_{\text{eff}}=nZ$ and $\varepsilon_{\text{eff}}=n/Z$.

## 3. Studied Samples and Experimental Setup

The functioning of the MM sensor has been verified using two types of materials, MSGs and FRPC.

### *3.1. Metallic Strip Gratings (MSGs)*

Metallic films on flexible substrates (polyimide or plastic) currently have superior mechanical properties compared to ones deposited on glass, in many aspects. Although the glass substrate is hard, the polyimide or plastic substrate is lighter, less expensive, flexible, and more suitable for use in small devices [77]. Other applications include protective coatings, EM shielding, and electric current conductors for microelectronic applications.

Two types of MSG were taken into consideration:

Flexible printed circuit with transparent polyimide support, 80 μm thickness, with silver conductive strip of 10 μm thickness, the width of the strips being $x_c$ = 0.6 mm and the width of the slits being $x_d$ = 0.4 mm;

Silver strips realized with polyimide support of 65 μm thickness made from a silver strip having 14 μm thickness, the width of the strips being $x_c$ = 1.2 mm and the width of the slits being $x_d$ = 0.8 mm.

For strips having $x_c$ = 1.2 mm, we have taken into account interruptions as well as non-alignments. Silver strips were realized by successive deposition of silver paste, using an adequate stencil by screen printed method. The adhesion of silver on polyimide has been done with a thin film of resin. The conductive silver paste made from microparticles with concentration >80%, density 10.49 g/cm$^3$, resistivity $1 \div 3 \times 10^{-5}$ Ω cm was used [78]. The silver paste has good adhesion and fast drying speed at room temperature. At frequencies around the value of 500 MHz, the permittivity of silver [79] is $\varepsilon_m = -48.8 + j \cdot 3.16$.

The polyimide as flexible substrate can be easily embedded in 3D structures, also ensuring retention of bioactive species in the developed structure. They satisfy just about all the requirements for electronics applications, being lightweight, flexible, and resistant to heat and chemicals [80]. The flexible MSG structures are presented in Figure 3a,b.

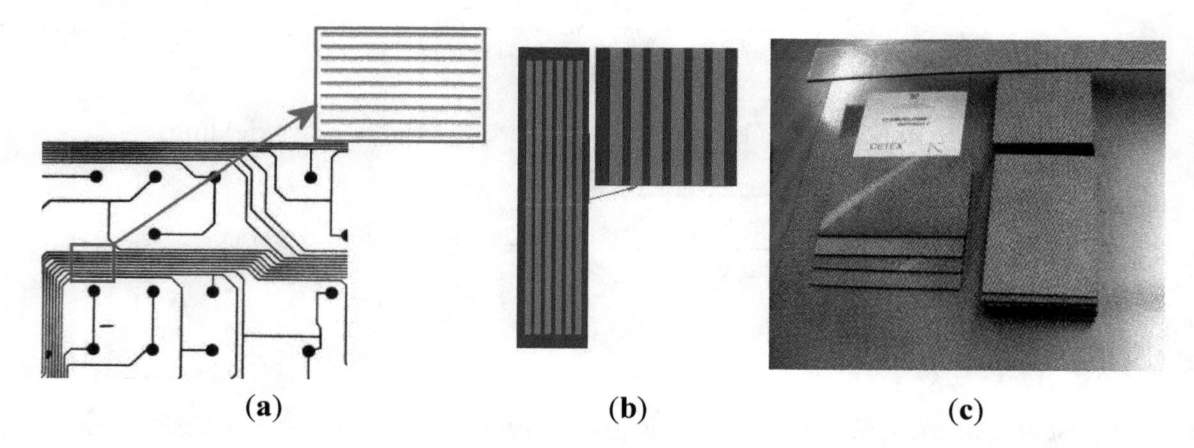


(a) (b) (c)

**Figure 3.** Studied samples: (**a**) MSGs from flexible printed circuit; (**b**) MSGs realized with polyimide support; (**c**) Plates from FRPC.

For the MSG having the width of strips $x_c$ = 0.6 mm, the study is focused on the appearance of abnormal and/or evanescent modes for the cases where various dielectric fluids fill the gaps between the strips, and also improvement of the EM images to obtain a better spatial resolution in order to exceed the limit imposed by diffraction.

### *3.2. Plates from FRPC Composite Materials*

The study involved quasi-isotropic FRPC plates produced by Tencate [35] (Almelo, The Netherlands), having 150 × 100 × 4.2 $mm^3$, containing 12 layers of 5 harness satin (5HS) carbon fibers woven with balanced woven fabric [81] (Figure 3c). The matrix is made of PPS, a thermoplastic polymer consisting of aromatic rings linked with sulfide moieties. It is resistant to chemical and thermal attack, and the amount of gas released due to matrix ignition is low. The carbon fibers are T300JB type (TORAYCA, Santa Ana, CA, USA); their volume ratio is 0.5 ± 0.03 and the density is 1460 $kg/m^3$.

Carbon fiber woven embedded into PPS offers strength of composite to impact. The FRPC has the transverse electric conductivity between 10 S/m and 100 S/m and longitudinal conductivity ranging between $5 \times 10^3$ S/m and $5 \times 10^4$ S/m and is paramagnetic, allowing eNDE. The samples were impacted with an energy of 8 J [59], in order to induce delaminations. The impacts are induced with FRACTOVIS PLUS 9350 CEAST (Instron, Nordwood, MA, USA), which allows the modification of impact energies as well as the temperature during impact. The conditions of the impact are designed according to ASTM D7136 [82] at a temperature of 20 °C. Under impact, the FRPCs suffer delamination, usually accompanied by a dent, deviation, and/or breaking of the carbon fibers. In all cases, a reduction in the space between fibers in the thickness direction appears and this causes an increase in fiber contact, leading to a decrease of electrical resistance in the thickness direction and modifying the local electrical conductivity both in the plane of the fibers and perpendicularly on fibers [83]. The energy absorbed by the composite serves as the plastic deformation of the composite in the contact zone, being dissipated through internal friction between the matrix's molecules, carbon fibers, and matrix-carbon fibers, as well as at the creation of delaminations. Typical records of force *vs.* time during impact can give information about the FRPC status (delaminated or not) [84].

The generation and detection of evanescent waves from slits/fibers has been made using a sensor with MM lens and the equipment presented in Figure 4. The rectangular frame used for the generation of the $TM_z$ polarized EM field has 35 × 70 mm dimensions, from a 1 mm diameter Cu wire. This represents the excitation part of the transducer. The lens is constructed from two CSRs with the large bases being placed face to face. The insulated conductor is a copper foil with 18 μm thickness laminated adhesiveless with a polyimide foil (LONGLITE™200, produced by Rogers Corporation (Connecticut, CT, USA)), in order to reduce the losses at high frequencies, with 12 μm thickness (Figure 2b). Each CSR has 1.25 turns, 20 mm diameter large base, 3.2 mm diameter small base, aperture angle of 20°, and 50 mm height. The reception coil with 1 mm diameter and one turn from 0.1 mm diameter Cu wire was placed in the focal plane of the lens. A grounded screen made from the same insulated conductor as the CSR, having a circular aperture with diameter $d$ = 100 μm, is placed in front of the lens (Figure 2a). The distance between screen aperture and the surface to be examined has been maintained at 20 ± 1 μm.

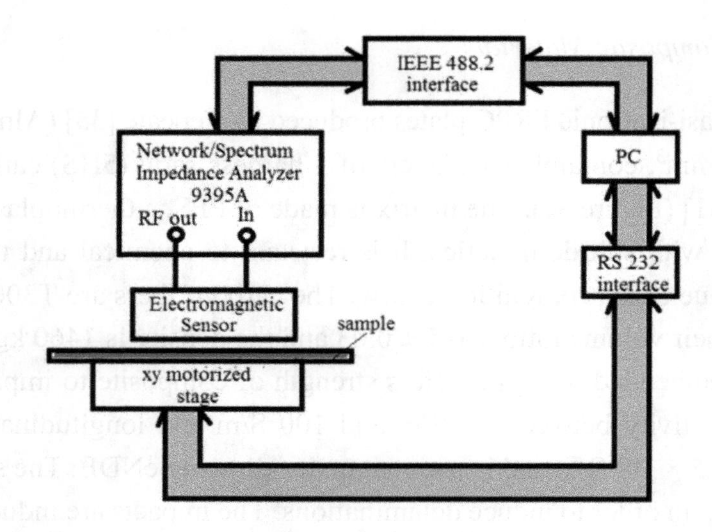


**Figure 4.** Experimental setup.

The EM sensor with MM lens, presented above, is connected to a Network/Spectrum/Impedance Analyzer type 4395A Agilent USA Analyzer (Agilent Technologies, Santa Clara, CA, USA). During the measurements, the transducer was maintained in fixed position and samples were displaced with a XY displacement system, type Newmark—Newmark Systems Inc. (Santa Margarita, CA, USA) that ensures the raster scanning in plane with ±10 μm precision and rotation with ± 2″.

The measurement system is commanded via PC through an RS232 interface (National Instruments, Mopac Expwy, Austin, TX, USA) for displacement system and IEEE 488.2 for Network/Spectrum/Impedance Analyzer (National Instruments, Mopac Expwy, Austin, TX, USA). The data acquisition and storage are made by software developed in Matlab 2012b (The MathWorks, Inc., Natick, MA, USA). The e.m.f. induced in the reception coil of the measurement system is the average of 10 measurements at the same point in order to reduce the effects of the white noise, the bandwidth of the analyzer being set to 10 Hz, also to diminish the noise level.

The measurements of *S* parameters were carried out with Agilent S Parameters Test kit 87511A (Agilent Technologies, Santa Clara, CA, USA) coupled to Agilent 4395A Analyzer.

## 4. Results and Discussion

The EM sensor used in eNDE must accomplish two roles:

Induce eddy current into the conductive material to be examined, and

Emphasize their flow modifications due to material degradation.

The simplest method to create time-variable magnetic fluxes that can induce eddy current into the material to be examined is represented by the coils circulated by alternative currents, by current impulses, or more alternative currents with different frequencies. The emission part of the EM sensor has this role.

To emphasize the induced eddy current and effect of material degradation over their propagation, sensors sensitive to variation of magnetic field can be used, in our case a sensor with MM lens. The theoretical and experimental study of eigenmodes [71] that appear in the studied MSGs open new domains of applications in the eNDE of stratified structures.

The method and the developed sensor can serve as the eNDE of conductive strips, in order to eventually detect interruptions as well as non-alignments of silver strips or short-circuits between the traces.

According to [59], in the case of MSG with silver strips and sub-wavelength features, excited with $TM_z$ polarized incident EM field, causing a single evanescent mode to appear in the space between strips but disappearing when water is inserted ($\varepsilon_{water} = 81$). This mode could be detected and visualized using the sensor described above.

For MSG with features compatible with the $\lambda$ value of incident EM field, $TM_z$ polarized; the structure presents known selective properties of transmission and reflection [85]. When an EM $TM_z$ polarized waves acts, at normal incidence, for MSG, according to Ref. [27], the reflection coefficient for a strip grating with $x_0 << \lambda_0$ is practically 1.

When MSG represents a layered structure having the features $x_c = 0.6$ mm, $x_d = 0.4$ mm, and thickness of 10 μm deposited on polyimide having a relative permittivity of 4.8, the study focuses on the appearance of abnormal and/or evanescent modes for the cases where dielectric fluids fill the gaps between the strip.

Using the sensors with MM lenses, it is experimentally confirmed that, in the space between the strips, evanescent and abnormal modes appear in the case of modes excited with a $TM_z$ polarized EM wave with $\lambda = 0.6$ m. The images present an increasing of the amplitude of the signal induced in the reception coil of the evanescent and abnormal modes created in slits at the scanning of a $1 \times 1$ mm$^2$ region with a 10 μm step in both directions. When the slits are filled with air, only the evanescent mode will be generated; the amplitude of the signal induced in the reception coil has the shape presented in Figure 5a.

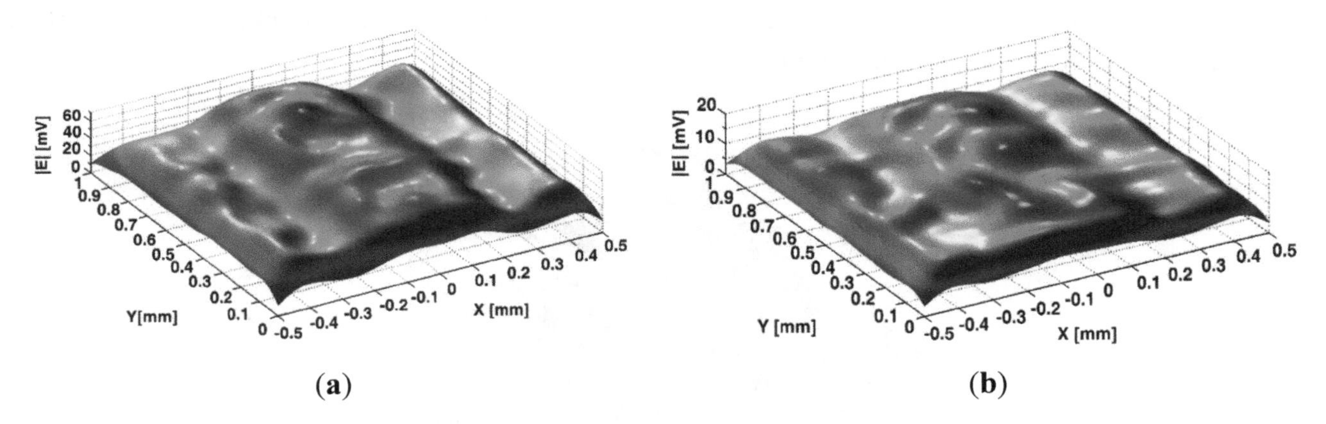


**Figure 5.** Image of evanescent and abnormal modes generated in slits for TMz polarized excitation at frequency of 476 MHz: (**a**) slits filled with air; (**b**) slits filled with isopropyl alcohol.

In the central zone of the slits the amplitude is maximum, followed by an accentuated decreasing towards the flanks of the metallic strips. The existence of a single evanescent mode, foreseen theoretically [7], is experimentally confirmed by the existence of a local maximum in the middle zone of the slits, with maximum amplitude on the middle of the slits, followed by an accentuated decreasing, symmetrically on the flanks of the strip. The two secondary maxima with smaller amplitude are at ±42 μm from the vertical strip wall.

In slits abnormal modes are generated, in the case of excitation with a $TM_z$ polarized wave, for large values of the liquid dielectric constants larger than 10, when $\varepsilon_d$ = 17.9, (isopropyl alcohol) (Figure 5b). Because the real component of the propagation constant $\beta_v$ for isopropyl alcohol is smaller ($\beta_v = 4.2 + i \cdot 104.321$) than the imaginary component, abnormal modes will be generated in slits; the EM image of these modes shows a similar behavior to the case of air in the slits. The amplitude of the signal is smaller and has a central maximum that is more flat (Figure 5b). This gives new ways for MSGs in sub-subwavelength regime to be used as sensors (as biosensors using evanescent modes generated space between strips and extremely low frequency plasmons).

In order to obtain a silver strip with different widths and thickness in the range of micrometers, different masks/stencils with different microstrip widths were used. Once a metal coating has been applied to a surface, it is critical to determine the adhesion properties of the deposit.

In the case of an MSG made from silver strip having 14 μm thickness, the width of strips where $x_c = 1.2$ mm and the width of slits is $x_d = 0.8$ mm is analyzed, considering that the wavelength of incident field is $\lambda = 0.6$ m (corresponding to a 500 MHz frequency).

A region of 16 × 16 $mm^2$ from MSG has been scanned with 0.1-mm steps in both directions. The scanning along the $x$ direction was done to correspond with a few periods of gratings, $x_0 = x_c + x_d$. The working frequency was 476 MHz.

For MSG excitation with EM field $TM_z$ polarized, the simulation was performed using XFDTD 6.3 software produced by REMCOM (State College, PA, USA) [86]. According to Ref. [79], the value of dielectric permittivity of silver is $\varepsilon_m = -48.8 + j \cdot 3.16$. Figure 6 shows the dependency of e.m.f. amplitude induced in the reception coil of a sensor on the scanning of the MSG taken into study, the image showing that this type of sensor correctly relies on conductive strips with 14 μm thickness and eventual non-adherence to support and/or interruptions.

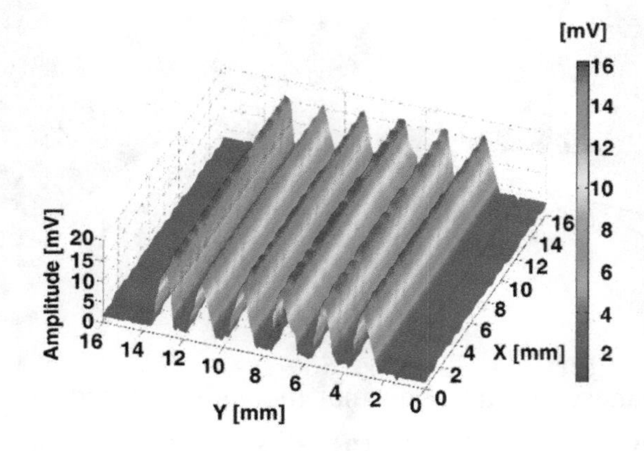


**Figure 6.** Amplitude of e.m.f. induced in the reception coil at the scanning of silver strip grating.

These results are in good concordance with theoretical estimations. They confirm a good adhesion of the silver paste on polyimide, as well as good alignment of the strips. It is known that the biosensing characteristic is strongly dependent on the deposition condition, which affects the physical properties of the thin film. Using the procedure and sensor described earlier for eNDE of MSG, interruption of a strip stopped the propagation of evanescent waves in the nearest slit so that the amplitude of e.m.f induced in the reception coil practically decreased to zero when the aperture's sensor was in the corresponding region of the slit. The roughness feature is emphasized by the propagation of surface polaritons [29].

For the second type of material taken into study (FRPC), the EM behavior of the composite was simulated by FDTD software; the samples were CAD designed following textiles features, and compared with eNDE tests. One cell of woven carbon fiber has been designed in the TexGenTextile Geometric modeler software (University of Nottingham, Nottingham, UK) (Figure 7a) and exported in CAD format in order to be used in FDTD software (XFDTD REMCOM, State College, PA, USA). The 5 × 5 tows were represented in different colors in order to easily follow their intersection when woven. The cell dimension in FDTD simulation is 0.04 mm, and the grid size is 146 × 146 × 45 cells. The perfectly matched layer boundary condition was applied at the grid boundaries.

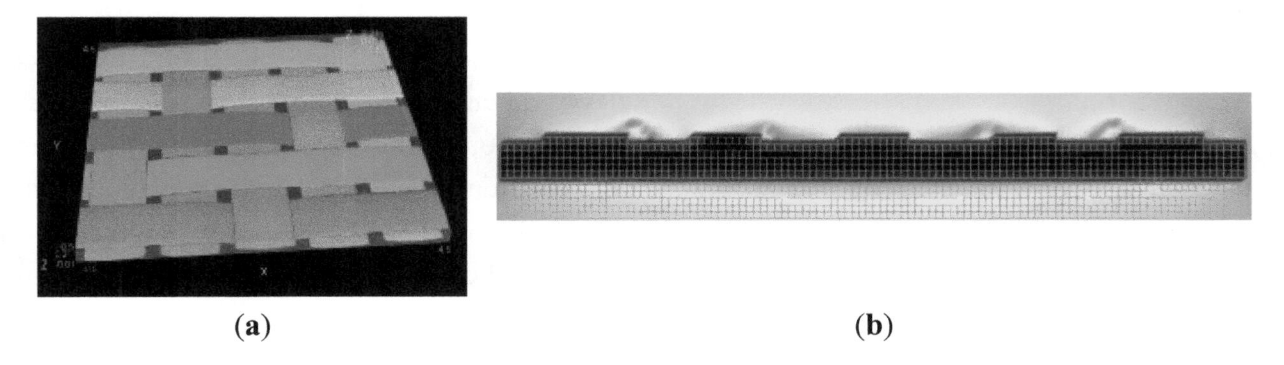

(a) (b)

**Figure 7.** Simulation of one woven carbon cell: (**a**) TexGen—different colors show the intersections in the woven 5HS; (**b**) XFDTD—$H_y$ propagation in a plane orthogonal at composite.

The result of the simulation presented in Figure 7b is a snapshot from a field sequence, showing the $H_y$ field progress at a particular slice of the geometry. In this case, the role of conductive strips is taken by carbon fibers, which act as MSG [71]; the apparition of evanescent waves can also be emphasized.

The detection of eventual delamination or the characterization of carbon fibers' woven structure using the eNDE method and the sensors with MM lens is an emerging nondestructive technique combining the advantages of conventional eddy current testing and evanescent wave detection, giving a higher resolution than a classical eddy current.

The results are shown in Figure 8, which presents a scan with 1-mm step of a 60 × 60 mm$^2$ region from the FRPC sample impacted with 8 J, scanned in both directions.

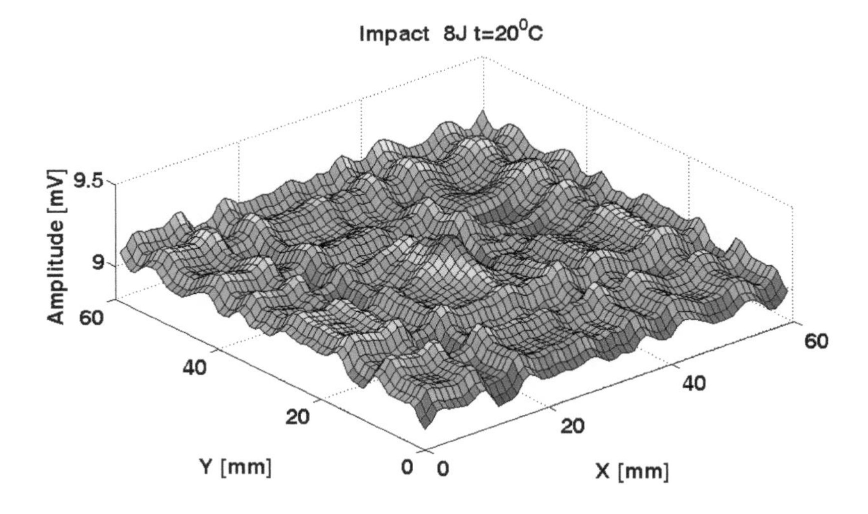


**Figure 8.** Sensor data with MM lens for 8 J impact.

It can be shown that the MM lens allow the enhancement of spatial resolution, with the layout of the woven being emphasized. The proposed method can thus be extended not only for the evaluation of MSG but also for the eNDE of FRPC in order to evaluate delaminations as well as the woven layout.

## 5. Conclusions

The structures' conductive grating allows, by extension, the estimation of results from eNDE of few real situations such as evaluation of metallic strips in printed circuit, MSG in the sub-wavelength regime as a biosensor, and FRPC.

In order to significantly enhance the spatial resolution of the EM method, the use of evanescent waves that can appear in slits, in the space between FRPC and on the edge of open microscopic cracks, is proposed. Special attention is granted to the MM lens based on CSR configurations, with an optimal frequency that assures the concentration of the incident electromagnetic field as well as the effective manipulation of the evanescent waves. The EM field TMz polarized can be created with a rectangular frame, having the plane perpendicular on the plane of MSG/FRPC and fed with an alternating current.

The use of evanescent waves and sensors with MM lens allows for the manipulation of evanescent waves to reach a spatial resolution of approximately λ/2000.

The performance of the EM sensors with MM lens is improved regarding sensitivity and spatial resolution by using the evanescent wave that can appear in the space between slits for structures excited with a polarized TMz plane wave.

**Acknowledgments**

This paper is partially supported by a grant from the Romanian Ministry of Education CNCS-UEFISCDI, project number PN-II-ID-PCE-2012-4-0437.

**Author Contributions**

Adriana Savin has analyzed the experimental data and wrote of the paper. Adriana Savin, Alina Bruma have contributed to development of the theoretical aspects of the EM sensor with MM lens and functioning principle. Adriana Savin and Rozina Steigmann have designed and realized the EM sensor with MM lens. Nicoleta Iftimie and Dagmar Faktorova performed the experiment and used the EM sensor with MM lens on different MSG in the sub-wavelength regime to involve them as biosensors, and on FRPC to evaluate delamination as well as the woven layout. All of the authors read and approved the final manuscript.

**References**

1. Bladel, V. *Electromagnetic Fields*, 2nd ed.; IEEE Press: Piscataway, NJ, USA, 2007.
2. Theodoulidis, T.P.; Kriezis, E.E. Impedance evaluation of rectangular coils for eddy current testing of planar media. *NDT E Int.* **2002**, *35*, 407–414.

3. Sablik, M.J.; Burkhardt, G.L.; Kwun, H.; Jiles, D.C. A model for the effect of stress on the low frequency harmonic content of the magnetic induction in ferromagnetic materials. *J. Appl. Phys.* **1988**, doi:10.1063/1.340609.
4. Harfield, N.; Yoshida, Y.; Bowler, J.R. Low-frequency perturbation theory in eddy-current non-destructive evaluation. *J. Appl. Phys.* **1996**, *80*, 4090–4100.
5. Theodoulidis, T.P.; Bowler, J.R. Eddy-current interaction of a long coil with a slot in a conductive plate. *IEEE Trans. Magn.* **2005**, *41*, 1238–1247.
6. Grimberg, R.; Udpa, L.; Savin, A.; Steigmann, R.; Palihovici, V.; Udpa, S.S. 2D eddy current sensor array. *NDT E Int.* **2006**, *39*, 264–271.
7. Grimberg, R.; Savin, A.; Steigmann, R. Electromagnetic imaging using evanescent waves. *NDT E Int.* **2012**, *46*, 70–76.
8. Auld, B.A.; Moulder, J.C. Review of advances in quantitative eddy current non-destructive evaluation. *J. Nondestruct. Eval.* **1999**, *18*, 3–36.
9. Li, Y.; Theodoulidis, T.; Tian, G.Y. Magnetic field based multi-frequency eddy current for multilayered specimen characterization. *IEEE Trans. Magn.* **2007**, *43*, 4010–4015.
10. Wang, L.; Xie, S.; Chen, Z.; Li, Y.; Wang, X.; Takagi, T. Reconstruction of stress corrosion cracks using signals of pulsed eddy current testing. *Nondestruct. Test Eva.* **2013**, *28*, 145–154.
11. He, Y.; Pan, M.; Luo, F.; Tian, G. Pulsed eddy current imaging and frequency spectrum analysis for hidden defect nondestructive testing and evaluation. *NDT E Int.* **2011**, *44*, 344–352.
12. Tai, C.C.; Yang, H.C.; Liang, D.S. Pulsed eddy current for metal surface cracks inspection: Theory and experiment. *AIP Conf. Proc.* **2002**, *21*, 388–395.
13. Wilson, J.W.; Tian, G.Y. Pulsed electromagnetic methods for defect detection and characterization. *NDT E Int.* **2007**, *40*, 275–283.
14. Mook, G.; Hesse, O.; Uchanin, V. Deep penetrating eddy currents and probes. *Mater. Test* **2007**, *49*, 258–264.
15. Wong, B.S. *Non-Destructive Testing-Theory, Practice and Industrial Applications*; Lap-Lambert Publishing GmbH KG: Saarbrücken, Germany, 2014.
16. Dodd, C.V.; Deeds, W.E. Analytical Solutions to Eddy-Current Probe-Coil Problems. *J. Appl. Phys.* **1968**, *39*, 2829–2838.
17. Grimberg, R.; Savin, A.; Radu, E.; Mihalache, O. Nondestructive evaluation of the severity of discontinuities in flat conductive materials by an eddy-current transducer with orthogonal coils. *IEEE Trans. Magn.* **2000**, *36*, 299–307.
18. Bowler, J.R.; Jenkins, S.A.; Sabbagh, L.D.; Sabbagh, H.A. Eddy-current probe impedance due to a volumetric flaw. *J. Appl. Phys.* **1991**, *70*, 1107–1114.
19. Bihan, Y.L. Study on the transformer equivalent circuit of eddy current nondestructive evaluation. *NDT E Int.* **2003**, *36*, 297–302.
20. Grbic, A.; Eleftheriades, G.V. Growing evanescent waves in negative-refractive-index transmission-line media. *Appl. Phys. Lett.* **2003**, *82*, 1815–1817.
21. Petit, R.L.; Botten, L.C. *Electromagnetic Theory of Gratings*; Springer-Verlag: Berlin, Germany, 1980.
22. Collin, R.E. *Field Theory of Guided Waves*; IEEE Press: New York, NY, USA, 1991.

23. Munk, B. *Frequency Selective Surfaces: Theory and Design*; John Wiley & Sons, Inc.: New York, NY, USA, 2000.
24. Balanis, C.A. *Antenna Theory: Analysis and Design*, 3rd ed.; John Wiley & Sons, Inc.: Hoboken, NJ, USA, 2005.
25. Brand, G.I. The strip grating as a circular polarizer. *Am. J. Phys.* **2003**, *71*, 452–457.
26. Arnold, M.D. An efficient solution for scattering by a perfectly conducting strip grating. *J. Electromagnet. Wave* **2006**, *20*, 891–900.
27. Peterson, A.F.; Ray, S.L.; Mittra, R. *Computational Methods for Electromagnetics*; IEEE Press: New York, NY, USA, 1998.
28. Porto, J.A.; Garcia-Vidal, F.J.; Pendry, J.B. Transmission resonances on metallic gratings with very narrow slits. *Phys. Rev. Lett.* **1999**, *83*, 2845–2849.
29. Kolomenski, A.; Kolomenskii, A.; Noel, J.; Peng, S.; Schuessler, H. Propagation length of surface plasmons in a metal film with roughness. *Appl. Opt.* **2009**, *48*, 5683–5691.
30. Grimberg, R. Electromagnetic metamaterials. *Mater. Sci. Eng. B* **2013**, doi:10.1016/j.mseb.2013.03.022.
31. Grimberg, R.; Savin, A.; Steigmann, R.; Serghiac, B.; Bruma, A. Electromagnetic non-destructive evaluation using metamaterials. *Insight* **2011**, *53*, 132–137.
32. Pilato, L.A.; Michno, M.J. *Advanced Composite*; Springer Verlag: Berlin, Germany, 1994.
33. Morgan, P. *Carbon Fibers and Their Composites*; CRC Press: Boca Raton, FL, USA, 2005.
34. Vieille, B.; Taleb, L. About the influence of temperature and matrix ductility on the behavior of carbon woven-ply PPS or epoxy laminates: Notched and unnotched laminates. *Compos. Sci. Technol.* **2011**, *71*, 998–1007.
35. TenCate Advanced Composites. Available online: http://www.tencate.com/ (accessed on 10 September 2015).
36. Kaw, A.K. *Mechanics of Composite Materials*, 2nd ed.; CRC Press: Boca Raton, FL, USA, 2006.
37. Standard AS 9100/2009. *Quality Systems-Aerospace-Model for Quality Assurance in Design, Development, Production, Installation and Servicing*; SAE: Warrendale, PA, USA, 1999.
38. Boller, C.; Meyendorf, N. State-of-the-art in Structural Health monitoring for aeronautics. In Proceedings of the International Symposium on NDT in Aerospace, Fürth/Bavaria, Germany, 3–5 December 2008; pp. 1–8.
39. Elmarakbi, A. Advanced Composite Materials for Automotive Applications: Structural Integrity and Crashworthiness; John Wiley & Sons Ltd.: Chichester, West Sussex, UK, 2013.
40. Boller, C. Next generation structural health monitoring and its integration aircraft design. *Int. J. Syst. Sci.* **2000**, *31*, 1333–1349.
41. Salvado, R.; Lopes, C.; Szojda, L.; Araújo, P.; Gorski, M.; Velez, F.J.; Krzywon, R. Carbon Fiber Epoxy Composites for Both Strengthening and Health Monitoring of Structures. *Sensors* **2015**, *15*, 10753–10770.
42. Kordatos, E.Z.; Aggelis, D.G.; Matikas, T.E. Monitoring mechanical damage in structural materials using complimentary NDE techniques based on thermography and acoustic emission. *Compos. Part B Eng.* **2012**, *43*, 2676–2686.

43. Adden, S.; Pfleiderer, K.; Solodov, I.; Horst, P.; Busse, G. Characterization of stiffness degradation caused by fatigue damage in textile composites using circumferential plate acoustic waves. *Compos. Sci. Technol.* **2008**, *68*, 1616–1623.
44. He, Y.; Tian, G.Y.; Pan, M.; Chen, D. Impact evaluation in carbon fiber reinforced plastic (CFRP) laminates using eddy current pulsed thermography. *Compos. Struct.* **2014**, *109*, 1–7.
45. Mian, A.; Han, X.; Islam, S.; Newaz, G. Fatigue damage detection in graphite/epoxy composites using sonic infrared imaging technique. *Compos. Sci. Technol.* **2004**, *64*, 657–666.
46. Vavouliotis, A.; Paipetis, A.; Kostopoulos, V. On the fatigue life prediction of CFRP laminates using the electrical resistance change method. *Compos. Sci. Technol.* **2011**, *71*, 630–642.
47. Park, J.M.; Kwon, D.J.; Wang, Z.J.; DeVries, K. Nondestructive sensing evaluation of surface modified single-carbon fiber reinforced epoxy composites by electrical resistivity measurement. *Compos. Part B Eng.* **2006**, *37*, 612–626.
48. Seon, G.; Makeev, A.; Cline, J.; Shonkwiler, B. Assessing 3D shear stress-strain properties of composites using Digital Image Correlation and finite element analysis based optimization, *Compos. Sci. Technol.* **2015**, *117*, 371–378.
49. Nikishkov, Y.; Airoldi, L.; Makeev, A. Measurement of voids in composites by X-ray Computed Tomography. *Compos. Sci. Technol.* **2013**, *89*, 89–97.
50. Schilling, P.J.; Karedla, B.R.; Tatiparthi, A.K.; Verges, M.A.; Herrington, P.D. X-ray computed microtomography of internal damage in fiber reinforced polymer matrix composites. *Compos. Sci. Technol.* **2015**, *65*, 2071–2078.
51. Ren, W.; Liu, J.; Tian, G.Y.; Gao, B.; Cheng, L.; Yang, H. Quantitative non-destructive evaluation method for impact damage using eddy current pulsed thermography. *Compos. Part B Eng.* **2013**, *54*, 169–179.
52. Hung, Y.Y.; Chen, Y.S.; Ng, S.P.; Liu, L.; Huang, Y.H.; Luk, B.L.; Ip, R.W.L.; Wu, C.M.L.; Chung, P.S. Review and comparison of shearography and active thermography for nondestructive evaluation. *Mater. Sci. Eng. R Rep.* **2009**, *64*, 73–112.
53. Solodov, I.; Döring, D.; Rheinfurth, M.; Busse, G. New Opportunities in Ultrasonic Characterization of Stiffness Anisotropy in Composite Materials. In *Nondestructive Testing of Materials and Structures*; Springer: Dordrecht, The Netherlands, 2013; Volume 6, pp. 599–604.
54. Kolkoori, S.; Hoehne, C.; Prager, J.; Rethmeier, M.; Kreutzbruck, M. Quantitative evaluation of ultrasonic C-scan image in acoustically homogeneous and layered anisotropic materials using three dimensional ray tracing method. *Ultrasonics* **2014**, *54*, 551–562.
55. Cacciola, M.; Calcagno, S.; Megali, G.; Pellicano, D.; Versaci, M.; Morabito, F.C. Eddy current modeling in composite materials. *PIERS Online* **2009**, *5*, 591–595.
56. Grimberg, R.; Savin, A.; Steigmann, R.; Bruma, A.; Barsanescu, P. Ultrasound and eddy current data fusion for evaluation of carbon-epoxy composite delaminations. *Insight* **2009**, *51*, 1–25.
57. Lamberti, A.; Chiesura, G.; Luyckx, G.; Degrieck, J.; Kaufmann, M.; Vanlanduit, S. Dynamic Strain Measurements on Automotive and Aeronautic Composite Components by Means of Embedded Fiber Bragg Grating Sensors. *Sensors* **2015**, *15*, 27174–27200.
58. Takeda, S.; Okabe, Y.; Takeda, N. Delamination detection in CFRP laminates with embedded small-diameter fiber Bragg grating sensors. *Compos. Part A* **2002**, *33*, 971–980.

59. Savin, A.; Steigmann, R.; Bruma, A.; Šturm, R. An Electromagnetic Sensor with a Metamaterial Lens for Nondestructive Evaluation of Composite Materials. *Sensors* **2015**, *15*, 15903–15920.
60. Grimberg, R.; Savin, A. Electromagnetic Transducer for Evaluation of Structure and Integrity of the Composite Materials with Polymer Matrix Reinforced with Carbon Fibers. Patent RO126245-A0, 2011.
61. Pendry, J.; Holden, A.J.; Robbins, D.J.; Stewart, W.J. Magnetism from conductors and enhanced non-linear phenomena. *IEEE Trans. Microw Theory Tech.* **1999**, *47*, 47–58.
62. Pendry, J.B. Negative Refraction makes a perfect lens. *Phys. Rev. Lett.* **2000**, *85*, 3966–3969.
63. Cai, W.; Chettiar, U.K.; Kildishev, A.V.; Shalaev, V.M. Optical cloaking with metamaterials. *Nat. Photonics* **2007**, *1*, 224–227.
64. Smith, D.R.; Pendry, J.B.; Wiltshire, M.C.K. Metamaterials and negative refractive index. *Science* **2004**, *305*, 788–792.
65. Shelby, R.A.; Smith, D.R.; Schultz, S. Experimental Verification of a Negative Index of Refraction. *Science* **2001**, *6*, 77–79.
66. Bai, Q.; Liu, C.; Chen, J.; Cheng, C.; Kang, M. Tunable slow light in semiconductor metamaterial in a broad terahertz regime. *J. Appl. Phys.* **2010**, doi:10.1063/1.3357291.
67. Wuttig, M.; Yamada, N. Phase-change materials for rewriteable data storage. *Nat. Mater.* **2007**, *6*, 824–832.
68. Engheta, N.; Ziolkowski, R.W. *Electromagnetic Metamaterials: Physics and Engineering Explorations*; John Wiley & Sons, Inc.: New York, NY, USA, 2006.
69. Zouhdi, S.; Sihvola, A.; Vinogradov, A.P. *Metamaterials and Plasmonics: Fundamentals, Modelling, Applications*; Springer: New York, NY, USA, 2008.
70. Veselago, V.G. The electrodynamics of substances with simultaneously negative values of $\epsilon$ and $\mu$. *Phys. Uspekhi* **1968**, *10*, 509–514.
71. Grimberg, R.; Tian, G.Y. High Frequency Electromagnetic Non-destructive Evaluation for High Spatial Resolution using Metamaterial. *Proc. R. Soc. A* **2012**, *468*, 3080–3099.
72. Born, M.; Wolf, E. *Principle of Optics*, 5th ed.; Pergamon Press: Oxford, UK, 1975.
73. Goodman, J.W. *Introduction to Fourier Optics*, 3rd ed.; Roberts and Company: Englewood, CO, USA, 2005.
74. Kong, J.A. *Electromagnetic Wave Theory*; EMW Publishing: Cambridge, MA, USA, 2000.
75. Shelby, R.A.; Smith, D.R.; Nemat-Nasser, S.C.; Schultz, S. Microwave transmission through a two-dimensional, isotropic, left-handed metamaterial. *Appl. Phys. Lett.* **2001**, *78*, 489–491.
76. Chen, X.; Grzegorczyk, T.M.; Wu, B.I.; Pacheco, J., Jr.; Kong, J.A. Robust method to retrieve the constitutive effective parameters of metamaterials. *Phys. Rev. E* **2004**, doi:10.1103/PhysRevE.70.016608.
77. Prentice, G.A.; Chen, K.S. Effects of current density on adhesion of copper electrodeposits to polyimide substrates. *J. Appl. Electrochem.* **1998**, *28*, 971–977.
78. SPI Supplies. Available Online: http://www.2spi.com/ (accessed on 10 October 2015).
79. Palik, E.D. *Handbook of Optical Constants of Solids*; Academic Press: London, UK, 1985.
80. Schlesinger, M. *Deposition on Nonconductors, Chapter 15 on Modern Electroplating*, 5th ed.; John Wiley & Sons: Hoboken, NJ, USA, 2010; pp. 413–420.

81. Akkerman, R. Laminate mechanics for balanced woven fabrics. *Compos. Part B Eng.* **2006**, *37*, 108–116.
82. Standard Test Method for Measuring the Damage Resistance of a Fiber-Reinforced Polymer Matrix Composite to a Drop-Weight Event; ASTM International: West Conshohocken, PA, USA; 2005.
83. Menana, H.; Féliachi, M. Electromagnetic characterization of the CFRPs anisotropic conductivity: Modeling and measurements. *Eur. Phys. J. Appl. Phys.* **2011**, doi:10.1051/epjap/2010100255.
84. Ullah, H.; Abdel-Wahab, A.A.; Harland, A.R.; Silberschmidt, V.V. Damage in woven CFRP laminates subjected to low velocity impacts. *J. Phys. Conf. Ser.* **2012**, doi:10.1088/1742-6596/382/1/012015.
85. Balanis, C. *Advanced Engineering Electromagnetics*; John Wiley & Sons: Hoboken, NJ, USA, 1989.
86. Kunz, K.S.; Luebbers, R.J. *The Finite Difference Time Domain Method for Electromagnetics*; CRC Press: Boca Raton, FL, USA, 1993.

# 14

# Correlation between Earthquakes and AE Monitoring of Historical Buildings in Seismic Areas

**Giuseppe Lacidogna *, Patrizia Cutugno, Gianni Niccolini, Stefano Invernizzi and Alberto Carpinteri**

Department of Structural, Geotechnical and Building Engineering, Politecnico di Torino, Corso Duca degli Abruzzi 24, 10129 Torino, Italy; E-Mails: patrizia.cutugno@polito.it (P.C.); gianni.niccolini@polito.it (G.N.); stefano.invernizzi@polito.it (S.I.); alberto.carpinteri@polito.it (A.C.)

† This paper is an extended version of paper published in the 6th International Conference on Emerging Technologies in Non-destructive Testing (ETNDT6), Brussels, Belgium, 27–29 May 2015.

* Author to whom correspondence should be addressed; E-Mail: giuseppe.lacidogna@polito.it

Academic Editors: Dimitrios G. Aggelis and Nathalie Godin

---

**Abstract:** In this contribution a new method for evaluating seismic risk in regional areas based on the acoustic emission (AE) technique is proposed. Most earthquakes have precursors, *i.e.*, phenomena of changes in the Earth's physical-chemical properties that take place prior to an earthquake. Acoustic emissions in materials and earthquakes in the Earth's crust, despite the fact that they take place on very different scales, are very similar phenomena; both are caused by a release of elastic energy from a source located in a medium. For the AE monitoring, two important constructions of Italian cultural heritage are considered: the chapel of the "Sacred Mountain of Varallo" and the "Asinelli Tower" of Bologna. They were monitored during earthquake sequences in their relative areas. By using the Grassberger-Procaccia algorithm, a statistical method of analysis was developed that detects AEs as earthquake precursors or aftershocks. Under certain conditions it was observed that AEs precede earthquakes. These considerations reinforce the idea that the AE monitoring can be considered an effective tool for earthquake risk evaluation.

**Keywords:** acoustic emission; structural monitoring; cultural heritage; earthquakes; seismic precursors

## 1. Introduction

A new method to evaluate seismic risk in regional areas is proposed as an attempt to preserve Italian historical and architectural cultural heritage. To this purpose, the spatial and temporal correlations between the acoustic emission (AE) data obtained from the monitoring sites and the earthquakes that have occurred in specific ranges of time and space are examined.

The predictive power and the non-invasive procedure of the AE technique (no external excitation is provided, since the source of energy is the damage process itself) can be exploited to preserve Italian cultural heritage, as historic buildings and monuments are exposed to seismic risk and, in general, to severe, long-term, cyclic loading conditions or harsh environmental conditions.

Buildings and structures exposed to the action of earthquakes with moderate magnitude, which are rather frequent in the central and southern regions of Italy, may undergo accelerated aging and deterioration. Such damage processes, often inaccessible to visual inspection, eventually lead to an increased vulnerability to the actions of major earthquakes, with catastrophic results [1].

The two regions considered in this paper surround the Italian cities of Varallo and Bologna, where the authors have recently carried out AE monitoring of important historical monuments for structural stability assessment. In particular, in the Italian Renaissance Architectural Complex of "The Sacred Mountain of Varallo" the structure of the Chapel XVII was analyzed, while in the City of Bologna the stability of the "Asinelli Tower", known as the highest leaning tower in Italy, was evaluated [2,3].

According to seismic analysis, several studies investigated spatial and temporal correlations of earthquake epicenters, involving, for example, the concepts of Omori's Law and fractal dimensions [4–6]. Other authors tried to study the complex phenomenon of seismicity using an approach that is able to analyze the spatial location and time occurrence in a combined way, without subjective a priori choices [7]. The present approach leads to a self-consistent analysis and visualization of both spatial and temporal correlations based on the definition of correlation integral [8].

Based on these considerations, we have tried not only to analyze the seismic activity in the considered regions, but also to correlate it with the AE activity detected during the structural monitoring. By adopting a modified Grassberger-Procaccia algorithm, we give the cumulative probability of the events' occurrence in a specific area, considering the AE records and the seismic events during the same period of time. In the modified integral the cumulative probabilities $C(r, \tau)$ are the function of the regional radius of interest, $r$, and of the time interval, $\tau$, both considering the peak of AEs as earthquake "precursors" or "aftershocks" [9].

## 2. Description of Chapel XVII and Asinelli Tower

In Varallo, placed in Piedmont in the province of Vercelli, there is the relevant Italian Renaissance Architectural Complex named "The Sacred Mountain of Varallo". It was built in the 15th century on a cliff above the town of Varallo and is composed of a basilica and 45 chapels.

Because of its high level of damage due to regional earthquakes that have occurred [2], we chose to consider Chapel XVII, known as the chapel of the "Transfiguration of the Christ on the Mount Tabor", for the AE analysis. This structure, having a circular plan, was built with stone masonry and mortar. The interior walls of the chapel are also equipped with some valuable frescoes (Figure 1).

**Figure 1.** Sacred Mountain of Varallo, Mount Tabor Chapel XVII.

In Bologna, Emilia-Romagna, there is another important masterpiece of Italian architectural and historical heritage, the "Asinelli Tower", which is also the symbol of the city. To study the damage evolution of the structure, we have recently analyzed the effect of repetitive and impulsive natural and anthropic events, such as earthquakes wind, or vehicle traffic, with the AE technique [3].

The Asinelli Tower was built in the early 12th century, and it rises to a height of 97.30 m above the ground. From the ground level, up to a height of 8.00 m, the tower is surrounded by an arcade built at the end of the 15th century. Studies conducted in the early 20th century revealed that the Asinelli Tower leaned westward by 2.25 m, and other recent studies have confirmed that its leaning is of 2.38 m, and it remains the tallest leaning tower in Italy (Figure 2).

**Figure 2.** The Asinelli and the adjacent Garisenda Towers in the city center of Bologna. The Asinelli tower is the tallest one on the right.

## 3. Acoustic Emission Monitoring

The AE monitoring is performed by analyzing the signals received from the transducers through a threshold detection device that counts the burst signals, which are greater than a certain voltage.

The adopted USAM (Unit for Synchronous Multichannel Acquisition) acquisition system consists of six pre-amplified AE sensors, equipped with six units of data storage, a central unit for synchronization, and a trigger threshold. The output voltage signals of the piezoelectric transducers (PZT) were filtered with a pass-band from 50 kHz to 800 kHz and a detection threshold of 100 μV. The obtained data from this device are the cumulative counting of each mechanical wave, the acquisition time, the measured amplitude in volts, the duration, and the number of oscillations across the threshold value for each AE signal [2,3]. Then, at the post-processing stage, we discarded signals with a duration shorter than 3 μs and containing less than three oscillations across the detection threshold in order to filter out electrical noise spikes.

### *3.1. Chapel XVII*

Firstly, we consider the analysis carried out in Chapel XVII of Varallo. The AE monitoring was led on the frescoed masonry on the north wall of the Chapel. The wall showed some damage: a vertical crack of about 3.00 m in length and fresco detachment. Four AE sensors were located around the vertical crack, while two were placed near the fresco detachment (Figure 3). The AE monitoring was conducted in two phases for a total duration of about 14 weeks. The first phase started on 9 May 2011 and finished on 16 June 2011; the second one was from 5 July 2011 to 5 September 2011. The monitoring results related to the chapel's structural integrity are reported in [2]. We interpreted the AE data considering the amplitude and time distribution of AE signals during the cracking phenomena. From this analysis, we found that the vertical crack, monitored on the north wall of the chapel, evolved during the acquisition period, while the process of detachment of the frescos was mainly related to the diffusion of moisture in the mortar substrate [2].

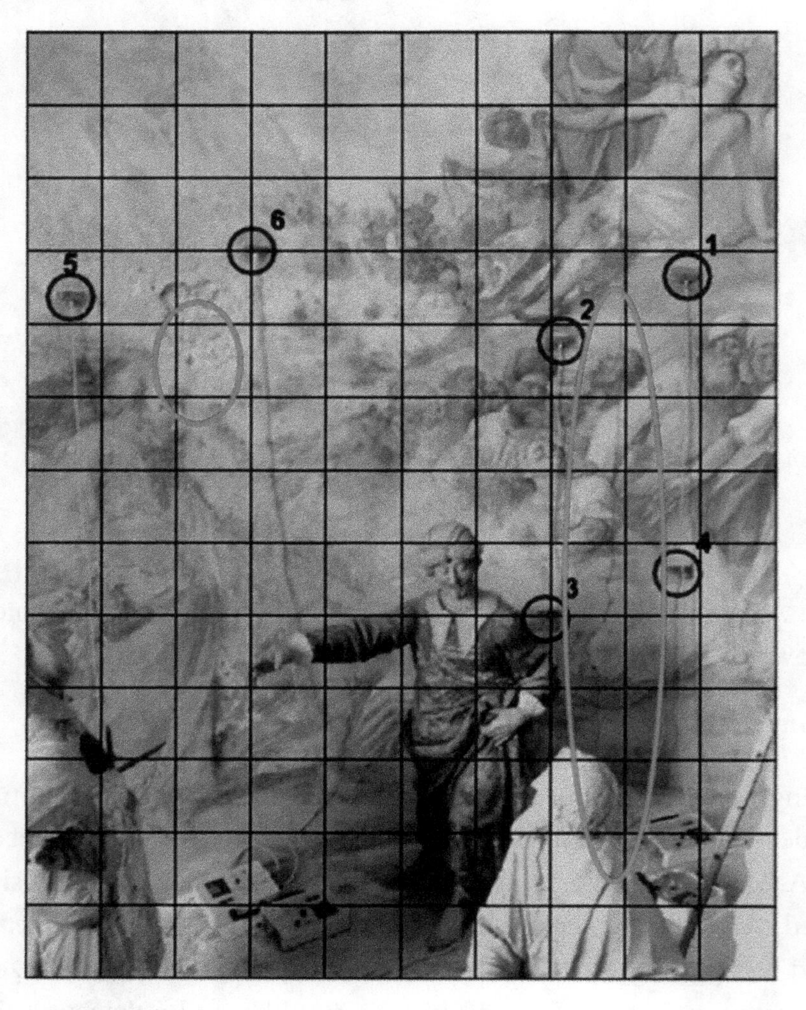


**Figure 3.** Chapel XVII. View of the monitored areas. Left side: sensors 5, 6, and the fresco detachment. Right side: sensors 1–4 and the vertical crack.

During the monitoring period, among all regional seismic events we considered 21 earthquakes with Richter magnitudes ≥1.2, having an epicenter within a radius from 60 to 100 km from Varallo. The strongest earthquake was a 4.3 magnitude event that occurred on 25 July 2011 at 12:31 p.m. in the Giaveno area (epicenter about 80 km from Varallo). Figure 4 displays the AE event rate, which counts simultaneous signal detection by multiple sensors as one event and the sequence of the earthquakes as functions of time, obtained during the monitoring period.

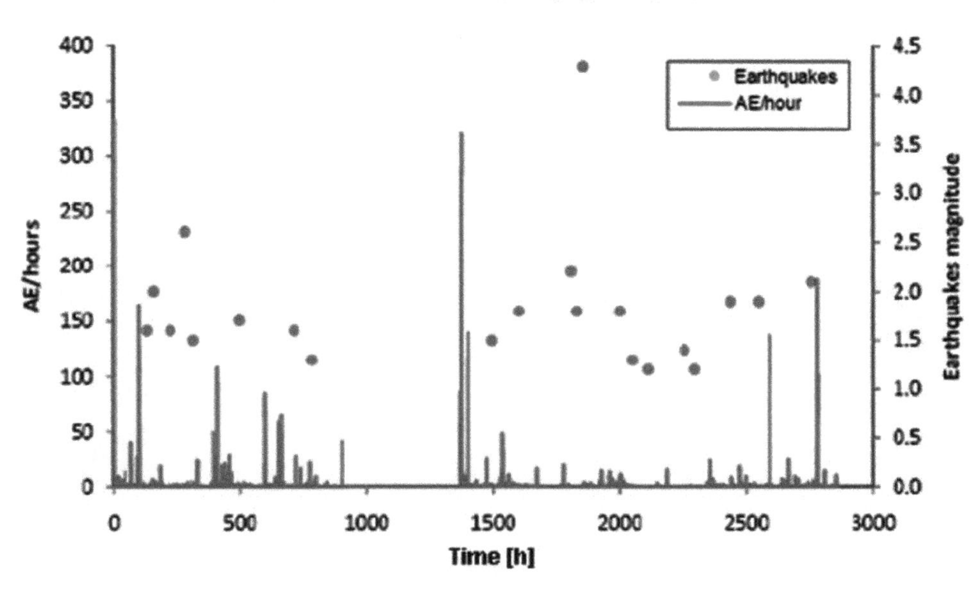


**Figure 4.** Chapel XVII: AE rate (blue chart) and nearby earthquake (red dots) occurrences as functions of time. The AE rate chart illustrates the number of AE events (averaged over 1 h), while seismic events are marked by points indicating occurrence time and Richter magnitude.

### *3.2. Asinelli Tower*

The AE activity on Asinelli Tower was examined in a significant zone for monitoring purposes. This was developed by attaching six piezoelectric sensors to the northeast corner of the tower at an average height of *ca.* 9.00 m above ground level, immediately above the terrace atop the arcade. In this area, the double-wall masonry has an average thickness of *ca.* 2.45 m (Figure 5). AE monitoring began on 23 September 2010 and ended on 10 January 2011, for a total duration of about 16 weeks. By monitoring a significant part of this tower, the incidence of vehicle traffic, seismic activity and wind action on the damage evolution within the structure were assessed [3].

In this case, among all regional seismic events, we considered 43 earthquakes with a Richter magnitude ≥1.2, having an epicenter within a radius from 25 to 100 km from Bologna, as the most likely to affect the tower's stability. The strongest earthquakes were the 4.1 magnitude event that occurred on 13 October 2010 at 11:43 p.m. in the Rimini area (epicenter about 100 km from Bologna) and the 3.4 magnitude event on 21 November 2010 at 4:10 p.m. in Modena Apennines (epicenter about 50 km from Bologna). The AE instantaneous rate (averaged over 1 h) and the sequence of the earthquakes obtained during the monitoring are displayed in Figure 6. Also, in this case a correlation between peaks of AE activity in the structure and regional seismicity can be observed (Figure 6).

The tower, in fact, as in another case investigated by the authors regarding the Medieval Towers of Alba in Italy, is very sensitive to earthquake motions [9,10].

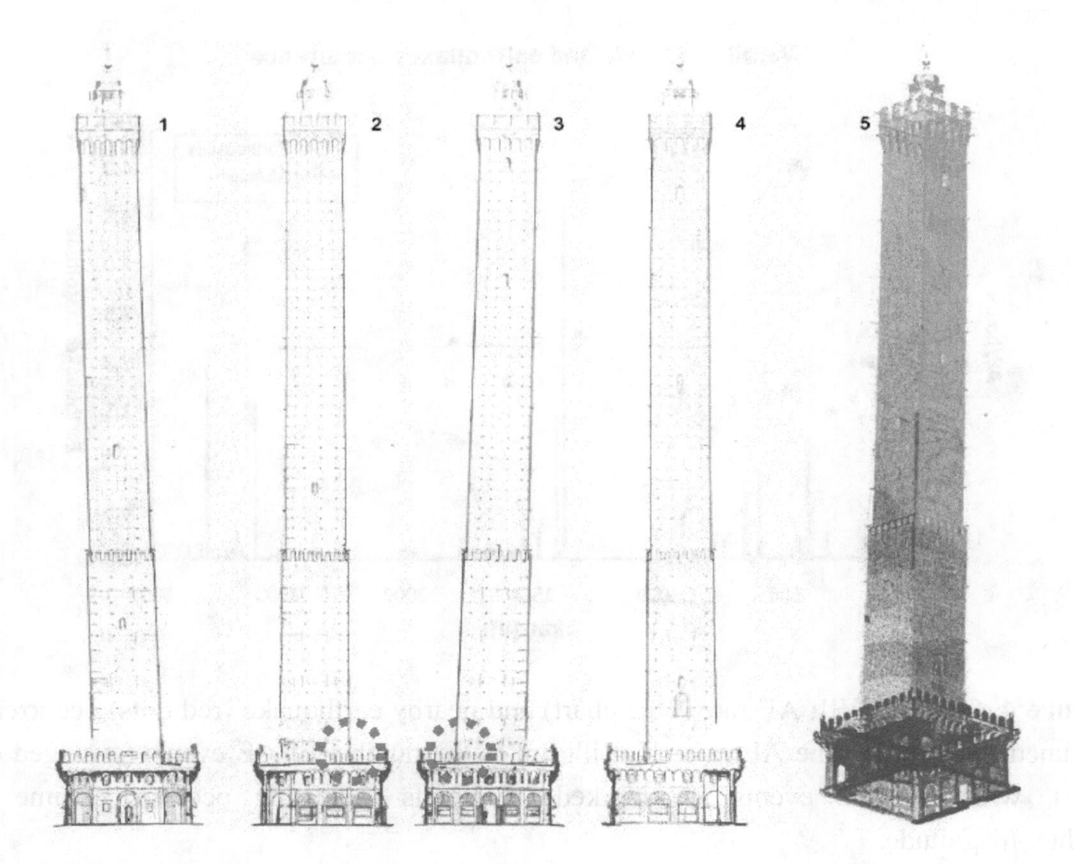


**Figure 5**. Front views and axonometric view of Asinelli Tower. Faces (**1**) South; (**2**) East; (**3**) North; (**4**) West; (**5**) Axonometric view. The AE transducers were applied to the northeast corner of the tower, in the zones marked with a circle.

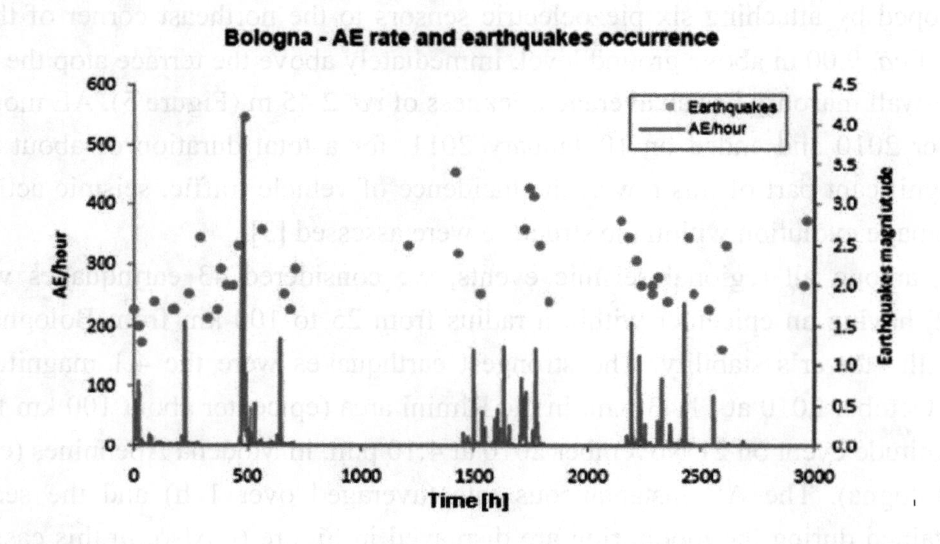


**Figure 6.** Asinelli Tower: AE rate (blue chart) and nearby earthquake (red dots) occurrences as functions of time. The AE rate chart illustrates the number of AE events (averaged over 1 h), while seismic events are marked by points indicating occurrence time and Richter magnitude.

## 4. Grassberger-Procaccia Algorithm

Acoustic emissions in materials and earthquakes in the crust are very similar phenomena though on very different scales. They both involve a sudden release of elastic energy from a source located in the medium: respectively, the tip of opening micro-cracks and the seismic hypocenter [11]. The aging and the damage of historical buildings due to the action of small and intermediate earthquakes, a situation very common in central and southern Italy, implies triggering of AE activity on these structures.

However, there appear to be some seismic events which follow AE bursts and, then, do not trigger structural AE activity. In fact, intense crises of crustal stress apparently cross large areas, as revealed by increased AE activity in several case histories from Italy [12], and precede the eventual occurrence of some earthquake within them.

This experimental evidence suggests that part of the AE activity from the structures might derive from precursive microseismic activity propagating across the ground-building foundation interface.

In this sense, AE monitoring provides twofold information: one concerning the structural damage and the other the amount of stress affecting the crust, in which the building foundation represents a sort of extended underground probe.

With this spirit we introduce a statistical approach to analyze spatial and temporal correlations between AE bursts from the structures and surrounding earthquakes, along the lines of similar investigations of the spatial and temporal correlations of earthquakes [2,3,9].

Seismicity levels of the Varallo and Bologna areas can be consistently compared by investigating the earthquakes' spatial distribution over equally sized regions (*i.e.*, circular regions with a radius of 100 km) centered at the two monitoring sites, during analogous monitoring periods (3000 h, as shown in Figures 4 and 6).

Quantification of this feature is possible through the notion of the fractal dimension of hypocenters, which is frequently estimated by the correlation integral because of its great reliability and sensitivity to small changes in clustering properties [13–15]. The correlation integral for hypocenter points embedded into a three-dimensional space is defined as follows:

$$C(r) = \frac{2}{N(N-1)} \sum_{k=1}^{N-1} \sum_{j=k+1}^{N} \theta\left(r - |x_k - x_j|\right) \tag{1}$$

where $N$ is the total number of seismic events, $x$ is the hypocentral coordinate vector, and $\Theta$ is the Heaviside step function ($\Theta(x) = 0$ if $x \leq 0$, $\Theta(x) = 1$ if $x > 0$). The correlation integral $C(r)$ gives the probability of finding another point in the sphere of radius $r$ centered at an arbitrary point. For a fractal population of points, $C(r)$ scales with $r$ as a power law, $D$ being the correlation dimension:

$$C(r) \sim r^D \tag{2}$$

We apply the correlation integral to the earthquake sequences under consideration, *i.e.*, the earthquakes with epicentral distance in the range of 60–100 km from the Varallo Chapel and 25–100 km from the Asinelli Tower, which occurred during the AE monitoring periods. The choice of the minimum magnitude threshold, $m_L \geq 1.2$, was driven by completeness criteria.

The seismic events were taken from the Italian Seismological Instrumental and Parametric Data-Base ISIDE [16].

Two patterns are evidenced in Figure 7a,b: shorter ranges, 3 km < $r$ < 10–20 km, are defined by a unique slope giving a relatively high value of the correlation dimension, $D = 1.6$ (Varallo) and $D = 2.2$ (Bologna); for distances over 10–20 km, the slope gives a value of $D = 0.7$ (Varallo) and $D = 0.9$ (Bologna). For short ranges, $D \approx 2$ corresponds to a distribution of events over a fault plane, such as the active faults in the region surrounding Bologna; the highest ranges, corresponding to a value of $D < 1$, suggest a dust-like setting of seismicity as in the region surrounding Varallo (see Figure 8).The higher dimension for Bologna apparently reflects the recent and active tectonics of the Emilia-Romagna in the northeastern Italy, with respect to the seismically quiet zone of Varallo [17].

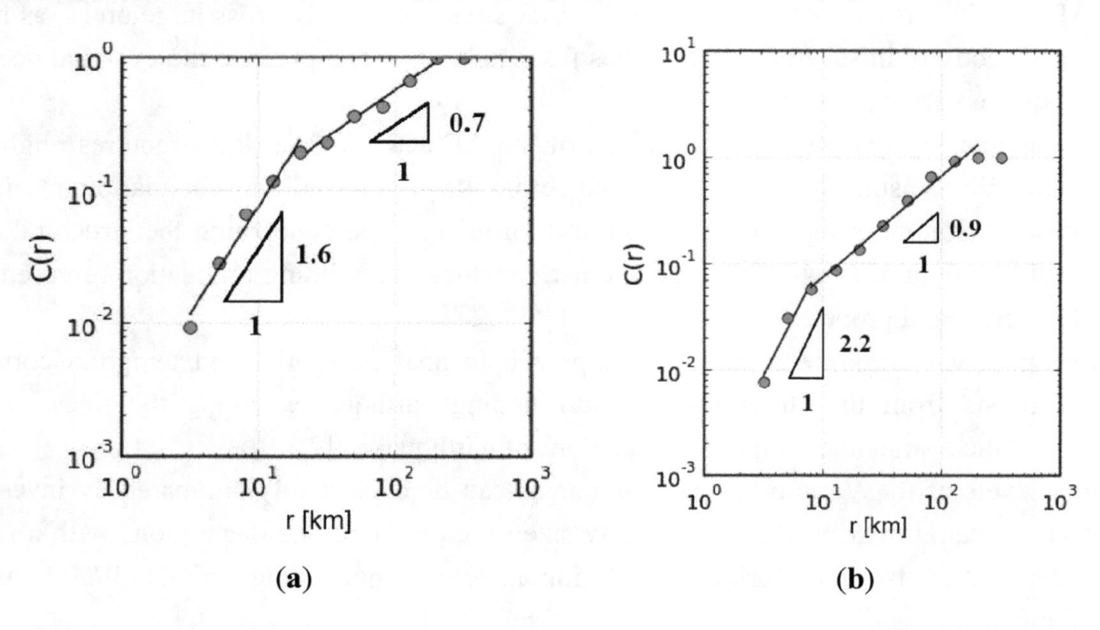


**Figure 7.** Correlation integral $C(r)$ *vs.* $r$ (km) and its slope $D$ in bi-logarithmic scale for Varallo regional seismicity (**a**) and Bologna regional seismicity (**b**).

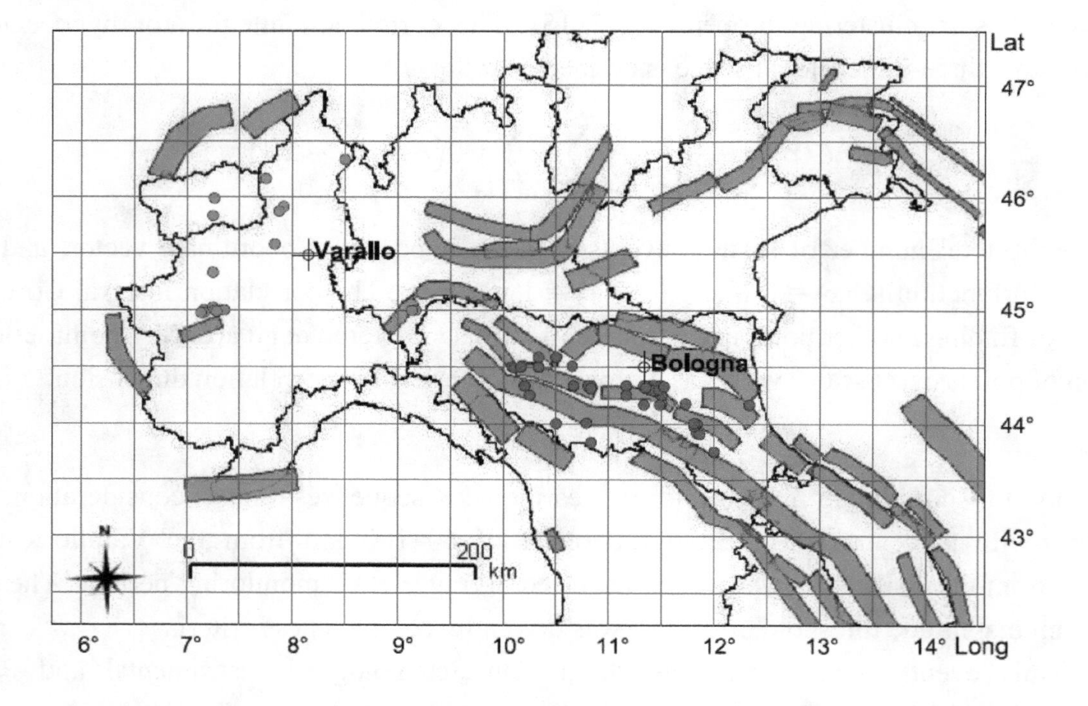


**Figure 8.** Map of Northern Italy depicting faults and epicenters.

## 5. Space-Time Correlation between AE and Seismic Events

Along the lines of studies on the space-time correlation among earthquakes based on a generalization of the Grassberger-Procaccia correlation integral [8], we calculate the degree of correlation between the AE time series and the sequence of nearby earthquakes recorded during the same period. No a priori assumptions on causal relationships between AE and seismic activities drive this approach, which is merely statistical. Only the choice of the time window and the region size is arbitrary.

The space-time combined correlation integral is defined as follows [7,9]:

$$C(r,\tau) = \frac{1}{N_{EQ}N_{AE}} \sum_{k=1}^{N_{EQ}} \sum_{j=1}^{N_{AE}} \theta(r - |x_k - x_0|) \cdot \theta(\tau - |t_k - t_j|) \tag{3}$$

where the index $j$ runs over all the $N_{AE}$ event bursts $\{x_0, t_j\}$, with $x_0$ being the coordinate vector of the monitoring site, while $k$ runs over all the $N_{EQ}$ seismic events $\{x_k, t_k\}$, with $x_k$ being the epicentral coordinate vector.

Since the double-sum in Equation (3) counts pairs formed by an AE event $(x_0, t_j)$ and a seismic event $(x_k, t_k)$ with mutual epicentral distance $|x_0 - x_k| \leq r$ and time intervals $|t_j - t_k| \leq \tau$ among all $N_{AE}$ and $N_{EQ}$ possible pairs, $C(r,\tau)$ can easily be interpreted as the probability of an AE burst and an earthquake occurring with an inter-distance $\leq r$ and a time interval $\leq \tau$.

If $C(r, \tau,)$ exhibits power-law behavior both in space and time variables, time and space fractal dimensions $D_t$ and $D_s$ can be defined similarly to Equation (2):

$$D_t(r,\tau) = \frac{\partial \log C(r,\tau)}{\partial \log \tau} \tag{4}$$

and

$$D_s(r,\tau) = \frac{\partial \log C(r,\tau)}{\partial \log r} \tag{5}$$

The time correlation dimension $D_t(r, \tau)$ characterizes the time-coupling of AE bursts to the earthquakes occurring up to a given distance $r$ from the AE monitoring site, whereas the space correlation dimension $D_s(r, \tau)$ characterizes the spatial distribution of nearby earthquakes with separation in time from AE bursts not exceeding a given $\tau$.

It is worth noting that Equation (5) identifies with Equation (2) if $\tau$ equals the whole time data span, as all earthquakes would be taken into account in the Grassberger-Procaccia integral $C(r, \tau)$.

Actually, the imposed condition $|t_j - t_k| \leq \tau$ does not specify the chronological order between the two types of event. On the other hand, the AE time series and the sequence of nearby earthquakes are two very closely interrelated sets in the time domain. Therefore, a given AE burst might be either due to structural damage triggered by a seismic event or due to precursive microseismic activity. In spite of intrinsic difficulties in high-frequency propagation across disjointed media (in particular at the ground-building foundation interface), AE bursts apparently indicate widespread crustal stress crises during the preparation of a seismic event [9,12], when part of the related deformation energy stored in the Earth's crust might be transferred to the building foundations.

It is interesting to carry out a probabilistic analysis of the available data considering AE events once as preceding an earthquake, *i.e.*, as potential seismic precursors, and next as following an earthquake, *i.e.*, as structural aftershocks. In this way, the obtained conditioned probability distributions can be compared in order to discover the prevailing trend. This analysis is performed applying the modified correlation integral [9]:

$$C^{\pm}(r,\tau) = \frac{1}{N_{EQ}N_{AE}} \sum_{k=1}^{N_{EQ}} \sum_{j=1}^{N_{AE}} \theta(r - |x_k - x_0|) \cdot \theta(\tau - |t_k - t_j|) \tag{6}$$

where "+" and "−" are used to account for AE events, respectively, as seismic precursors and as aftershocks. $C^+ (r, \tau)$, for example, gives the occurrence probability of an AE burst followed by an earthquake in the next interval $\tau$ and within a radius $r$ of the monitoring site.

We applied the modified integral (Equation 6) to investigate the space-time correlation between the two above-mentioned regional seismic sequences and the AE time series from the Asinelli Tower and the Varallo Chapel (see Figures 4 and 6).

The AE signals originating from damage sources definitely localized in the building's masonry are filtered out. In this way, we can preliminarily distinguish between environmental contributions due to crustal trembling (external source) and structural damage contributions (inner sources localized by means of triangulation techniques [18]). Then, by applying the modified Grassberger-Procaccia correlation integral to the remaining AE data series and to the earthquake sequence, we obtained the cumulative probabilities $C^{\pm}(r, \tau)$ as functions of time $C^{\pm}{}_r(\tau)$ for different values of range $r$. This provides two-dimensional plots, which are easier to read than a three-dimensional representation of $C^{\pm}(r, \tau)$.

Figure 9a,b show that $C^+{}_r > C^-{}_r$ for all considered values of range $r$, suggesting that AE bursts are more likely to precede earthquakes than to follow them. The interpretation of this evidence is that the monitored structures behave as receptors of microseismic precursive activity during the preparation of a seismic event, *i.e.*, as sensitive earthquake receptors.

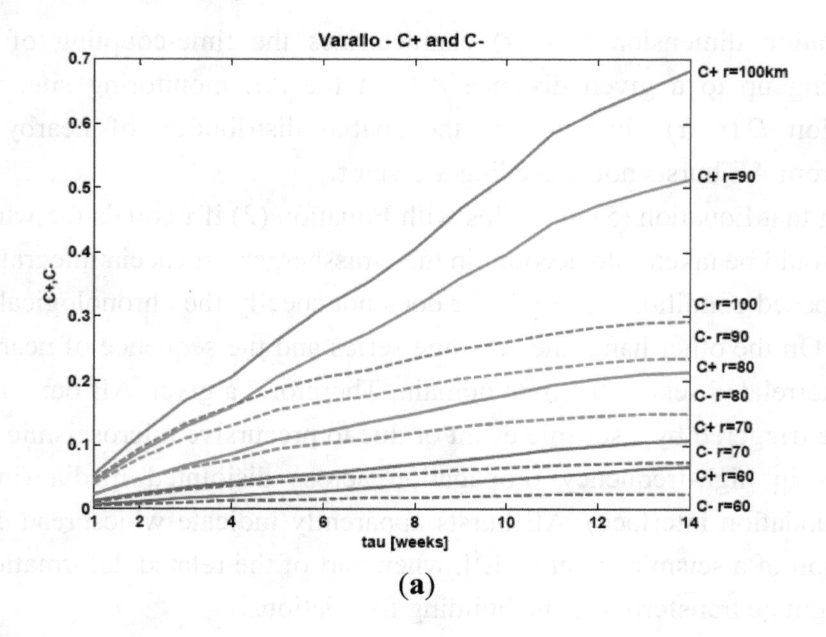


(a)

**Figure 9.** *Cont.*

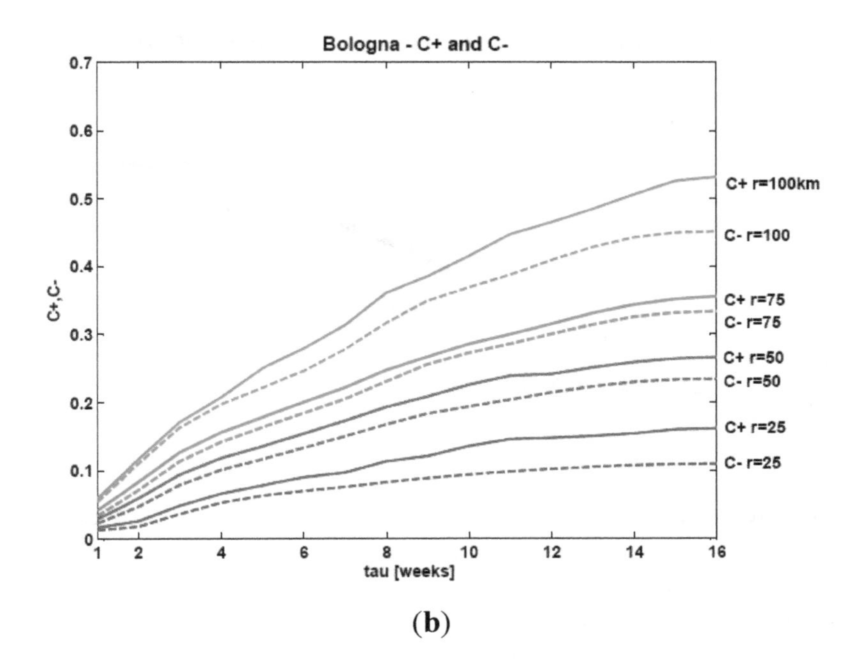


**(b)**

**Figure 9**. Modified correlation integrals $C^{\pm}{}_{r}(\tau)$, both considering AEs as earthquake "precursors" (+) and as "damage aftershocks" (−), plotted as functions of the time separation $\tau$ for different values of the spatial range $r$ for: (**a**) Chapel XVII and area around the city of Varallo; (**b**) Asinelli Tower and area around the city of Bologna.

## 6. AE Clustering in Time as a Seismic Precursor or an Aftershock

Since the cumulative probabilities $C^{\pm}{}_{r}(\tau)$ are represented in a time domain ranging from 1 to 14–16 weeks, shorter values of $\tau$ are further investigated.

In particular, time coupling of regional seismicity to AE activity from Chapel XVII and the Asinelli Tower is analyzed in terms of $D_t$ for $\tau$ ranging from 3 to 24 h after an earthquake occurrence, and from 10 min to 24 h before an earthquake occurrence.

The varying local slope $D_t$ of Log $C^{-}$ vs. Log $\tau$ reveals that the AE activity following seismic events is not equally probable over the time (see Figure 10a,b). For short time delays, $\tau$ = 3–10 min, high values of the time fractal dimension, $D_t$ = 1.38 and 1.24, indicate tight coupling of the AE activity to the earthquakes. In other words, AE bursts following an earthquake are more likely to occur within $\tau$ = 3–10 min after the seismic event. Then, such short time delays suggest a triggering action exerted by nearby earthquakes on damage processes of Chapel XVII and the Asinelli Tower. Contrarily, for $\tau$ ranging from $1.0 \times 10^3$ to $1.440 \times 10^3$ min, the obtained low values of $D_t$, 0.48 and 0.45, suggest that the effects of nearby earthquakes on the structural damage evolution disappear after 24 h.

In regards to the AE events preceding earthquakes, *i.e.*, considering AE as "seismic precursors", the linear bi-logarithmic plots of $C^{+}$ ($D_t$ = 1.06 and 0.94) shown in Figure 11 describe a uniform probability density of finding AE events prior to an earthquake for a wide range of time intervals, up to $\tau = 1.440 \times 10^3$ min. In other words, we observe a constant AE activity in the 24 h preceding an earthquake. The absence of accelerating precursive activity in the presented case studies confirms the need for further investigation, possibly with the aid of electromagnetic seismic precursors.

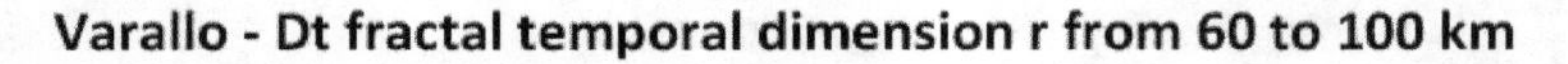


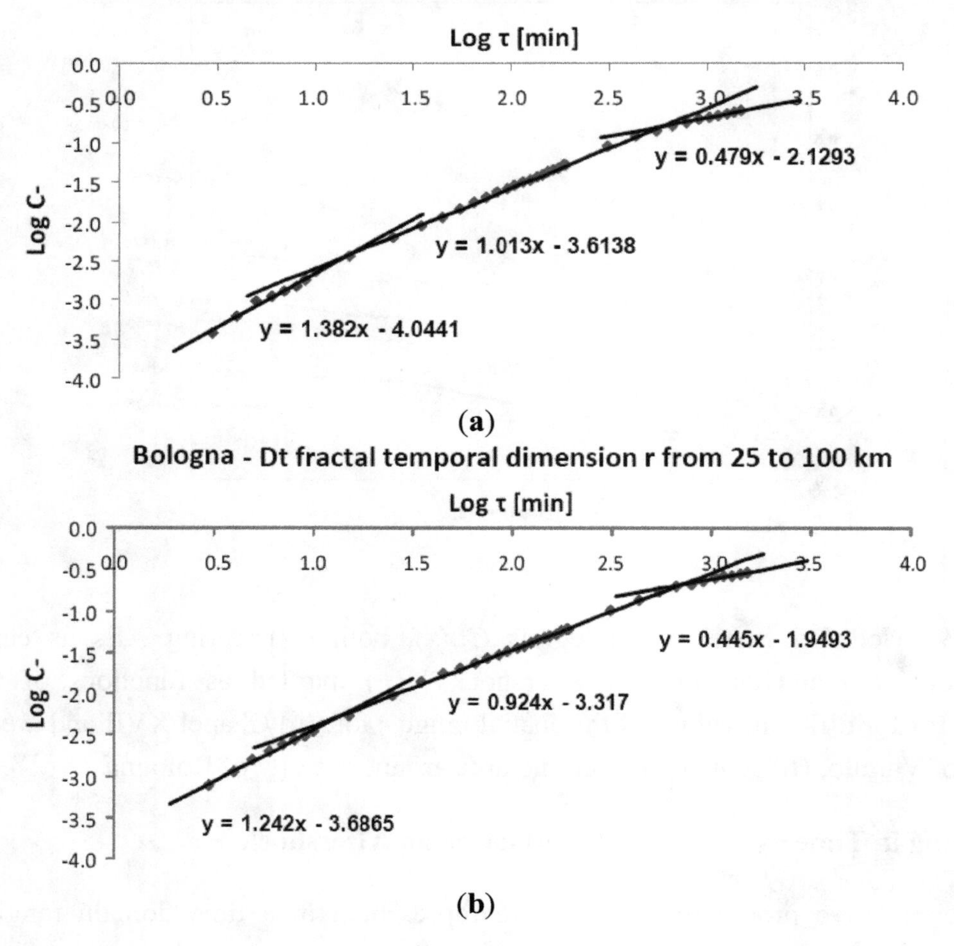


**Figure 10.** Analysis of time-clustering features of AE events considered as "seismic aftershocks" by the time correlation dimension $D_t$, as a local slope of Log $C^-$ *vs.* Log τ. The time range for τ is 3 min to 24 h. The selected seismic events are all those considered during the defined monitoring periods. (**a**) Chapel XVII (Varallo); (**b**) Asinelli Tower (Bologna).

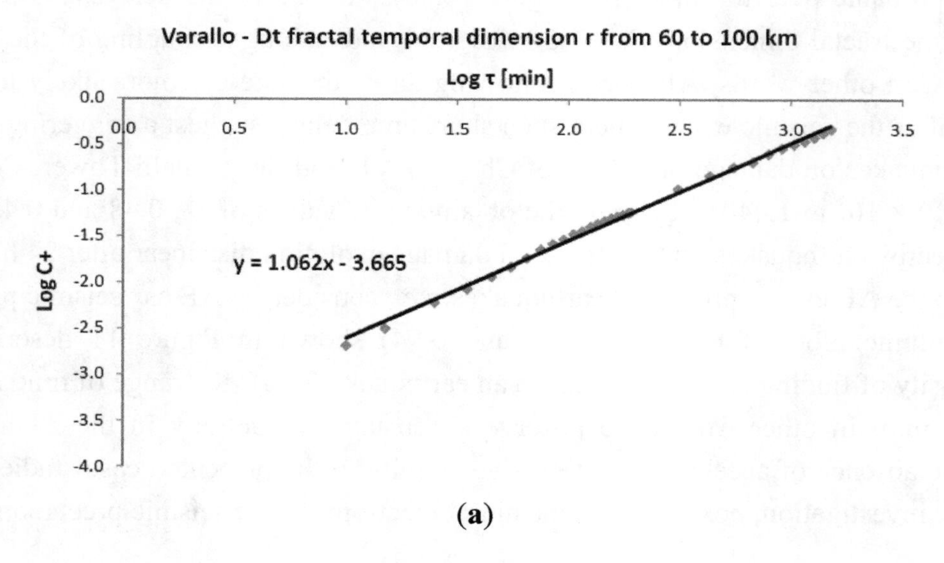


(**a**)

**Figure 11.** *Cont.*

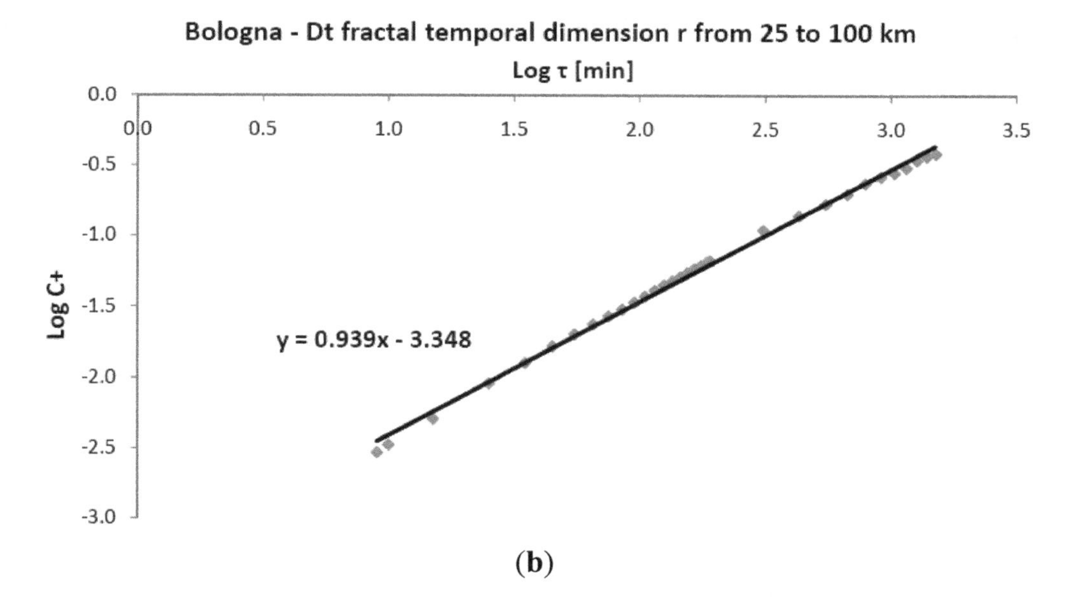


(b)

**Figure 11.** Analysis of time-clustering features of AE events considered "seismic precursors" by the time correlation dimension $D_t$, as the local slope of Log $C^+$ *vs.* Log τ. The time range for τ is 10 min to 24 h. The selected seismic events are all those considered during the defined monitoring periods. (**a**) Chapel XVII (Varallo); (**b**) Asinelli Tower (Bologna).

## 7. Conclusions

In order to assess the seismic hazard in different areas, it is suggested to observe the spatial and temporal correlation of seismic events and monitored AEs. It emerges that AEs behave as earthquake precursors. A statistical method of analysis is proposed, based on the Grassberger-Procaccia integral. It gives the cumulative probability of the events' occurrence in a specific area, considering the AE records and the seismic events during the same period of time. Two important constructions of Italian cultural heritage were considered: the "Sacred Mountain of Varallo" and "Asinelli Tower". Given the interesting premises that emerged from this study, the Grassberger-Procaccia integral could also be applied more extensively in several monitoring sites belonging to different seismic regions, not only for defining the seismic hazard, but also to handle it as an effective tool for earthquake risk mitigation.

However, the authors are aware that a rigorous investigation to ascertain the existence of crustal stress crises crossing areas of a few hundred kilometers' radius during the preparation of a seismic event would require simultaneous and numerous operations of AE monitoring sites, adequately placed in the territory.

## Acknowledgments

E. de Filippis, Director of the Piedmont Sacred Mountains Institute, is gratefully acknowledged. The authors thank the Municipality of Bologna and Eng. R. Pisani for having allowed the study on the Asinelli Tower.

## Author Contributions

A.C. supervised the research; P.C. and S.I. performed the numerical simulations; G.N. carried out AE and NDT tests; G.L. wrote the paper with discussions from all the authors.

## References

1. Niccolini, G.; Carpinteri, A.; Lacidogna, G.; Manuello, A. Acoustic emission monitoring of the Syracuse Athena Temple: Scale invariance in the timing of ruptures. *Phys. Rev. Lett.* **2011**, doi:10.1103/PhysRevLett.106.108503.
2. Carpinteri, A.; Lacidogna, G.; Invernizzi, S.; Accornero, F. The Sacred Mountain of Varallo in Italy: Seismic risk assessment by Acoustic Emission and structural numerical models. *Sci. World J.* **2013**, *2013*, 1–10.
3. Carpinteri, A.; Lacidogna, G.; Manuello, A.; Niccolini, G. A study on the structural stability of the Asinelli Tower in Bologna. *Struct. Control HLTH* **2015**, doi:10.1002/stc.1804.
4. Bak, P.; Christensen, K.; Danon, L.; Scanlon, T. Unified scaling laws for earthquakes. *Phys. Rev. Lett.* **2002**, *88*, 178501–178504.
5. Parson, T. Global Omori law decay of triggered earthquakes: Large aftershocks outside the classical aftershock zone. *J. Geophys. Res.* **2002**, *107*, 2199–2218.
6. Corral, A. Long-term clustering, scaling and universality in the temporal occurrence of earthquakes. *Phys. Rev. Lett.* **2004**, doi:10.1103/PhysRevLett.92.108501.
7. Tosi, P.; de Rubeis, V.; Loreto, V.; Pietronero, L. Space-time combined correlation integral and earthquake interactions. *Ann. Geophys.* **2004**, *47*, 1–6.
8. Grassberger, P.; Procaccia, I. Characterization of strange attractors. *Phys. Rev. Lett.* **1983**, *50*, 346–349.
9. Carpinteri, A.; Lacidogna, G.; Niccolini, G. Acoustic emission monitoring of medieval towers considered as sensitive earthquake receptors. *Nat. Hazard Earth Syst. Sci.* **2007**, *7*, 251–261.
10. Carpinteri, A.; Lacidogna, G. Structural monitoring and integrity assessment of medieval towers. *J. Struct. Eng.* **2006**, *132*, 1681–1690.
11. Scholz, C.H. The frequency-magnitude relation of microfracturing in rock and its relation to earthquakes. *Bull. Seismol. Soc. Am.* **1968**, *58*, 399–415.
12. Gregori, G.P.; Paparo, G. Acoustic emission (AE). A diagnostic tool for environmental sciences and for non destructive tests (with a potential application to gravitational antennas). In *Meteorological and Geophysical Fluid Dynamics*; Schroeder, W., Ed.; Science Edition: Bremen, Germany, 2004; pp. 166–204.
13. Kagan, Y.Y.; Knopoff, L. Spatial distribution of earthquakes: The two-point correlation function. *Geoph. J. Roy. Astron. Soc.* **1980**, *62*, 303–320.
14. Hirata, T. A correlation between the *b*-value and the fractal dimension of earthquakes. *J. Geophys. Res.* **1986**, *94*, 7507–7514.

15. Carpinteri, A.; Lacidogna, G.; Niccolini, G.; Puzzi, S. Morphological fractal dimension *versus* power-law exponent in the scaling of damaged media. *Int. J. Damage Mech.* **2009**, *18*, 259–282.
16. Italian Seismological Instrumental and Parametric Data-Base (ISIDE). Available online: http://iside.rm.ingv.it/iside/standard/result.jsp?rst=1&page=EVENTS#result (accessed on 5 December 2015).
17. Burrato, P.; Vannoli, P.; Fracassi, U.; Basili, R.; Valensise, G. Is blind faulting truly invisible? Tectonic-controlled drainage evolution in the epicentral area of the May 2012, Emilia-Romagna earthquake sequence (northern Italy). *Ann. Geophys. Italy* **2012**, *55*, 525–531.
18. Carpinteri, A.; Xu, J.; Lacidogna, G.; Manuello, A. Reliable onset time determination and source location of acoustic emissions in concrete structures. *Cem. Concr. Compos*. **2012**, *34*, 529–537.

# Permissions

All chapters in this book were first published in MDPI; hereby published with permission under the Creative Commons Attribution License or equivalent. Every chapter published in this book has been scrutinized by our experts. Their significance has been extensively debated. The topics covered herein carry significant findings which will fuel the growth of the discipline. They may even be implemented as practical applications or may be referred to as a beginning point for another development.

The contributors of this book come from diverse backgrounds, making this book a truly international effort. This book will bring forth new frontiers with its revolutionizing research information and detailed analysis of the nascent developments around the world.

We would like to thank all the contributing authors for lending their expertise to make the book truly unique. They have played a crucial role in the development of this book. Without their invaluable contributions this book wouldn't have been possible. They have made vital efforts to compile up to date information on the varied aspects of this subject to make this book a valuable addition to the collection of many professionals and students.

This book was conceptualized with the vision of imparting up-to-date information and advanced data in this field. To ensure the same, a matchless editorial board was set up. Every individual on the board went through rigorous rounds of assessment to prove their worth. After which they invested a large part of their time researching and compiling the most relevant data for our readers.

The editorial board has been involved in producing this book since its inception. They have spent rigorous hours researching and exploring the diverse topics which have resulted in the successful publishing of this book. They have passed on their knowledge of decades through this book. To expedite this challenging task, the publisher supported the team at every step. A small team of assistant editors was also appointed to further simplify the editing procedure and attain best results for the readers.

Apart from the editorial board, the designing team has also invested a significant amount of their time in understanding the subject and creating the most relevant covers. They scrutinized every image to scout for the most suitable representation of the subject and create an appropriate cover for the book.

The publishing team has been an ardent support to the editorial, designing and production team. Their endless efforts to recruit the best for this project, has resulted in the accomplishment of this book. They are a veteran in the field of academics and their pool of knowledge is as vast as their experience in printing. Their expertise and guidance has proved useful at every step. Their uncompromising quality standards have made this book an exceptional effort. Their encouragement from time to time has been an inspiration for everyone.

The publisher and the editorial board hope that this book will prove to be a valuable piece of knowledge for researchers, students, practitioners and scholars across the globe.

# List of Contributors

**Fedor V. Shugaev, Dmitri Y. Cherkasov and Oxana A. Solenaya**
Faculty of Physics, M.V. Lomonosov Moscow State University, GSP-1, Leninskiye Gory, Moscow 119991, Russia

**Jae-Won Lee**
Underwater Communication/Detection Research Center, Kyungpook National University, Daegu 702-701, Korea

**Ho-Shin Cho**
College of IT Engineering, Kyungpook National University, Daegu 702-701, Korea

**Jun Liu**
School of Electrical Engineering, Xi'an Jiaotong University, Xi'an 710049, China

**Jiuqiang Han, Hongqiang Lv and Bing Li**
School of Electronic and Information Engineering, Xi'an Jiaotong University, Xi'an 710049, China

**Yan Chen and Jiyan Dai**
The Hong Kong Polytechnic University, Shenzhen Research Institute, Shenzhen 518057, China
Department of Applied Physics, The Hong Kong Polytechnic University, Hong Kong, China

**Kai Mei, Chi-Man Wong and Helen Lai Wa Chan**
Department of Applied Physics, The Hong Kong Polytechnic University, Hong Kong, China

**Dunmin Lin**
College of Chemistry and Materials Science, Sichuan Normal University, Chengdu 610066, China

**Helder Puga and Joaquim Barbosa**
Centre for Micro-Electro Mechanical Systems (CMEMS), University of Minho, Campus of Azurém, 4800-058 Guimarães, Portugal

**Vitor Carneiro and Vanessa Vieira**
Department of Mechanical Engineering, University of Minho, Campus of Azurém, 4800-058 Guimarães, Portugal

**Zhixin Zhang, Ji Liang, Daihua Zhang, Wei Pang and Hao Zhang**
State Key Laboratory of Precision Measuring Technology and Instruments, Tianjin University, Tianjin 300072, China

**Fangqian Xu**
Zhejiang University of Media and Communications, 998 Xueyuan Street Higher Education Zone Xia Sha, Zhejiang 310018, China

**Wen Wang, Xiuting Shao, Xinlu Liu and Yong Liang**
State Key Laboratory of Acoustics, Institute of Acoustics, Chinese Academy of Sciences, No. 21, BeiSiHuan West Road, Beijing 100190, China

**Chang Sun, Weidong Zhang and Yibo Ai**
National Center for Materials Service Safety, University of Science and Technology Beijing, Beijing 100083, China

**Hongbo Que**
Qishuyan Institute Co. Ltd., China South Locomotive & Rolling Stock Corporation Limited, Changzhou 213011, China

**Chiara Scrosati and Fabio Scamoni**
Construction Technologies Institute, National Research Council of Italy, via Lombardia 49, 20098 San Giuliano Milanese (MI), Italy

**Ernst Niederleithinger, Julia Wolf, Frank Mielentz, Herbert Wiggenhauser and Stephan Pirskawetz**
BAM Federal Institute for Materials Research and Testing, Berlin 12200, Germany

**Angélica Díaz, María T. Casas and Jordi Puiggalí**
Chemical Engineering Department, Polytechnic University of Catalonia, Av. Diagonal 647, Barcelona E-08028, Spain

**Maria Kogia, Liang Cheng, Vassilios Kappatos, Wamadeva Balachandran and Cem Selcuk**
Brunel Innovation Centre (BIC), Brunel University, Cambridge CB21 2AL, UK

**Abbas Mohimi and Tat-Hean Gan**
Brunel Innovation Centre (BIC), Brunel University, Cambridge CB21 2AL, UK
TWI Ltd., Granta Park, Great Abington, Cambridge CB21 6AL, UK

**Adriana Savin and Nicoleta Iftimie**
Nondestructive Testing Department, National Institute of R&D for Technical Physics, Iasi 700050, Romania

**Alina Bruma**
CRISMAT Laboratory, Ecole Nationale Superieure d'Ingenieurs de Caen, Universite de Caen Basse Normandie, 6 Blvd Marechal Juin, Caen 14050, France

**Rozina Steigmann**
Nondestructive Testing Department, National Institute of R&D for Technical Physics, Iasi 700050, Romania
Faculty of Physics, University Al.I. Cuza, 11 Carol I Blvd, Iasi 700506, Romania

**Dagmar Faktorova**
Faculty of Electrical Engineering, University of Žilina, Univerzitná 1, Žilina 010 26, Slovakia

**Giuseppe Lacidogna, Patrizia Cutugno, Gianni Niccolini, Stefano Invernizzi and Alberto Carpinteri**
Department of Structural, Geotechnical and Building Engineering, Politecnico di Torino, Corso Duca degli Abruzzi 24, 10129 Torino, Italy

# Index

Printed in the USA
CPSIA information can be obtained
at www.ICGtesting.com
JSHW051358091023
49903JS00006B/193

9 781639 897452